Polymerase Chain Reaction
Theory and Technology

AF904091

Authors:
Mark A. Behlke, Integrated DNA Technologies, Coralville, Iowa, USA
Kornelia Berghof-Jäger, BIOTECON Diagnostics, Potsdam, Germany
Tom Brown, University of Southampton, UK
Stephen A. Bustin, Anglia Ruskin University, Chelmsford, UK
Nigel Cook, Jorvik Food Safety Services, York, UK
Martin D'Agostino, Campden BRI, UK
Joseph R. Dobosy, Integrated DNA Technologies, Coralville, Iowa, USA
Trine Hansen, Technical University of Denmark, Søborg, Denmark
Marta Hernández, ITACyL, Valladolid, Spain
Jeffrey Hoorfar, Technical University of Denmark, Søborg, Denmark
Jim Huggett, University of Surrey, Guildford, UK
Waqar Hussain, The Binding Site, Birmingham, UK
Mathilde H. Josefsen, Technical University of Denmark, Søborg, Denmark
Tomáš Kuchta, Food Research Institute, Bratislava, Slovakia
Martin A. Lee, Fluorogenics Limited, Porton Down, UK
Dario L. Leslie, Defence Science and Technology Laboratory, Porton Down, UK
Charlotta Löfström, Agrifood and Bioscience, RISE Research Institute, Sweden
Tania Nolan, The Gene Team, Bury St. Edmunds, UK
Richard Owczarzy, Integrated DNA Technologies, Coralville, Iowa, USA
Michael W. Pfaffl, Technical University Munich, Germany
Eyjólfur Reynisson, Alvotech, Reykjavík, Iceland
Gabriel A. de Ridder, Onderstepoort Biological Products, Pretoria, South Africa
David Rodríguez-Lázaro, ITACyL, Valladolid, Spain
Scott D. Rose, Integrated DNA Technologies, Coralville, Iowa, USA
Chaminda Salgado, GlaxoSmithKline, UK
Nick A. Saunders, Health Protection Agency, London, UK
David J. Squirrell, Enigma Diagnostics, Salisbury, UK
Maureen B. Taylor, University of Pretoria, South Africa
Sara Zaccara, Università degli Studi di Trento, Italy

Caister Academic Press **www.caister.com**

Caister Academic Press
Norfolk, UK

www.caister.com

British Library Cataloguing-in-Publication Data
A catalogue record for this book is available from the British Library

ISBN: 978-1-912530-24-3 (book)
ISBN: 978-1-912530-25-0 (ebook)

Description or mention of instrumentation, software, or other products in this book does not imply endorsement by the author or publisher. The author and publisher do not assume responsibility for the validity of any products or procedures mentioned or described in this book or for the consequences of their use.

Contents

Current books of interest

- SUMOylation and Ubiquitination: Current and Emerging Concepts 2019
- Avian Virology: Current Research and Future Trends 2019
- Pathogenic Streptococci: From Genomics to Systems Biology and Control 2019
- Insect Molecular Virology: Advances and Emerging Trends 2019
- Methylotrophs and Methylotroph Communities 2019
- Prions: Current Progress in Advanced Research (Second Edition) 2019
- Microbiota: Current Research and Emerging Trends 2019
- Microbial Ecology: Current Advances from Genomics, Metagenomics and Other Omics 2019
- Porcine Viruses: From Pathogenesis to Strategies for Control 2019
- Lactobacillus Genomics and Metabolic Engineering 2019
- Cyanobacteria: Signaling and Regulation Systems 2018
- Viruses of Microorganisms 2018
- Protozoan Parasitism: From Omics to Prevention and Control 2018
- Genes, Genetics and Transgenics for Virus Resistance in Plants 2018
- Plant-Microbe Interactions in the Rhizosphere 2018
- DNA Tumour Viruses: Virology, Pathogenesis and Vaccines 2018
- Pathogenic Escherichia coli: Evolution, Omics, Detection and Control 2018
- Postgraduate Handbook: A Comprehensive Guide for PhD and Master's Students and their Supervisors 2018
- Enteroviruses: Omics, Molecular Biology, and Control 2018
- Molecular Biology of Kinetoplastid Parasites 2018
- Bacterial Evasion of the Host Immune System 2017
- Illustrated Dictionary of Parasitology in the Post-Genomic Era 2017
- Next-generation Sequencing and Bioinformatics for Plant Science 2017
- Brewing Microbiology: Current Research, Omics and Microbial Ecology 2017
- Metagenomics: Current Advances and Emerging Concepts 2017
- The CRISPR/Cas System: Emerging Technology and Application 2017
- Bacillus: Cellular and Molecular Biology (Third edition) 2017
- Cyanobacteria: Omics and Manipulation 2017
- Foot-and-Mouth Disease Virus: Current Research and Emerging Trends 2017
- Staphylococcus: Genetics and Physiology 2016
- Microbial Biodegradation: From Omics to Function and Application 2016
- Influenza: Current Research 2016
- Chloroplasts: Current Research and Future Trends 2016
- The Bacteriocins: Current Knowledge and Future Prospects 2016
- Brain-eating Amoebae: Biology and Pathogenesis of Naegleria fowleri 2016

Full details at: www.caister.com

DOI: https://doi.org/10.21775/9781912530243.01

Introduction to the Real-time Polymerase Chain Reaction

1

David Rodríguez-Lázaro and Marta Hernández

Abstract

Food safety and quality control programmes are increasingly applied throughout the production food chain in order to guarantee added value products as well as to minimize the risk of infection for the consumer. The development of real-time PCR has represented one of the most significant advances in food diagnostics as it provides rapid, reliable and quantitative results. These aspects become increasingly important for the agricultural and food industry. Different strategies for real-time PCR diagnostics have been developed including unspecific detection independent of the target sequence using fluorescent dyes such as SYBR Green, or by sequence-specific fluorescent oligonucleotide probes such as TaqMan probes or molecular beacons.

Introduction

Kleppe *et al.* described the principles of polymerase chain reaction (PCR) for first time in 1971 (Kleppe *et al.*, 1971). But it was in 1985, with the introduction of a thermostable DNA polymerase, that Saiki and collaborators working in Cetus Corporation published the first experimental data using this technique (Saiki *et al.*, 1985, 1988; Mullis *et al.*, 1986). The discovery of PCR is attributed to a collaborator of Saiki, Kary B. Mullis, winner of the Nobel Prize in Chemistry in 1993. Since then, PCR has been used in more than 275,000 scientific publications (search 'PCR' in pubmed.com). This technique has been applied in different areas due to its versatility, specificity and sensitivity. Accordingly, PCR has been successfully used for microorganism identification (Allmann *et al.*, 1995; Candrian, 1995; Olsen *et al.*, 1995; Rodríguez-Lázaro *et al.*, 2007), for the detection of ingredients of food products, e.g. cereals (Allmann *et al.*, 1993), vegetables (Jankiewicz *et al.*, 1997), and different animal (Zimmerman *et al.*, 1998b) and fish species (Hübner *et al.*, 1997).

The principle of PCR

PCR is a simple, versatile, sensitive, specific and reproducible assay (Saiki *et al.*, 1988; Cha *et al.*, 1993). It consists of an exponential amplification of a DNA fragment, and its principle is based on the mechanism of DNA replication *in vivo*: dsDNA is denatured to ssDNA, duplicated, and this process is repeated along the reaction according to the following formula (Higuchi *et al.*, 1993; Innis *et al.*, 1999).

$$C = C_0(1+E)^n$$

where C is the final amount of DNA, C_0 is the initial amount of DNA, E is efficiency ($E = 10^{-1/s} - 1$; if $E = 1$ then $s = -3.3219$), n is the number of cycles and s is the slope of the exponential phase.

During the denaturation step, the dsDNA melts opening up to ssDNA, and all enzymatic reactions stop (i.e. the extension from a previous cycle). To DNA denaturation, the temperature is usually raised to 93–96°C, breaking the H-bonds and thus increasing the number of non-paired bases. The temperature at which half of the dsDNA is single-stranded is known as the melting temperature, T_m. The type of solvent, the salt concentration and the pH influence the denaturation process. The concentration of G/C and T/A can also affect the T_m value. G/C-rich DNA sequences have higher T_m values compared with those T/A rich. The second phase, i.e. annealing of primers to ssDNA, takes place at temperatures closer to their T_m (usually 55–65°C) and is named as temperature of annealing (T_a). The oligonucleotides used as primers typically consist of relatively short sequences (15–25 nt) complementary to recognition sites, flanking the segment of target DNA to be amplified. Once the temperature is reduced, the two complementary ssDNA chains tend to rehybridize into a dsDNA molecule. In this phase, ionic bonds are constantly formed and broken between the single-stranded primer and the single-stranded template. If primers adequately anneal to the template, the ionic bond is strong enough between the template and the primer to stabilize the nascent double stranded structure and allow the polymerase to attach and begin copying the template. The extension phase is carried out across the target sequence by using a heat-stable DNA polymerase in the presence of dNTPs and $MgCl_2$, resulting in a duplication of the starting target material. This enzyme has 5′→3′ DNA polymerase activity, i.e. it adds dNTPs from 5′ to 3′, reading the template from 3′ to 5′. When the primers have been extended a few bases, they posses a stronger ionic attraction to the template, which reduces the probability of unbinding. The duration of the extension step can be increased if the region of DNA to be amplified is long (> 1000 bp); however, for the majority of PCR experiments 1 min is sufficient to get a complete extension.

After each cycle, the newly synthesized DNA strands can serve as template in the next cycle. As shown in Fig. 1.1, the major product of this exponential reaction is a segment of ds-DNA whose termini are defined by the 5′ termini of the 2 primers and whose length is defined by the distance between the primers. The products of a successful first round of amplification are heterogeneously sized DNA molecules, whose lengths may exceed the distance between the binding sites of the two primers. In the second round, these molecules generate DNA strands of defined length that will accumulate in an exponential fashion in later rounds of amplification and will form the dominant products of the reaction.

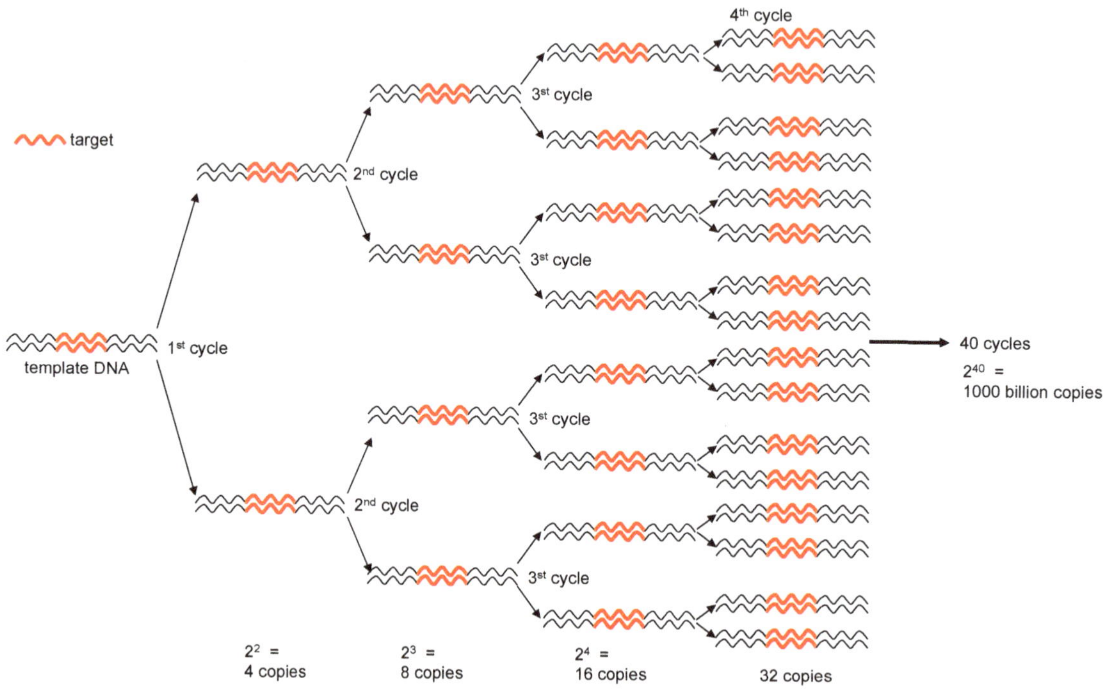

Figure 1.1 The exponential amplification of DNA in PCR.

Thus, amplification, as a final number of copies of the target sequence, is expressed by the following equation:

$$(2^n - 2n)x$$

where n is the number of cycles, $2n$ is the first product obtained after the first cycle and second products obtained after the second cycle with undefined length and x is the number of copies of the original template.

Potentially, after 30 PCR cycles there will be around 2^{30}-fold amplification, assuming 100% efficiency during each cycle. The efficiency of a PCR will vary from template to template and according to the degree of optimization that has been carried out.

Multiplex PCR

Whereas standard PCR usually uses one pair of primers to amplify a specific sequence, multiplex PCR allows the simultaneous amplification of more than one target sequence in a single reaction by using a set of primers (Elnifro *et al.*, 2000; Higuchi *et al.*, 1992; Wittwer *et al.*, 2001). This saves considerable time and effort, and decreases the number of reactions that need to be performed to detect the desired targets in the sample. The presence of many PCR primers in a single tube can cause some problems, such as the increased formation of misprimed PCR products, 'primer dimers', and the amplification discrimination of longer DNA fragments (Atlas and Bej, 1994; Higuchi *et al.*, 1992). For multiplex PCR, primers should be chosen with similar annealing temperatures. The lengths of amplified products should be similar; as large differences in the lengths of the target DNAs may favour the amplification of shorter targets over the longer ones, which results in differential yields of amplified products.

Real-time PCR

The development of real-time (q-) PCR represents a significant advance in many molecular techniques involving nucleic acids analysis. qPCR allows monitoring of the synthesis of new amplicon molecules during the PCR (i.e., in real time). Data is therefore collected throughout the PCR process, i.e. monitoring of the process of amplification in real time by using fluorescence, and not only at the end of the reaction (as occurs in conventional PCR) (Heid *et al.*, 1996; Higuchi *et al.*, 1993; Nazarenko *et al.*, 1997). The qPCR results consist of amplification curves (Fig. 1.2), that can be used to quantify the initial amounts

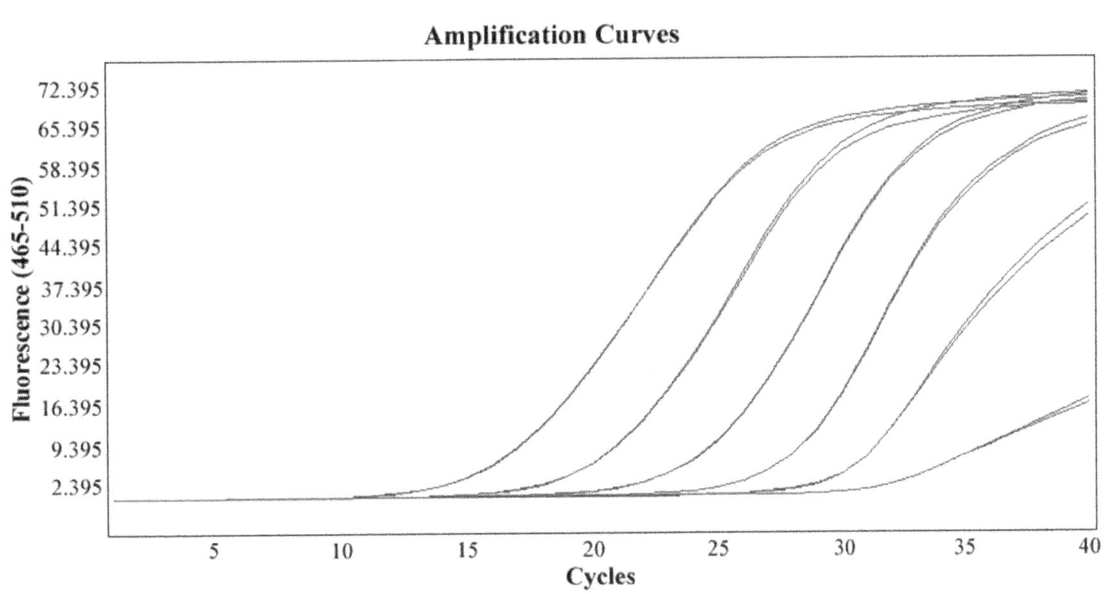

Figure 1.2 Amplification curves in semilogarithmic view obtained from serial dilutions of a target DNA. Inset, regression curve obtained from C_T values.

of template DNA molecules with high precision over a wide range of concentrations (Schmittgen *et al.*, 2000). Major advantages of qPCR are the closed-tube format (that avoids risks of carry-over contamination), fast and easy to perform analysis, the extremely wide dynamic range of quantification (more than eight orders of magnitude) (Heid *et al.*, 1996) and the significantly higher reliability and sensitivity of the results compared to conventional PCR. qPCR can be used for many different purposes like conventional PCR, particularly for quantifying nucleic acids and for genotyping. Since the fist scientific work published in 1996 (Heid *et al.*, 1996), the number of publications where qPCR is used has increased nearly exponentially (Fig. 1.3).

Chemistries or detection formats

The fluorescence that is monitored along all the qPCR process can be detected by a unspecific detection strategy independent of the target sequence, e.g. through fluorescent dyes that have special fluorescent properties when bound to dsDNA, or by sequence-specific fluorescent oligonucleotide probes, i.e. a sequence-specific strategy.

Unspecific detection formats

Ethidium bromide was the first dye used for this purpose (Higuchi *et al.*, 1993; Le Pecq and Paoletti, 1966; Wittwer *et al.*, 1997). Other intercalating dyes such as YO-PRO-1, BEBO or SYBR Green I have since been used (Bengtsson *et al.*, 2003; Ishiguro *et al.*, 1995; Tseng *et al.*, 1997). SYBR Green I is the most frequently used dsDNA-specific dye in qPCR. It is an asymmetrical cyanine dye, structurally related to the dsDNA-specific dyes YOYO-1 and TOTO-1 (Jin *et al.*, 1994; Singer *et al.*, 1994). Its binding affinity is more than 100 times higher than that of ethidium bromide. SYBR Green I largely binds sequence independently to the minor groove of ds-DNA. It can be excited with blue light with a wavelength of 480 nm, and its emission spectrum is comparable to that of fluorescein with a maximum at 520 nm and a quantum yield of 0.8 (Jin *et al.*, 1994). The fluorescence of the bound dye is above 1000-fold higher than that of the free dye and, therefore, is well suited for monitoring the product accumulation during PCR (Wittwer *et al.*, 1997; Morrison *et al.*, 1998). When monitored in real time, this results in an increase in the fluorescence signal that can be observed during the polymerization step, and that falls off when the DNA is denatured (Fig. 1.4). Consequently, fluorescence measurements have to be performed at

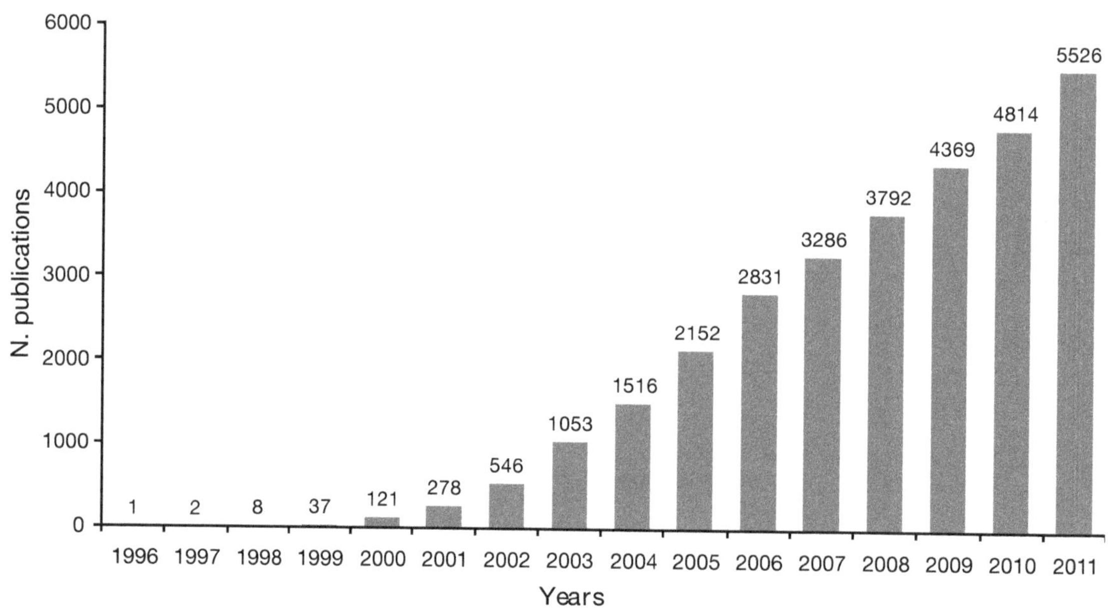

Figure 1.3 Result of a PubMed search using 'real-time PCR', sorted by year of publication.

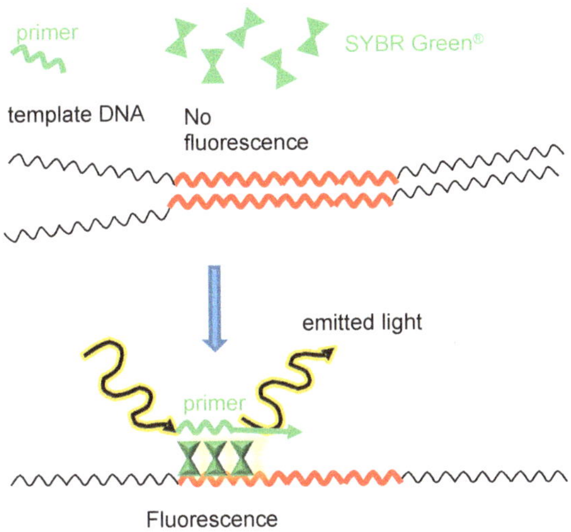

Figure 1.4 Principle of detection using Sybr Green.

the end of the elongation step of every PCR cycle. This method obviates the need for target-specific fluorescent probes, and hence it can be used with any pair of primers for any target, making its use less expensive (Giulietti *et al.*, 2001). However, its major disadvantage is that specificity is determined entirely by the primers and thus the risk of amplifying non-specific PCR products has to be considered during optimization (Simpson *et al.*, 2000). However, PCR product verification can be achieved at the end-point by plotting fluorescence as a function of temperature to generate a melting curve of the amplicon (Ririe *et al.*, 1997).

The *AmpliFluor system* is an unspecific detection system developed by Intergen co. AmpliFluor technology uses a universal energy-transfer hairpin primer (UniPrimer) which emits a fluorescent signal when unfolded during its incorporation into an amplification product. The UniPrimer contains a 18-nucleotide sequence (Z sequence: 5′-act gaa cct gac cgt aca-3′) at its 3′ end, that is also present at the 5′ end of one of the target-specific primers so that it anneals to the PCR product and acts as universal PCR primer. In the first step, the forward primer is extended. This extended product serves as template for the reverse primer in the second step. In the end, the polymerase opens the hairpin structure and a double-stranded PCR product is formed in which reporter and quencher are separated (Nazarenko *et al.*, 1997) (Fig. 1.5).

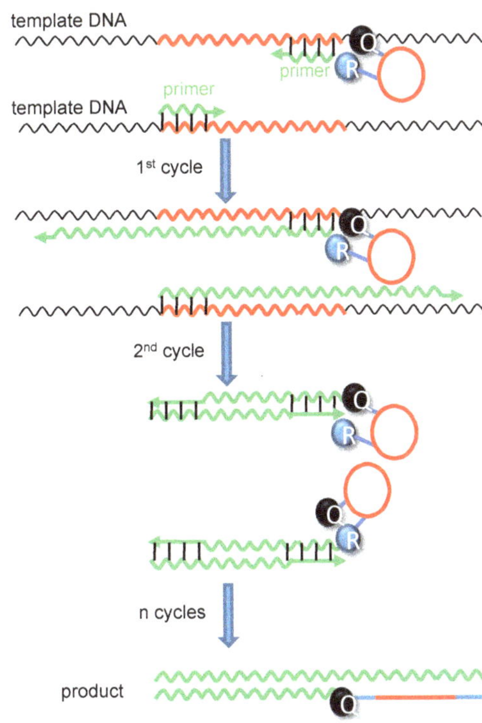

Figure 1.5 Principle of detection using AmpliFluor system. Adapted from Whitcombe *et al.* (1999).

Sequence-specific fluorescent oligonucleotide probes

There are different types of specific-sequence fluorescent probes, and they can be classified into two major groups, *hydrolysis probes* and *hybridization probes*, both types being homologous to the internal region amplified by the two primers. The fluorescence signal intensity can be related to the amount of PCR product (i) by a product-dependent decrease of the quench of a reporter fluorophore or (ii) by an increase of the fluorescence resonance energy transfer (FRET) from a donor to an acceptor fluorophore. FRET, also called Förster transfer, is the radiationless transfer of excitation energy by dipole–dipole interaction between fluorophores with overlapping emission and excitation spectra. The FRET and the quench efficiency are strongly dependent on the distance between the fluorophores (Förster, 1948; Clegg, 1992). Therefore, the PCR-product-dependent change in the distance between the fluorophores is used to generate the sequence-specific signals.

Several different formats can be used. In principle, all of them could function by a decrease of quench or an increase of FRET; in practice, most formats are based on a decrease of quench. The most commonly used fluorescent reporter dyes are FAM, TET (tetrachloro-6-car-boxyfluorescein), JOE (2,7-dimethoxy-4,5-dichloro-6-carboxy-fluorescein) or HEX (hexacholoro-6-carboxyfluorescein), and the most frequently used quenchers are TAMRA, DABCYL and Black Hole Quencher (BHQ) (Table 1.1).

Sequence-specific probes allow multiplexing (Bernard *et al.*, 1998; Dupont *et al.*, 2002; Wittwer *et al.*, 2001) and easy identification of point mutations (Giesendorf *et al.*, 1998; Lay and Wittwer, 1997; Lyon, 2001). A common drawback of probe systems that use the decrease-of-quench mechanism is unwanted generation of a signal due to probe destruction (e.g. by unintentional hydrolysis of the probes by the *Taq* DNA polymerase) or by formation of secondary structures of the probes that lead to a decrease in quench (Wilhem and Pingoud, 2003).

Hydrolysis probes

The hydrolysis probes are cleaved when hybridized by the 5′–3 exonuclease activity of particular DNA polymerases (Holland *et al.*, 1991) during the elongation phase of primers, yielding a real time measurable fluorescence emission directly proportional to the concentration of the target sequence (Heid *et al.*, 1996). It usually utilizes either *Taq* or *Tth* polymerase, but any enzyme with equivalent 5′–3′ exonuclease activity properties (e.g. *Tfl*) can be used (Gut *et al.* 1999). The best known hydrolysis probes are TaqMan® probes and TaqMan® MGB (minor groove binder) probes, both developed by Applied Biosystems.

A *TaqMan probe* is an oligonucleotide double-labelled with a reporter fluorophore at the 5′ end (reporter dye) and with a quencher internally or at the 3′ end (quencher dye). In addition, the probes must be blocked at their 3′- end to prevent the extension during the annealing step. The TaqMan assay (Fig. 1.6) uses three oligonucleotides. Two conventional primers allow amplification of the product, to which the TaqMan probe will anneal. The quencher dye absorbs the fluorescence of the reporter dye due to its proximity, which permits FRET. When the correct amplicon is amplified, the probe can hybridize to the target after the

Table 1.1 Excitation and emission spectre of different fluorophores

Fluorophores	Excitation (nm)	Emission (nm)
DABCYL	453	–
6-FAM	492	515
SYBR Green I	497	520
TET	521	536
JOE	527	548
HEX	535	556
TAMRA	555	580
ROX	575	602
VIC	538	554

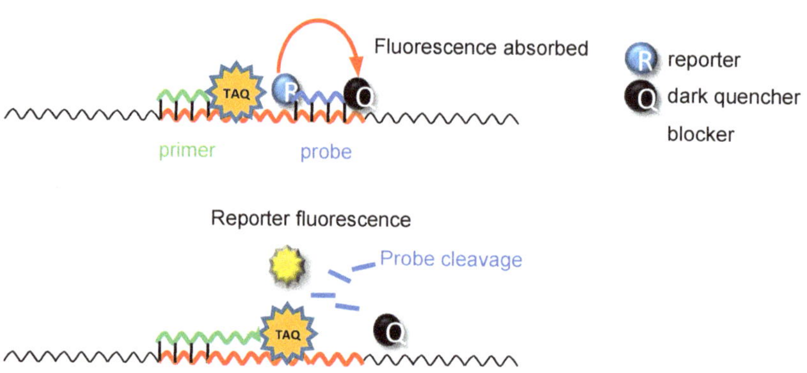

Figure 1.6 Principle of detection using TaqMan probes. 5′–3′ polymerase and exonuclease activity of the Taq DNA polymerase (TAQ). R, reporter; Q, quencher.

denaturation step. It remains hybridized while the polymerase extends the primers until it reaches the probe. Then, it displaces its 5′ end to hold it in a forked structure. The enzyme continues to move from the now free end to the bifurcation of the duplex, where cleavage takes place (Lyamichev *et al.* 1993). The quencher is hence released from the fluorophore, which now fluoresces after excitation (Heid *et al.*, 1996; Holland *et al.*, 1991; Gibson *et al.*, 1996). As the polymerase will cleave the probe only while it remains hybridized to its complementary strand, the temperature conditions of the polymerization phase of the PCR must be adjusted to ensure probe binding. Most probes have a T_m of around 70°C; therefore, the TaqMan system uses a combined annealing and polymerization step at 60–63°C. This ensures that the probe remains bound to its target during the primer extension step. It also ensures maximum 5′–3′ exonuclease activity of the *Taq* and *Tth* DNA polymerases (Tombline *et al.* 1996).

The *TaqMan MGB probes* are similar to TaqMan probes. They contain a non-fluorescent quencher (NFQ) and an oligopeptide at the 3′ end. This oligopeptide is a DNA minor groove binder (MGB), with very high affinity for the minor groove of AT-rich double-stranded DNA (Afonina *et al.*, 1997). Addition of the MGB ligand significantly enhances duplex stability. The shorter the probe, the greater the MGB contribution to the overall duplex stability: with 12- to 18-bp oligonucleotides, it raises the T_m from 44–56°C up to 66–70°C (Kutyavin *et al.*, 2000). This allows designing suitable probes in sequences such as those rich in AT, in which conventional TaqMan probes require an excessive length.

Hybridization probes

In contrast to hydrolysis probes, hybridization probes are not hydrolysed during PCR. The fluorescence is generated by a change in its secondary structure during the hybridization phase, which results in an increase of the distance separating the reporter and the quencher dyes. The most relevant hybridization probes are those containing hairpins (Molecular Beacons, Scorpion primers, etc.), and FRET hybridization probes.

Molecular beacons form a stem-and-loop structure through complementary sequences on the 5′ and 3′ ends of the probe (Tyagi and Kramer 1996). The loop portion is complementary to the target nucleic acid (Tan *et al.*, 2000). A reporter and a quencher fluorophore are attached one at the end of each arm. The quencher is a non-fluorescent chromophore that dissipates the energy it receives from the fluorophore as heat. The fluorescence is quenched when the probe is in a hairpin-like structure (stem-and-loop structure) due to the proximity between quencher and fluorophore allowing FRET (Cardullo *et al.*, 1998; Stryer, 1978). In the presence of a complementary sequence, designed internal to the primer binding sites, the probe undergoes a conformational transition that forces the stem apart and results in the formation of a probe/target hybrid that is more stable than the former stem (Bonnet *et al.*, 1999). This conformational change separates the fluorophore from the quencher and consequently

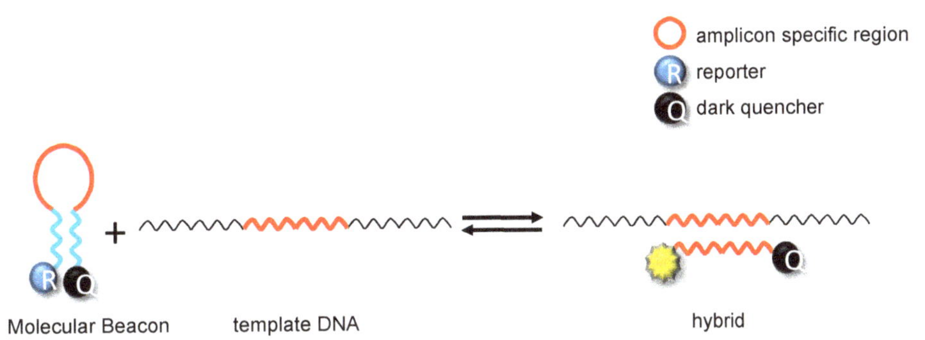

Figure 1.7 Principle of detection using Molecular Beacons. During hybridization PCR step, the fluorophore and quencher components of molecular beacons become spatially separated and the fluorescence is generated.

FRET no longer occurs, thus increasing reporter fluorescence emission (Kostrikis *et al.*, 1998; Tyagi *et al.*, 1998) (Fig. 1.7). Molecular beacons have been reported to be significantly more specific than conventional oligonucleotide probes of equivalent length, due to the presence of a stem structure (Bonnet *et al.*, 1999; Kaboev *et al.*, 2000; Tyagi *et al.*, 1998). The main drawback of molecular beacons is associated with its design as the fluorescence yield is very sensitive to the hybridization conditions.

FRET probes or *hybridization probes* use four oligonucleotides, i.e. two primers and two sequence-specific probes (Wittwer *et al.*, 1997). Each probe has a single label: either a donor fluorophore at the 3′-end or an acceptor fluorophore at the 5′-end. The emission spectrum of the donor fluorophore overlaps the excitation spectrum of the acceptor fluorophore. The FRET probes must be blocked at their 3′-end to prevent the extension during the annealing step. The two probes hybridize to the target sequences in a head-to-tail arrangement, thus bringing the two dyes close (typically 1–5 nucleotides distant), allowing FRET. During PCR, only the donor fluorophore is excited. In solution, only background fluorescence is emitted by the donor. During annealing, the two probes hybridize adjacently to their target sequence and thus the excitation energy is transferred by FRET from the donor dye in one of the probes to the acceptor dye in the other probe (Fig. 1.8), allowing the acceptor dye to dissipate fluorescence at a different wavelength (Bernard and Wittwer, 2000). The use of two independent probes results in high specificity and flexibility for probe design. Furthermore, as the probes are not hydrolysed, fluorescence is reversible and allows the generation of melting curves.

Scorpion primers are structurally and functionally related to molecular beacons, but serve as primers in the PCR. They consist of a probe sequence linked to the 5′ end of a primer via a non-amplifiable stopper moiety (Whitcombe *et al.*, 1999). The probe presents a fluorophore linked at the 5′-end and a quencher at the 3′-end, and is held in a hairpin loop structure by complementary sequences on its 5′- and 3′-ends. This configuration brings the fluorophore in close proximity with the quencher and avoids fluorescence similarly to Molecular Beacons. In addition, the probe sequence is complementary to an internal region of the sequence extended by the adjacent primer. In the first step, the primer is extended, yielding a single-stranded template for the reverse primer in the second step. Upon hybridization, the hairpin is opened, producing a physical separation of the fluorophore and quencher such that increases in signal are observed (Fig. 1.9). In contrast to the sunrise primers, the reverse extension is blocked by a hexethylene glycol group. This ensures that

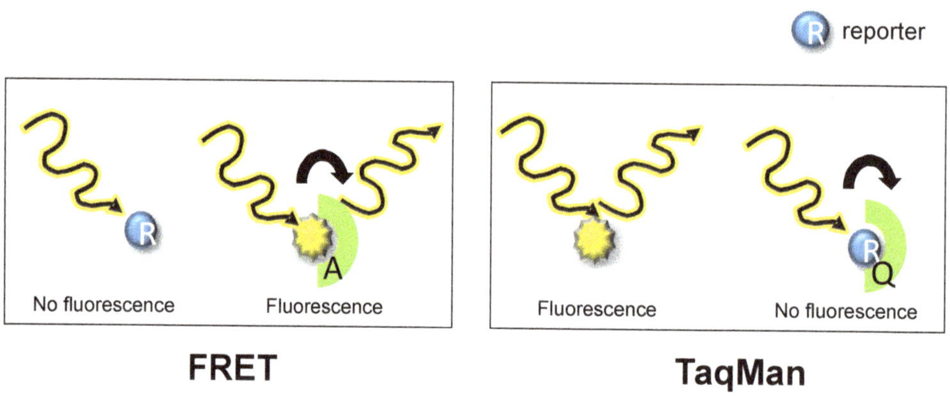

Figure 1.8 Differences between detection using FRET probes and TaqMan probes. A. Energy transfer in FRET probes (the e⁻ acceptor emits fluorescence when placed close to the e- donor. B. Energy transfer in hydrolysis probes (TaqMan) (the e⁻ donor emits fluorescence when the distance to the acceptor or quencher is higher than 10 nm. A, acceptor; D, donor; and Q, quencher.

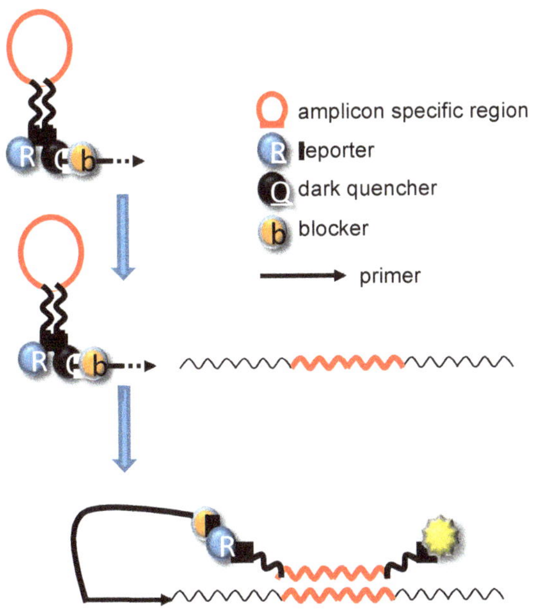

Figure 1.9 Principle of detection using Scorpion primers. (A) A Scorpion primer carries a 5′ extension comprising a probe element, a pair of self-complementary stem sequences, and a fluorophore/quencher pair. The extension is 'protected' from copying by the inclusion of a blocking HEG monomer. B. After a round of PCR extension from the primer, a newly synthesized target region is now attached to the same strand as the probe. C. Following a second round of denaturation and annealing, the probe and target hybridize. Adapted from Whitcombe *et al.* (1999).

the reporter of the scorpion primer remains quenched in unspecific products like primer dimers (Whitcombe *et al.*, 1998, 1999).

The *light-up technology* utilizes a nucleic acid analogue instead of natural DNA as sequence recognizing element (Isacsson *et al.*, 2000). Light-up probes are peptide nucleic acids (PNAs) (Nielsen *et al.*, 1991) that use thiazole orange, a derivative of cyanine, as reporter fluorophore (Svanvik *et al.*, 2000a). It forms sequence-specific complexes with DNA and RNA which are more stable than double-stranded natural nucleic acids (Iyer *et al.*, 1995; Egholm *et al.*, 1993). These features are attributed mainly to the charge-neutral nature of PNA, which eliminates the electrostatic repulsion between the hybridizing strands. The probe has low fluorescence when free in solution, however

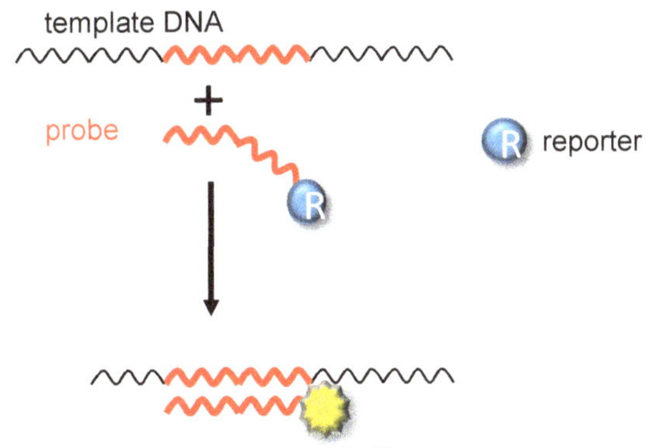

Figure 1.10 Principle of detection using light-up probes.

they show increased fluorescence intensity upon hybridization with DNA (Fig. 1.10).

Real-time PCR design

There are already defined prerequisites to design TaqMan oligonucleotides, and most of them can be also applied to other probes. They are summarized in Table 1.2.

When a qPCR assay is designed, the most important parameters are the amplicon length and the T_m of the primers and probe. The optimal amplicon length should be less than 150 bp, but it is advisable to reduce the length below 80 bp. However, amplicons up to 300 bp amplify efficiently. Shorter amplicons amplify more efficiently than longer ones and are more tolerant to suboptimal reaction conditions. This is because they are more likely to be denatured during the 92–95°C PCR step, allowing the probes and primers to compete more effectively for binding to their complementary targets. As the extension rate of *Taq* polymerase is between 30 and 70 bases per second (Jeffreys *et al.*, 1988), polymerization times as short as 5 s are sufficient to replicate such amplicon, making amplification of artefacts less likely and reducing the time of the assay. Primers are generally used in the 50–300 nM range. Higher concentrations may promote mispriming and accumulation of non-specific products, and lower concentrations may lead to primer exhaustion, although target copy numbers will have been calculated well before.

Table 1.2 Requisites for the design of oligonucleotides for RTi-PCR assays

Parameter	Primers	Probes
GC content (%)	30–80%	30–80%
Calculated T_m	50–60°C, always >55°C as UNG works at 50°C T_m of the primers should not differ >2°C	68–70°C (5–10°C above primer T_m)
Runs of identical nucleotides	Maximum 3 (no Gs!!)	Maximum 3 (no Gs!!)
Sequence length	Minimum 15 bp (15–30 bp)	Maximum. 30 bp
Amplicon length	The shorter the better. With TaqMan probes, 50–150 bp	
Distance forward primer to probe	Maximum 50 bp	
Primer dimers, hairpin loops	Avoid	Avoid
3′- instability (3′- rule)	Primers only. Maximum two Gs/Cs in the last 5 bp	
Autoquenching		No G at 5′-end
G–C ratio		Select the strand with C>G

Non-specific priming can be minimized by selecting primers that have only one or two G/Cs within the 3′ last five nucleotides. A relative instability at the 3′ ends makes primers less likely to hybridize transiently causing non-specific extension.

The T_a of the probe is also a critical parameter. Since amplification primers are extended as soon as they bind to their targets, the hybridization target sequence is rapidly masked with newly synthesized DNA. Therefore, the T_a of the probes must be significantly greater (approximately 10°C) than that of the primers. The presence of G at the 5′ end of the probe is to be avoided, because it slightly quenches the reporter signal, even after probe cleavage. Furthermore, the probe should contain more Cs than Gs; if this is not the case, the antisense probe should be used. The probe should never overlap with, or be complementary to either of the primers. The optimum concentration of fluorogenic probes will vary with the type of probe, as it depends on background fluorescence: quenching of hydrolysis probes is often below 100%, and thus they produce background fluorescence levels higher than molecular beacons and FRET probes.

An additional requirement of the FRET system is that the target sequence should be located towards the 3′ end of the amplicon, close to the reverse primer-binding site on the opposite strand. As fluorescence is measured during the annealing phase, this arrangement allows the qPCR device to measure fluorescence before the polymerase displaces the probes. Paired probes used for quantification should have equal $T_{a,}$ whereas for mutational analysis the T_m of the sensor probe should be lower than that of the anchor probe, thus ensuring that the sensor probe controls the generation of the fluorescent signal.

Quantification analysis

As indicated above, the fluorescence emission during qPCR is proportional to the synthesized DNA, and can be can be visualized as an amplification plot (Fig. 1.2). Typically, an amplification curve presents three different phases (Fig. 1.11). The first is called the *initiation phase*, it occurs during the first PCR cycles where the emitted fluorescence can not be distinguished from the baseline. During the *exponential* or *log phase* there is an exponential increase in fluorescence, before the plateau phase is reached. In this last phase, the reagents are exhausted, and no increase in fluorescence is observed. Only in the exponential phase, quantification is possible.

A fluorophore (so-called passive reference) may be present in the PCR buffer in order to well-to-well normalize the fluorescence (Table

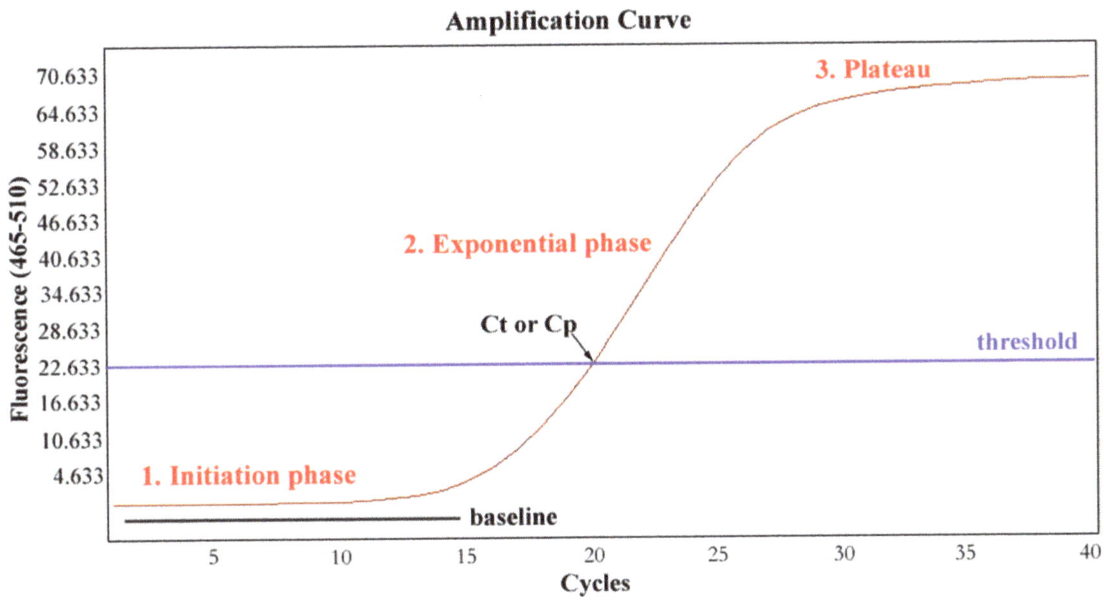

Figure 1.11 Phases of a PCR amplification curve. Red: amplification curve of a positive sample. Blue: threshold. Black: baseline cycles.

1.1). This helps to overcome the effects of pipetting errors. This normalization is accomplished by dividing the emission intensity of the reporter dye by the emission intensity of the passive reference to obtain a ratio, i.e. Rn (normalized reporter) for any given reaction. At any given cycle, Rn^+ is the Rn value of a reaction containing all components including the template and Rn^- is the Rn value of an unreacted sample obtained from the early cycles of the PCR, i.e. those prior to a detectable increase in fluorescence, or from a reaction not containing template. The ΔRn is the difference between the Rn at the end point (Rn^+) and at the starting point (Rn^-). It reliably indicates the magnitude of the signal generated by the given set of PCR conditions, and is proportional to the DNA amount during the exponential phase. The fluorescence emitted in the first cycles is used to calculate the baseline. A threshold is established at the fluorescence value of the average standard deviation of Rn for the baseline cycles, multiplied by an adjustable factor (usually 10 times). Alternatively, it can be established by the operator in order to compare different qPCR experiments. The C_P (crossing point, PCR cycle) or C_T (threshold cycle) value is the cycle at which fluorescence achieves a defined threshold. It corresponds to the cycle at which a statistically significant increase in fluorescence (ΔRn or non-normalized) is first detected. This concept is the basics for accurate and reproducible quantification using qPCR; the number of cycles needed for the amplification-associated fluorescence to reach a specific threshold level of detection (the C_T or C_P value) is inversely correlated to the amount of nucleic acid that was in the original sample (Walker, 2002). This value always is in the exponential phase of amplification, when amplification is most efficient, and therefore quantification is least affected by reaction-limiting conditions.

The quantity of DNA at the start of the PCR can then be determined by interpolation of the resulting C_T or C_p value in a linear standard curve of values obtained from serially diluted known-amount standards (Fig. 1.12). This standard curve correlates the emitted fluorescence (C_T or C_P value) with the initial concentration of the standards used and the final result is achieved by interpolation of the produced fluorescence (C_T or C_P value) during the amplification of the sample in this standard curve. In practice, such curves are linear over more than five orders of magnitude.

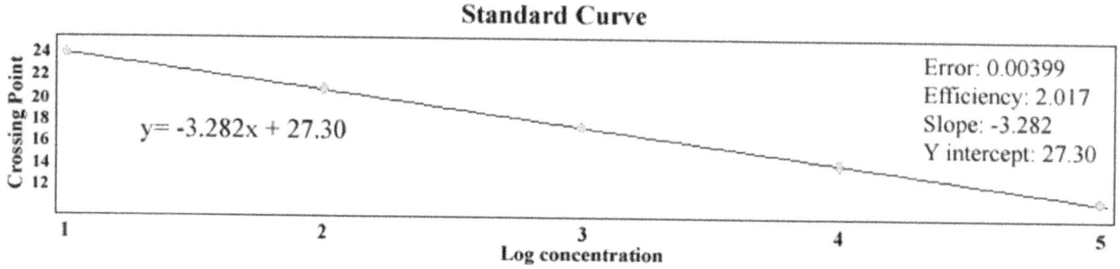

Figure 1.12 Standard curve built with DNA standards of known concentration (Fig. 1.2). *x*-axis, initial amount of DNA (number of copies, ng, etc.); *y*-axis, C_T values.

Diagnostic real-time PCR parameters

An international standard guideline (ISO 16140:2003) was prepared by CEN/TC275 in collaboration with the Technical Committee ISO/TC34 to establish the general principle and technical procedure for the validation of alternative methods in the field of microbiological analysis of food, animal feeding stuffs, and environmental and veterinary samples. Many of the performance criteria are well recognized and accepted by numerous internationally recognized organizations for standardization (ISO, EMEA, Codex Alimentarius, etc.). There are a series of concepts that define an analytical method and that are applied to diagnostic qPCR. The main concepts are: linearity, accuracy, precision, sensitivity and specificity.

Linearity can be defined as the ability of the method, when used with a given matrix, to give results that are in proportion to the amount of analyte present in the sample; that is, an increase in analyte corresponds to a proportional increase in the result (Anonymous, 2003). In the same manner, linearity has been defined by Paoletti and Weighardt (2002) as the proportionality of the signal to the amount of reference material, demonstrated by the calculation of a regression line with an adequate statistical method. The European Commission, in the particular case of genetically modified organisms (GMO), has established that the linearity within the range of applicability of the method should be not less than 98% (Paoletti and Weighardt, 2002).

The accuracy describes the veracity of the test results (Skog and Leary, 1992), and can be defined as closeness of agreement between a test result and the accepted reference value (Anonymous, 1993; Paoletti and Weighardt, 2002; Hoorfar and Cook, 2003). The precision describes the reproducibility of the test results (Skog and Leary, 1992), and can be defined as the closeness of agreement between independent test results obtained under stipulated conditions of repeatability and reproducibility (Anonymous, 1993; Thompson *et al.*, 2002).

Sensitivity is the ability of the alternative method to distinguish two different amounts of analyte (as measured by the reference method) within a given matrix, at a specified average value, or over the whole measurement range; that is, the minimal quantity variation which gives a significant variation of the measurement signal (Anonymous, 2003). The sensitivity differs from detection limit because it is calculated for each value of the measurement range. From a qualitative approach (i.e. for qualitative methods), the sensitivity can also be defined as the ability of the alternative method to detect the analyte when it is detected by the reference method (Anonymous, 2003; Hoorfar and Cook, 2003). The limit of detection is the smallest critical level detected with a probability $(1 - \beta)$, which has to be well over 50%, for example 95% (Anonymous, 2003) or 99% (Hoorfar and Cook, 2003; Knutsson *et al.*, 2002). The limit of detection has been also described as the smallest amount or concentration of analyte in the test that can be distinguished from zero (Anonymous, 1997; IUPAC, 1995; Thompson *et al.*, 2002); the smallest number of culturable target microorganisms necessary to create a positive test response (Hoorfar and Cook, 2003), or the minimum level at which the analyte can reliably be detected with a probability of 95%

(Paoletti and Weighardt, 2002). The limit of quantification is the smallest amount of analyte which can be measured and quantified with defined precision and accuracy by the method under validation (Anonymous, 2003). The Association of Analytical Communities (AOAC) defines the quantification limit for quantitative methods as: $LOQ = 10SD_0$ (Coleman *et al.*, 1997).

The specificity is the degree to which a method is affected by the other components present in a multi-component sample (Anonymous, 2003; Malorny *et al.*, 2003b). In other words, the ability of a method to measure a given analyte within the sample without interference from non-target components or background noise (Anonymous, 2003). For food microbiology diagnostics other three related concepts have been defined: selectivity, inclusivity and exclusivity. Selectivity is defined as a measure of the degree of non-interference in the presence of non-target analytes (Anonymous, 2003). A method is selective if it can be used to detect the analyte and that a guarantee can be provided that the detected signal can only be a product by the specific analyte. Inclusivity is defined as the ability of an alternative method to detect the target analyte from a wide range of strains, and exclusivity as the lack of interference from a relevant range of non-target strains of the alternative method.

Quality control of the real-time PCR

The qPCR is an instrumental technique, and thus can produce false negative and false positive results. Contamination remains an issue for diagnostic laboratories (Pellet *et al.*, 1999; Scherczinger *et al.*, 1999). The main causes of production of false positive results are the accidental contamination of the samples or the reagents with positive samples (cross-contamination) or with amplification products and plasmid clones (carry-over contamination). To minimize these risks, several practices should be applied (Dieffenbach *et al.*, 1995; Hernández, 2003; Mifflin, 2003). For example, it is advisable to establish space and time separation of pre- and post-amplification activities (Heinrich, 1991; Yap *et al.*, 1994), in particular to separate pre-PCR activities from recombinant DNA activities, as well as to define two different pre-PCR areas: for the preparation of the master mix and for the addition the nucleic acid template to the reaction mixture. Each working area must have its own separate set of equipment, including pipettes, reagents, pipette tips, racks, centrifuges, vortex, lab coats, etc., and lab coats and gloves should be worn at all times, and gloves should be changed frequently. In addition, all reagents should be prepared in large volumes and aliquoted into single-use volumes for storage and free from contaminating nucleic acids and/or nucleases. Other actions to prevent or reduce the risk of PCR contaminations are the use of UV light before carrying out the experiments in the sample preparation lab and in the cabinet where the amplification mixture is prepared, and samples should be manipulated with specialized barrier or positive-displacement pipettes, which prevent the carry-over of aerosols created during pipetting. Finally, the use of uracil-DNA-glycosylase (UDG; also called uracil-*N*-glycosylase or UNG) is recommended, as it can eliminate the contamination arising from PCR products (Longo *et al.*, 1990; Thornton *et al.*, 1992). During amplification, dTTP is substituted with dUTP, so the DNA product will contain dU instead of dT. Subsequent PCR runs are programmed to include a 50°C incubation step with UDG, which will mediate cleavage of any contaminating DNA strands containing dU.

In addition to the prevention measures for avoiding contamination, a suite of controls must be included in order to correctly interpret the results of a qPCR (see Chapters 2 and 4).

Application of real-time PCR for the detection of food-borne pathogens

Bacterial detection and control are two key aspects in food microbiology (Stewart, 1997). Microbiological quality control programmes are increasingly applied throughout food chain production in order to minimize the risk of infection for the consumer. Thus, the availability of reliable, rapid and accepted test systems to detect the presence or absence, or even the degree of contamination of pathogens, becomes increasingly important for the agricultural and food industry. Such systems would also find a place within the framework of legislative control measures. During the last 15 years, PCR has significantly contributed to this field.

Traditional and standardized analysis of food for the presence of microorganisms relies on the enrichment and isolation of presumptive colonies of bacteria on agar media, using approved diagnostic artificial media. This is generally followed by biochemical and/or serological identification. Traditional methods of detection, although reliable and efficient, require several days to weeks before results are produced. Furthermore, phenotypic properties by which the bacteria are identified may not be always expressed; and when expressed, they may be difficult to interpret and classify. Another disadvantage of traditional methods is that cells which are viable but otherwise non-culturable cannot be detected, e.g. some stressed *Campylobacter* spp. (Rollins and Colwell, 1986; Tholozan *et al.*, 1999). Thus, introduction of amplification techniques in microbial diagnostics has been established in research laboratories as a valuable alternative to traditional detection methods. The most extensively used amplification method is PCR, but other methods have also been developed such as NASBA, the Q-beta replicase amplification system and the ligase chain reaction (LCR), although until now, they have had limited practical relevance for food monitoring and control. Speed, excellent detection limit, selectivity, specificity, sensitivity and potential for automation are among the most important advantages of qPCR. These advantages compared to traditional detection methods might well encourage end-users to adopt amplification techniques in routine testing for food-borne pathogens.

References

Afonina, I., Zivarts, M., Kutyavin, I., Lukhtanov, E., Gamper, H., and Meyer, R.B. (1997). Efficient priming of PCR with short oligonucleotides conjugated to a minor groove binder. Nucleic Acids Res. *25*, 2657–2660.

Allmann, M., Candrian, U., Hofelein, C., and Luthy, J. (1993). Polymerase chain reaction (PCR), a possible alternative to immunochemical methods assuring safety and quality of food. Detection of wheat contamination in non-wheat food products. Z. Lebensm. Unters. Forsch. *196*, 248–251.

Allmann, M., Hofelein, C., Koppel, E., Luthy, J., Meyer, R., Niederhauser, C., Wegmuller. B., and Candrian, U. (1995). Polymerase chain reaction (PCR) for detection of pathogenic microorganisms in bacteriological monitoring of dairy products. Res. Microbiol. *146*, 85–97.

Anonymous (1993). ISO 3534-1,1993 Statistics. Vocabulary and symbols. Part 1, Probability and general statistical terms. International Organization for Standardization, Geneva, Switzerland.

Anonymous (1997). ISO 11843-1,1997 Capacity of detection. International Organization for Standardization, Geneva, Switzerland.

Anonymous (2003). ISO 16140,2003, Microbiology of food and animal feeding stuffs – Protocol for the validation of alternative methods. International Organization for Standardization, Geneva, Switzerland.

Atlas, R.M., and Bej, A.K. (1994). Polymerase Chain Reaction. In, Gerhardt, P., Murrey, R.G.E., Wood, W.A., and Krieg, N.R. (eds.). Method for general and molecular bacteriology. (Washington. D.C., USA: American Society for Microbiology), pp. 418–435.

Bengtsson, M., Karlsson, H.J., Westman, G., and Kubista, M. (2003). A new minor groove binding asymmetric cyanine reporter dye for real-time PCR. Nucleic Acids Res. *31*, e45.

Bernard, P.S., and Wittwer, C.T. (2000). Homogeneous amplification and variant detection by fluorescent hybridization probes. Clin. Chem. *46*, 147–148.

Bernard, P.S., Ajioka, R.S., Kushner, J.P., and Wittwer, C.T. (1998). Homogeneous multiplex genotyping of hemochromatosis mutations with fluorescent hybridization probes. Am. J. Pathol. *153*, 1055–1061.

Bonnet, G., Tyagi, S., Libchaber, A., and Kramer, F.R. (1999). Thermodynamic basis of the enhanced specificity of structured DNA probes. Proc. Natl. Acad. Sci. USA. *96*, 6171–6176.

Candrian, U. (1995). Polymerase Chain Reaction in food microbiology. J. Microbiol. Methods *23*, 89–103.

Cardullo, R.A., Agrawal, S., Flores, C., Zamecnik, P.C., and Wolf, D.E. (1988). Detection of nucleic acid hybridization by nonradiative fluorescence resonance energy transfer. Proc. Natl. Acad. Sci. USA 85, 8790–8794.

Cha, R.S., and Tilly, W.G. (1993).. Specificity, efficiency and fidelity of PCR. PCR Methods Appl. *3*, 518–529.

Clegg, R.M. (1992). Fluorescence resonance energy transfer and nucleic acids. Methods Enzymol. *211*, 353–388.

Dieffenbach, C.W., Lowe, T.M.J., and Dveksler, G.S. (1995). General concepts for PCR primers design. In, Dieffenbach, C.W., and Dveksler, G.S. (eds.) PCR primers, a laboratory manual. (Cold Springer Harbour, USA: Cold Springer Harbour Laboratory Press), pp. 133–142.

Dupont, M., Goldsborough, A., Levayer, T., Savare, J., Rey, J.M., Rossi, J.F., Demaille, J., and Lavabre-Bertrand, T. (2002). Multiplex fluorescent RT-PCR to quantify leukemic fusion transcripts. Biotechniques *33*, 158–164.

Egholm, M., Buchardt, O., Christensen, L., Behrens, K., Freier, S.M., Driver, D.A., Berg, R.H., Kim, S.K., Nordén, B., and Nielsen, P.E. (1993). PNA hybridizes to complementary oligonucleotides obeying the Watson-Crick hydrogen-bonding rules. Nature *365*, 566–568.

Elnifro, E.M., Ashshi, A.M., Cooper, R.J., and Klapper, P.E. (2000). Multiplex PCR, optimization and application in diagnostic virology. Clin. Microbiol. Rev. *13*, 559–570.

Förster, T. (1948). Zwischen molekulare energiewanderung und fluoreszenz. Ann. Phys. *6*, 55–75.

Gibson, U.E., Heid, C.A., and Williams, P.M. (1996). A novel method for real time quantitative RT-PCR. Genome Res. *6*, 995–1001.

Giesendorf, B.A., Vet, J.A., Tyagi, S., Mensink, E.J., Trijbels, F.J., and Blom, H.J. (1998). Molecular beacons, a new approach for semi-automated mutation analysis. Clin Chem. *44*, 482–486.

Giulietti, A., Overbergh, L., Valckx, D., Decallonne, B., Bouillon, R., and Mathieu, C. (2001). An overview of real-time quantitative PCR, applications to quantify cytokine gene expression. Methods. *25*, 386–401.

Gut, M., Leutenegger, C.M., Huder, J.B., Pedersen, N.C., and Lutz, H. (1999). One-tube fluorogenic reverse transcription-polymerase chain reaction for the quantitation of feline coronaviruses. J. Virol. Methods. *77*, 37–46.

Heid, C.A., Stevens, J., Livak, K.J., and Williams, P.M. (1996). Real time quantitative PCR. Genome Res. *6*, 986–994.

Heinrich, M. (1991). PCR carry-over. Biotech Forum Europe. Int. J. Biotechnol. Heft. *10*, 594–597.

Hernández, M. (2003). Detección, identificación y cuantificación de organismos modificados genéticamente (OMG) en alimentos mediante PCR a tiempo real. Tesis Doctoral. Universidad de León, Spain.

Higuchi, R., Dollinger, G., Walsh, P.S., and Griffith, R. (1992). Simultaneous amplification and detection of specific DNA sequences. Biotechnology (N.Y.) *10*, 413–417.

Higuchi, R., Fockler, C., Dollinger, G., and Watson, R. (1993). Kinetic PCR analysis, real-time monitoring of DNA amplification reactions. Biotechnology (N.Y.) *11*, 1026–1030.

Holland, P.M., Abramson, R.D., Watson, R., and Gelfand, D.H. (1991). Detection of specific polymerase chain reaction product by utilizing the 5′–3′ exonuclease activity of *Thermus aquaticus* DNA polymerase. Proc. Natl. Acad. Sci. USA *88*, 7276–7280.

Hübner, P., Burgener, M., and Lüthy, J. (1997).. Application of molecular biology for the identification of fish. In, Amadò, R., and Battaglia, R. (eds.) Authenticity and Adulteration of food, the analytical approach. Proceedings of European Food Chem. Congress IX, 24–26 September, Interlaken, Switzerland, pp. 49–54.

Innis, M.A., Gelfand, D.H., and Sninsky, J.J. (1999). PCR Applications, protocols for functional genomics. (San Diego, USA: Academic Press).

Isacsson, J., Cao, H., Ohlsson, L., Nordgren, S., Svanvik, N., Westman, G., Kubista, M., Sjoback, R., and Sehlstedt, U. (2000). Rapid and specific detection of PCR products using light-up probes. Mol. Cell. Probes *14*, 321–328.

Ishiguro, T., Saitoh, J., Yawata, H., Yamagishi, H., Iwasaki, S., and Mitoma, Y. (1995). Homogeneous quantitative assay of hepatitis C virus RNA by polymerase chain reaction in the presence of a fluorescent intercalater. Anal. Biochem. *229*, 207–213.

IUPAC (1995). Nomenclature in evaluation of analytical methods, including quantification and detection capabilities. IUPAC Recommendation. Pure Appl. Chem. *67*, 1699–1723.

Iyer, M., Norton, J.C., and Corey, D.R. (1995). Accelerated hybridization of oligonucleotides to duplex DNA. J. Biol. Chem. *270*, 14712–14717.

Jankiewicz, A., Hübner, P., Bögl, K.W., Dehne, L.I., Vieths, S., Baltes, W., and Lüthy, J. (1997). Celery allergy, PCR as a tool for the detection of trace amounts of celery in processed foods. In, Amadò, R., and Battaglia, R. (eds.) Authenticity and Adulteration of food, the analytical approach. Proceedings of European Food Chem. Congress IX, 24–26September, Interlaken, Switzerland, pp. 131–136.

Jeffreys, A.J., Wilson, V., Neumann, R., and Keyte, J. (1988). Amplification of human minisatellites by the polymerase chain reaction, towards DNA fingerprinting of single cells. Nucleic Acids Res. *16*, 10953–10971.

Jin, X., Yue, S., Wells, K.S., and Singer, V.L. (1994). SYBR Greeen I: a new fluorescent dye optimized for detection of picogram amounts of DNA in gels. Biophys. J. *66*, A159.

Kaboev, O.K., Luchkina, L.A., Tretiakov, A.N., and Bahrmand, A.R. (2000). PCR hot start using primers with the structure of molecular beacons (hairpin-like structure). Nucleic Acids Res. *28*, E94.

Knutsson, R., Löfström, C., Grage, H., Hoorfar, J., and Rådström, P. (2002). Modeling of 5′ nuclease real-time responses for optimization of a high-throughput enrichment PCR procedure for *Salmonella enterica*. J. Clin. Microbiol. *40*, 50–62.

Kostrikis, L.G., Tyagi, S., Mhlanga, M.M., Ho, D.D., and Kramer, F.R. (1998). Spectral genotyping of human alleles. Science *279*, 1228–1229.

Kutyavin, I.V., Afonina, I.A., Mills, A., Gorn, V.V., Lukhtanov, E.A., Belousov, E.S., Singer, M.J., Walburger, D.K., Lokhov, S.G., Gall, A.A., *et al.* (2000). 3′-minor groove binder-DNA probes increase sequence specificity at PCR extension temperatures. Nucleic Acids Res. *28*, 655–661.

Lay, M.J., and Wittwer, C.T. (1997). Real-time fluorescence genotyping of factor V Leiden during rapid-cycle PCR. Clin Chem. *43*, 2262–2267.

Le Pecq, J.B., and Paoletti, C. (1966). A new fluorometric method for RNA and DNA determination. Anal. Biochem. *17*, 100–107.

Longo, M.C., Berninger, M.S., and Hartley, J.L. (1990). Use of uracil DNA glycosylase to control carry-over contamination in polymerase chain reactions. Gene 93,125–128.

Lyamichev, V., Brow, M.A., and Dahlberg, J.E. (1993). Structure-specific endonucleolytic cleavage of nucleic acids by eubacterial DNA polymerases. Science *260*, 778–783.

Lyon, E. (2001). Mutation detection using fluorescent hybridization probes and melting curve analysis. Exp. Rev. Mol. Diagn. *1*, 92–101.

Malorny, B., Tassios, P.T., Rådström, P., Cook, N., Wagner, M., and Hoorfar, J. (2003). Standardization of diagnostic PCR for the detection of food-borne pathogens. Int. J. Food Microbiol. *83*, 39–48.

Miffin, T.E. (2003). Setting up a PCR laboratory. In, Dieffenbach, C.W., and Dveksler, G.S. (eds.) PCR Primers, A Laboratory Manual. (Cold Springer Harbour, USA: Cold Springer Harbour Laboratory Press), pp. 5–14.

Morrison, T.B., Weis, J.J., and Wittwer, C.T. (1998). Quantification of low-copy transcripts by continuous SYBR Green I monitoring during amplification. Biotechniques *24*, 954–962.

Mullis, K., Faloona, F., Scharf, S., Saiki, R., Horn, G., and Erlich, H. (1986). Specific enzymatic amplification of DNA *in vitro*: the polymerase chain reaction. Cold Spring Harbor Symp. Quant. Biol. *51*, 263–273.

Nazarenko, I., Bhatnagar, S.K., and Hohman, R.J. (1997). A closed tube format for amplification and detection of DNA based on energy transfer. Nucleic Acids Res. *25*, 2516–2521.

Nielsen, P.E., Egholm, M., Berg, R.H., and Buchardt, O. (1991). Sequence-selective recognition of DNA by strand displacement with a thymine-substituted polyamide. Science *254*, 1497–1500.

Olsen, J.E., Aabo, S., Hill, W., Notermans, S., Wernars, K., Granum, P.E., Popovic, T., Rasmussen, H.N., and Olsvik, Ø. (1995). Probes and polymerase chain reaction for detection of food-borne bacterial pathogens. Int. J. Food Microbiol. *28*, 1–78.

Paoletti, C., and Wighardt, F. (2002). Definition of pre-validation performance requirements. 4th meeting of the European Network of GMO Laboratories (ENGL), 29–30 April, 2002. (Ispra, Italy: EU Joint Research Centre).

Pellett, P.E., Spira, T.J., Bagasra, O., Boshoff, C., Corey, L., de Lellis, L., Huang, M.L., Lin, J.C., Matthews, S., Monini, P., *et al.* (1999). Multicenter comparison of PCR assays for detection of human herpesvirus 8 DNA in semen. J. Clin. Microbiol. *37*, 1298–1301.

Ririe, K.M., Rasmussen, R.P., and Wittwer, C.T. (1997). Product differentiation by analysis of DNA melting curves during the polymerase chain reaction. Anal. Biochem. *245*, 154–160.

Rodríguez-Lázaro, D., Lombard, B., Smith, H., Rzezutka, A., D'Agostino, M., Helmuth, R., Schroeter, A., Malorny, B., Miko, A., Guerra, B., *et al.* (2007). Trends in analytical methodology in food safety and quality: monitoring microorganisms and genetically modified organisms. trends food sci. Technol. *18*, 306–319.

Rollins, D.M., and Colwell, R.R. (1986). Viable but non-culturable stage of *Campylobacter jejuni* and its role in survival in the natural aquatic environment. Appl. Environ. Microbiol. *52*, 531– 538.

Saiki, R.K., Scharf, S., Faloona, F., Mullis, K.B., Horn, G.T., Erlich, H.A., and Arnheim, N. (1985). Enzymatic amplification of beta-globin genomic sequences and restriction site analysis for diagnosis of sickle cell anemia. Science *230*, 1350–1354.

Saiki, R.K., Gelfand, D.H., Stoffel, S., Scharf, S.J., Higuchi, R., Horn, G.T., Mullis, K.B., and Erlich, H.A. (1988).. Primer-directed enzymatic amplification of DNA with a thermostable DNA polymerase. Science *239*, 487–491.

Scherczinger, C.A., Ladd, C., Bourke, M.T., Adamowicz, M.S., Johannes, P.M., Scherczinger, R., Beesley, T., and Lee, H.C. (1999). A systematic analysis of PCR contamination. J. Forensic. Sci. *44*, 1042–1045.

Schmittgen, T.D., Zakrajsek, B.A., Mills, A.G., Gorn, V., Singer, M.J., and Reed, M.W. (2000). Quantitative reverse transcription-polymerase chain reaction to study mRNA decay, comparison of endpoint and real-time methods. Anal. Biochem. *285*, 194–204.

Simpson, D.A.C., Feeney, S., Boyle, C., and Stitt, A.W. (2000). Retinal VEGF mRNA measured by SYBR green I fluorescence, A versatile approach to quantitative PCR. Mol. Vis. *6*, 178–183.

Singer, V.L., Jin, X., Ryan, D., and Yue, S. (1994). Biomed. Products *19*, 72–86.

Skoog, D.A., and Leary, J.J. (1992). Principles of Instrumental Analysis. (London, UK, Saunders College Publishing).

Stewart, D., and Gendel, S.M. (1998). Specificity of the BAX polymerase chain reaction system for detection of the food-borne pathogen *Listeria monocytogenes*. J. AOAC Int. *81*, 817–822.

Stryer, L. (1978). Fluorescence energy transfer as a spectroscopic ruler. Annu. Rev. Biochem. *47*, 819–846.

Svanvik, N., Stahlberg, A., Sehlstedt, U., Sjoback, R., and Kubista, M. (2000). Detection of PCR products in real time using light-up probes. Anal. Biochem. *287*,179–182.

Tan, W., Fang, X., Li, J., and Liu, X. (2000). Molecular beacons, a novel DNA probe for nucleic acid and protein studies. Chemistry *6*, 1107–1111.

Tholozan, J.L., Cappelier, J.M., Tissier, J.P., Delattre, G., and Federighi, M. (1999). Physiological characterization of viable-but-nonculturable *Campylobacter jejuni* cells. Appl. Environ. Microbiol. *65*, 1110–1116.

Thompson, M., Ellison, S.L.R., and Wood, R. (2002). Harmonized guidelines for single-laboratory validation of methods of analysis (IUAPC Technical Report). Pure Appl. Chem. *74*, 835–855.

Thornton, C.G., Hartley, J.L., and Rashtchian, A. (1992). utilizing uracil DNA glycosylase to control carry-over contamination in PCR, characterization of residual UDG activity following thermal cycling. BioTechniques *13*, 180–183.

Tombline, G., Bellizzi, D., and Sgaramella, V. (1996). Heterogeneity of primer extension products in asymmetric PCR is due both to cleavage by a structure-specific exo/endonuclease activity of DNA polymerases and to premature stops. Proc. Natl. Acad. Sci. U.S.A. 93, 2724–2728.

Tseng, S.Y., Macool, D., Elliott, V., Tice, G., Jackson, R., Barbour, M., and Amorese, D. (1997). An homogeneous fluorescence polymerase chain reaction assay to identify *Salmonella*. Anal. Biochem. *245*, 207–212.

Tyagi, S., and Kramer, F.R. (1996). Molecular Beacons, probes that fluoresce upon hybridization. Nat. Biotechnol. *14*, 303–308.

Tyagi, S., Bratu, D.P., and Kramer, F.R. (1998). Multicolour molecular beacons for allele discrimination. Nature Biotechnol. *16*, 49–53.

Walker, N. (2002). A technique whose time has come. Science *296*, 557–559.

Whitcombe, D., Brownie, J., Gillard, H.L., McKechnie, D., Theaker, J., Newton, C.R., and Little, S. (1998). A homogeneous fluorescence assay for PCR amplicons, its application to real-time, single-tube genotyping. Clin. Chem. *44*, 918–923.

Whitcombe, D., Theaker, J., Guy, S.P., Brown, T., and Little, S. (1999). Detection of PCR products using self-probing amplicons and fluorescence. Nat. Biotechnol. *17*, 804–807.

Wilhelm, J., and Pingoud, A. (2003). Real-time polymerase chain reaction. Chembiochemistry *4*, 1120–1128.

Wittwer, C.T., Herrmann, M.G., Moss, A.A., and Rasmussen, R.P. (1997). Continuous fluorescence monitoring of rapid cycle DNA amplification. Biotechniques *22*, 130–138.

Wittwer, C.T., Herrmann, M.G., Gundry, C.N., and Elenitoba-Johnson, K.S. (2001). Real-time multiplex PCR assays. Methods *4*, 430–442.

Yap, E.P.H., Lo, Y., Y.M.O., Fleming, K.A., and McGee, J.O.D. (1994). False-positives and contamination in PCR. En H.G. Griffin, A.M. Griffin (eds.). PCR Technology, current innovations, (Boca Raton, USA, CRC Press).

Zimmermann, A., Lüthy, J., and Pauli, U. (1998). Quantitative and qualitative evaluation of nine different extraction methods for nucleic acids on soya bean food samples. Z. Lebensm. Unters. Forsch. A. *207*, 81–90.

DOI: https://doi.org/10.21775/9781912530243.02

Principles of the Real-time Polymerase Chain Reaction

2

Stephen A. Bustin, Sara Zaccara and Tania Nolan

Abstract

The real-time, fluorescence-based, quantitative polymerase chain reaction (qPCR) has become the benchmark technology for the detection of nucleic acids in every area of microbiology, biomedical research, biotechnology and in forensic applications. Unlike conventional ('legacy') PCR, which is a qualitative end-point assay, qPCR allows accurate quantification of amplified DNA in real time, during the exponential phase of the reaction. The cost of instruments and reagents is well within reach of individual laboratories, assays are easy to perform, capable of high throughput and combine high sensitivity with reliable specificity. It is possible to achieve accurate and biologically meaningful quantification if meticulous attention is paid to the details of every step of the qPCR assay, starting with sample selection, acquisition and handling through to assay design, validation and optimization. The growing awareness of the need for standardization, quality control and the significant problems associated with inadequate reporting of the assay has resulted in the publication of guidelines for minimum information for the publication of qPCR experiments (MIQE; Bustin *et al.*, 2009).

Introduction

The capacity of the polymerase chain reaction (PCR) to amplify many million-fold any known DNA fragment from complex genetic material, in a short time (Saiki *et al.*, 1985), has revolutionized all areas of the life sciences, making it the most widely used molecular technology today. However, conventional PCR has several disadvantages; it is labour-intensive, not easily automated or adapted for high-throughput applications, and quantification of nucleic acids is arduous. This is because amplification typically reaches a yield plateau where the PCR product is no longer being doubled at each cycle owing to reagent constraints and end product inhibition. That plateau will differ for each assay owing to the different reaction kinetics for each sample, but since conventional PCR is an endpoint assay, samples are typically analysed after reaching this linear phase of the reaction. Consequently, post-PCR gel electrophoresis analysis shows a broadly similar amount of product DNA, independent of the initial amount of template. This makes quantification by legacy PCR difficult since it requires the establishment of additional empirical quantification parameters, e.g. competitive PCR, that vary with each assay and are tedious to reproduce.

The introduction of novel reagents, detection chemistries and instrumentation platforms has modified legacy PCR to create a technology, real-time, fluorescence-based quantitative PCR (qPCR) that addresses these issues. The use of qPCR allows monitoring of the progress of PCR in real time using fluorescent reporter molecules. In this way parallel and immediate amplification, detection and quantification of PCR products is possible, firmly establishing qPCR as the method of choice for the detection of nucleic acids (Lu *et al.*, 2008; Le Guillou-Guillemette and Lunel-Fabiani, 2009; Lopez *et al.*, 2009; Smith and Osborn, 2009; Zhang *et al.*, 2009). The features that make qPCR so universally popular are:

- Speed: assay reaction times are typically

measured in tens of minutes, but can be even shorter.

- Convenience: it is a homogeneous assay, hence there is no need for post-amplification processing.
- Simplicity: the assay requires target, buffer, two primers, a fluorescent reporter and a single enzyme.
- Sensitivity: single copy targets can be detected, if not quantified.
- Specificity: a well-designed assay is specific for a single target, but mismatch-tolerant assays are easily designed.
- Robustness: a wide range of reaction conditions will yield results.
- High throughput: thousands of assays can be carried out on a single run.
- Quantification: accuracy typically covers a wide dynamic range (8–11 orders of magnitude).
- Familiarity: PCR has been around since 1985 and its advantages and disadvantages are well understood.
- Cost: all assay reagents are inexpensive; together with the trend towards smaller reaction volumes, the costs per assay can be very low.

Three essential innovations underlie this technology:

1. A procedure had to be established that allowed the concurrent amplification and detection of specific DNA sequences. The original arrangement involved the addition of a DNA intercalator, ethidium bromide (EtBr) to the PCR (Higuchi *et al.*, 1992) together with the use of a thermal cycler that has an ultraviolet light source and a computer-controlled CCD camera (Higuchi *et al.*, 1993). A plot of increased fluorescence – caused by EtBr binding to newly synthesized DNA – against cycle number allowed the calculation of the quantity of starting material. The drawback of this approach was that this measurement could not easily distinguish between specific and non-specific products.
2. The second development contributed that specificity by adding a separate oligonucleotide probe, labelled with ^{32}P at the 5′ end and designed to hybridize within the target amplicon during the amplification. Successful amplification of the target sequence generated a substrate suitable for the double-strand-specific 5′–3′ exonuclease activity of Taq DNA polymerase (Holland *et al.*, 1991) and resulted in cleavage of the probe. However, measurement of the ^{32}P-labelled cleavage products was neither simple nor practical, requiring their separation by thin layer chromatography.
3. The invention of fluorogenic probes (Lee *et al.*, 1993) made it possible to combine the above innovations in a fluorescence-based hydrolysis probe assay, also known as 5′-nuclease or 'TaqMan' assay, the first practical qPCR method (Heid *et al.*, 1996).

Taken together, these features make qPCR the molecular-enabling technology *par excellence* (Bustin, 2008c) and have transformed the approach to PCR-based quantification of nucleic acids (Bustin, 2004) and, excitingly with recent modifications, for proteins (Reuter *et al.*, 2009).

Principles and basics

Assay characteristics

The output of a qPCR is the 'amplification plot', which is used to quantify DNA targets and provides key information with regards to the quality of the qPCR. The amplification plot is characterized by a series of phases; the background noise of the detection system prior to sufficient amplification signal is recorded as the baseline, this is followed by a period of exponential amplification where the detected signal is correlated to increase in DNA concentration until the plateau phase is reached (Fig. 1.1). All qPCR assays are characterized by the quantification cycle (Cq; Bustin *et al.*, 2009) (also known as the threshold cycle or C_t), the fractional PCR cycle used for quantification. The Cq defines the cycle at which amplification of a PCR product is first detected and is inversely correlated to the logarithm of the initial copy number. There are different ways of determining a Cq, but in general data are collected from the first PCR cycles, where

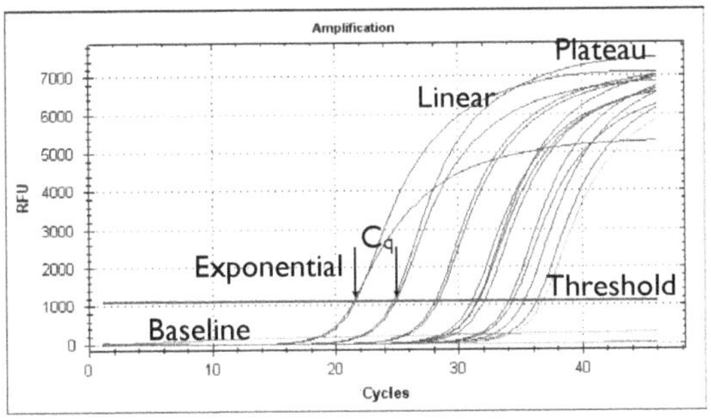

Figure 1.1 qPCR phases. *The baseline region:* the initial cycles of qPCR during which amplification proceeds but there is no change in fluorescence signal as it is below the detection threshold of the instrument. *The exponential phase:* during this phase the doubling of product at every cycle (assuming 100% reaction efficiency) creates exponential signal growth. The quantification cycle, Cq, is used for determining initial target copy numbers. *The linear phase:* reaction components are becoming limited, resulting in reduced amplification reaction efficiency and the signal no longer growing exponentially. *The plateau phase:* this phase, also known as end-point detection, is where traditional PCR products are detected. Reaction components have been exhausted and no more fluorescence is generated. Since each sample is characterized by different reaction kinetics, each reaction plateau at different points. This can be seen from the replicates that have similar Cqs but widely different endpoint fluorescence values.

there is little change in fluorescence signal, and are used to calculate the background fluorescence and the standard deviation. These are subtracted from subsequent results when setting the baseline, resulting in an normalized fluorescence emission intensity. Most systems use a threshold method to define the Cq: this involves identifying the intersection point between the amplification plot and the fractional cycle where fluorescence levels are significantly above the baseline within the exponential region of the amplification curve (Fig. 1.1). The fewer cycles it takes to reach a detectable level of fluorescence, the greater the initial copy number. However, although most analysis software packages have automated threshold settings, the choice of threshold can be made by the operator and introduces a subjective element into real-time quantification. This has led to the development of alternative methods, e.g. sigmoidal curve fitting using non-linear regression (Rutledge and Stewart, 2008), that analyse the curve shape using a variety of algorithms that predict the cycle at which a positive signal occurs. Regardless of method used, the Cq is dependent on the starting template copy number, the efficiency of the PCR amplification and the background fluorescence.

A well-optimized assay will be linear over a range of at least six logs of template concentration, with efficiency close to 100% and high reproducibility between technical replicates (Fig. 1.2A). Importantly, the linearity of the serial dilution of template defines the dynamic range of detection for the assay and only data within these limits are acceptable for experimental samples. Amplification efficiency is best determined by generating a standard curve using serial dilutions of a template and determining the slope from the linear regression of a plot of Cq (y-axis) versus log template quantity (Fig. 1.2B). The template may be any suitable material such as cDNA, genomic DNA, PCR product or a synthetic DNA oligo manufactured for the purpose. Analysis of the standard curve assumes that perfect doubling (or consistent amplification) occurs with each amplification cycle. Thus, the spacing of the fluorescence curves will be determined by the equation 2n = dilution factor, where n is the number of cycles between curves. For example, with a 10-fold serial dilution of DNA, 2n = 10. Therefore, n = 3.32 and the Cq values should be separated by approximately 3.32 cycles.

Amplification efficiency, *E*, is calculated from the slope of the standard curve using the formula $E = 10^{(-1/\text{slope})}$, which is usually converted into a percentage efficiency [% Efficiency = (E − 1) × 100%]. In a perfect scenario where doubling occurs at each cycle, $E = 10^{(-1/-3.32)} = 2$ and the % efficiency = (2 − 1) × 100% = 100%. In practice, amplification efficiency will be around 90–105% or result in a gradient of around −3.2 to −3.5. It is important to note that the presence of inhibitors can result in an apparent increase in efficiency. This is because samples with the highest concentration of template also have the highest level of inhibitors, and therefore display a greater lag between Cq values than samples with lower template concentrations and lower levels of inhibitors. As a result, the absolute value of the slope decreases and the calculated efficiency appears to increase.

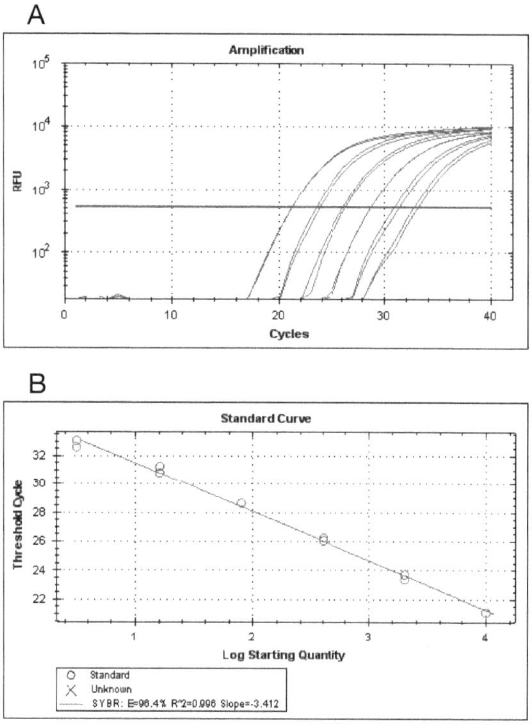

Figure 1.2 (A) Wide dynamic range over several orders of magnitude of template starting concentration. A 10-fold serial dilution of human total RNA/DNA was amplified using iQ supermix (BioRad) and incorporation of SYBR Green I dye. Duplicates of each reaction were monitored (RFU, relative fluorescence units). (B) Standard curve. The closer the slope is to –3.323, the closer the amplification efficiency is to the 100%, showing good reproducibility, with the R^2 being 0.996.

Similarly, inaccurate pipetting can lead to data suggestive of higher or lower efficiency and so it is important to assess the correlation coefficient in order to determine the linearity and reproducibility of the assay.

During assay evaluation and optimization, technical replicates of each template dilution should be processed in parallel in order to establish that the assay is reproducible. A stable assay will demonstrate a correlation coefficient of $R^2 > 0.99$ over at least six logs and with three replicates.

Detection chemistries

A wide range of chemistries utilizing fluorescent reporter molecules has been developed to permit the monitoring of PCR assays in real time. The simplest chemistries use either non-specific fluorescent dyes, e.g. SYBR Green I or EvaGreen®, that bind to any double-stranded DNA or primers labelled with a fluorophore that emit light only when incorporated into amplification products. More complex chemistries make use of template-specific fluorescent probes (e.g. Molecular beacons or hydrolysis probes) using single or dual probes with appropriately positioned fluorescent reporters and quenchers. Although less common, others use only a reporter and make use of the quenching properties of the DNA.

DNA-binding dyes

SYBR Green I dye is the most widely used double-strand DNA-specific reporter. In solution, unbound dye exhibits very little fluorescence; during amplification, fluorescence is dramatically enhanced when the dye is bound to the nascent double-stranded DNA (Morrison *et al.*, 1998) (Fig. 1.3A). When monitored in real-time, this results in an increase in the fluorescence signal that can be observed during the polymerization step, and that decreases when the DNA is denatured. Consequently, fluorescence measurements are performed at the end of the elongation step of every PCR cycle to monitor the increasing amount of amplified DNA. Specificity is dependent entirely on the primers, but the PCR product can be verified by plotting fluorescence as a function of temperature to generate a melting curve of the amplicon (Ririe *et al.*, 1997). This is done by slowly elevating the temperature above the amplicon's melting temperature (T_m) and measuring the fluorescence. Since the T_m of the amplicon depends markedly on its nucleotide composition and length, it is possible to identify the signal obtained from the correct product. A characteristic melting peak at the amplicon's melting temperature (T_m) will distinguish it from amplification artefacts that generally melt at lower temperatures in broader peaks (Fig. 1.3B). A potential drawback is that multiple dye molecules bind to a single amplified molecule and consequently the amount of signal generated, following excitation, is dependent on the mass of double-stranded DNA produced in the reaction. Assuming the same amplification efficiencies, amplification of a longer product will generate

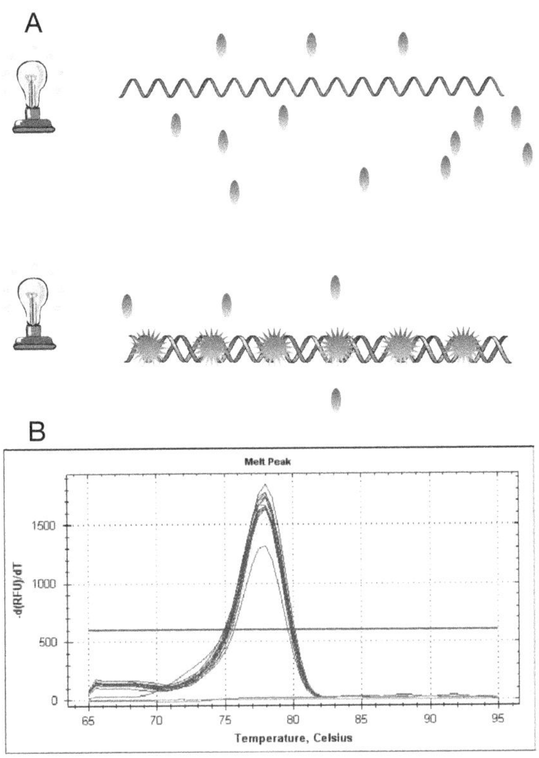

Figure 1.3 (A) The simplest and most economical detection system uses the double-stranded DNA-specific dye SYBR Green I. SYBR Green I dye binds double-stranded DNA and emits light upon excitation. Thus, as a PCR product accumulates, fluorescence increases. The main advantages are that it is simple to use and inexpensive. The disadvantage is that SYBR Green I will bind to any double-stranded DNA in the reaction, including primer-dimers and other non-specific reaction products, which results in an overestimation of the target concentration. (B) Melting curve analysis. The melting temperature of a DNA double helix depends on its base composition and its length. All PCR amplicons for a particular primer pair should have the same melting temperature as shown here; if not, there are likely to be contamination, primer dimers or other problems. If the peaks are not similar, problems with contamination, mispriming or primer-dimer artefacts are indicated.

more signal than a shorter one. If amplification efficiencies are different, quantification can be somewhat inaccurate. This is in contrast to the use of fluorogenic probes, in which the signal from a single fluorophore is observed for each amplified molecule synthesized, regardless of its length.

Probe-based assays

There are a number of chemistries available for probe-based detection of DNA targets; all make use of the same principle: a fluorescent signal is only generated if the amplicon-specific probe hybridizes to its complementary target. Depending on the chemistry used, the increase in fluorescence may be reported at the annealing step when the probe is hybridized (e.g. Molecular beacons) or at the end of the polymerization step when probe hydrolysis has resulted in the physical separation of fluorophore and quencher (e.g. hydrolysis probes).

The hydrolysis probe assay (also known as 5′-nuclease or 'TaqMan') is the most widely used probe-based assay (Gibson *et al.*, 1996) (Fig. 1.4). As its success relies on the 5′-nuclease activity of the DNA polymerase required to hydrolyse the hybridization probe bound to its target amplicon (hence its alternative name 5′-nuclease assay), the correct choice of DNA polymerase is important (Kreuzer *et al.*, 2000). Since the polymerase will cleave the probe only while it remains hybridized to its complementary strand, the temperature conditions of the polymerization phase of the PCR must be adjusted to ensure probe binding. Most probes have a T_m of around 70°C; therefore, the hydrolysis probe method uses a combined annealing and polymerization step at 60–62°C, thus ensuring that the probe remains bound to its target during the primer extension step. It also ensures maximum 5′–3′ exonuclease activity of the DNA polymerases.

Hybridization probe chemistry uses two probes to maximize specificity (Wittwer *et al.*, 1997) (Fig. 1.5). One of the probes carries at its 3′-end a fluorescein donor, which emits green fluorescent light when excited by a light source. Its emission spectrum overlaps the excitation spectrum of an acceptor fluorophore that is attached to the 5′-end of the second probe. This probe must be blocked at its 3′-end to prevent its extension during the annealing step. Excitation of the donor results in fluorescence resonance energy transfer (FRET) to the acceptor and the emission of red fluorescent light. In solution, the two dyes are apart, and since the energy transfer depends on the spacing between the two dye molecules, only background fluorescence is emitted

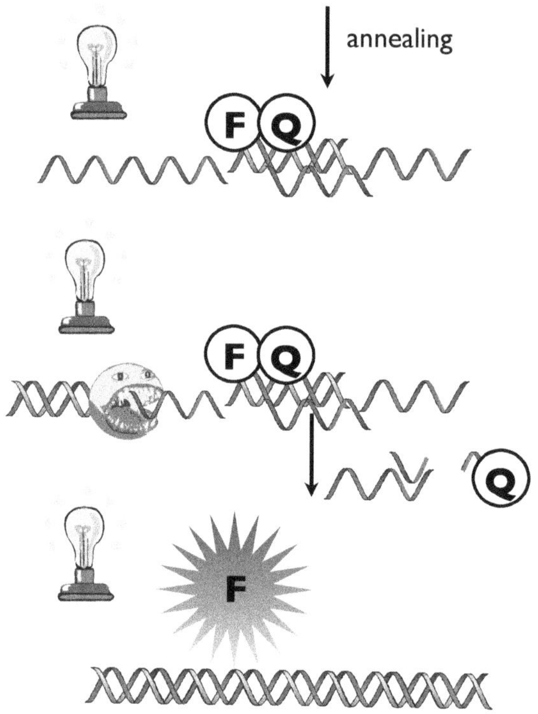

Figure 1.4 Hydrolysis probes. If the specific target sequence is present, the probe anneals downstream from one of the primer sites. During the extension phase of PCR, it is displaced and cleaved by the 5′ nuclease activity of Taq DNA polymerase. This cleavage of the probe separates the reporter dye from quencher dye, resulting in an increased reporter dye signal. Primer extension then continues to the end of the template strand and additional reporter dye molecules are cleaved from their respective probes with each cycle. This generates an increase in fluorescence intensity that is proportional to the amount of PCR amplicon produced.

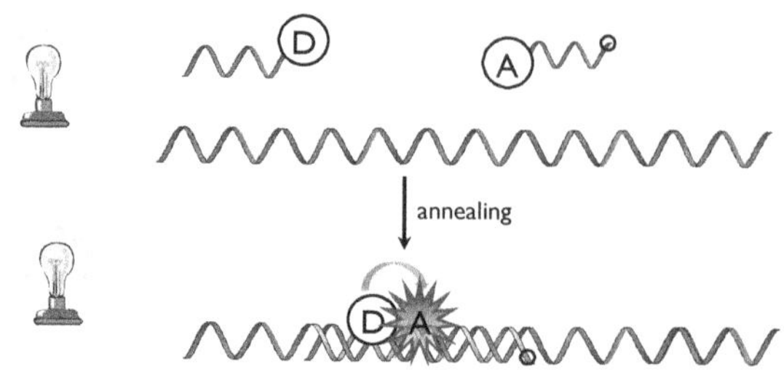

Figure 1.5 Hybridization probes. Hybridization probes are designed as a pair of oligonucleotides with one labelled at its 3′-end with a donor, and the other labelled at its 5′-end with an acceptor fluorphore. As FRET decreases with the sixth power of distance, hybridization probes have to be designed to hybridize to adjacent regions of the template DNA, separated by 1–5 nucleotides. During the denaturation step, PCR template, primers and probes are single-stranded and separate. At the annealing temperature, primers and probes bind to their specific target regions and the donor dye comes into close proximity of the acceptor dye. Energy emitted from the donor dye excites the acceptor dye, which now emits light that is detected and recorded. At the PCR melting temperature, the probes melt off and DNA is ready for the next cycle.

by the donor. Following the denaturation step, both probes hybridize to their target sequence in a head-to-tail arrangement during the annealing step. This brings the two dyes in close proximity to one another and the fluorescein can transfer its energy at high efficiency. The intensity of the light of longer wavelength emitted by the second dye is measured, with increasing amounts of measured fluorescence being proportional to the amount of DNA synthesized during the PCR. A fluorescent signal is detected only as a result of two independent probes hybridizing to their correct target sequence. This increases specificity and generates additional flexibility for probe design. Furthermore, since the probes are not hydrolysed, fluorescence is reversible and allows the generation of melting curves.

Molecular beacons are DNA hybridization probes that form a stem-and-loop structure; the loop portion of the molecule is complementary to the target nucleic acid molecule and the stem is formed by the annealing of complementary arm sequences on the ends of the probe sequence (Tyagi and Kramer, 1996) (Fig. 1.6). A fluorescent marker is attached to the end of one arm and a quencher is attached to the end of the other arm. In solution, free Molecular beacons adopt a hairpin structure and the stem keeps the arms in close proximity, resulting in efficient quenching of the fluorophore. When Molecular beacons encounter a complementary target at the annealing temperature, they undergo a conformational transition that forces the stem apart and results in the formation of a probe/target hybrid that is longer and more stable than the stem (Bonnet *et al.*, 1999). This separates the fluorophore and the quencher, leading to the emission of fluorescence, which can be detected, whereas any free Molecular

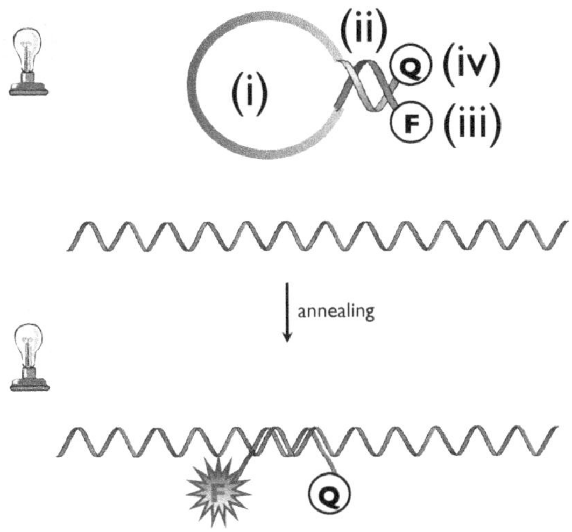

Figure 1.6 Molecular beacons. A Molecular beacon consists of four parts: (i) a loop that is complementary to the target sequence; (ii) a stem that is typically 7 bp long with sequences that are complementary to each other; (iii) a fluorophore that fluoresces in the presence of a complementary target and (iv) a quencher that prevents the fluorophore from emitting light when the Molecular beacon is in a closed loop shape. In the absence of a complementary target sequence, the Molecular beacon remains closed and there is no appreciable fluorescence. When the probe sequence in the loop hybridizes to its target, a conformational reorganization occurs that separates the quencher from the fluorophore, restoring fluorescence which is monitored and reported during each annealing step.

beacons remain closed and non-fluorescent. If the target DNA sequence does not exactly match the sequence of these conventional Molecular beacons, hybridization and fluorescence will not occur, as the thermodynamic properties of the Molecular beacons favour the formation of a hairpin form rather than continued hybridization to a less than perfectly matched target sequence. Conversely, more recently introduced 'sloppy' Molecular beacons possess relatively long probe sequences, enabling them to form hybrids with amplicons despite the presence of mismatched base pairs (El-Hajj *et al.*, 2009). This is useful for screening of families of microorganisms that have some sequence divergence. Furthermore, by using multiple sloppy Molecular beacons with different probe sequences and labelled with different fluorophores, different T_m values can be determined simultaneously, allowing potential screening of numerous microorganisms in a single assay (Chakravorty *et al.*, 2010).

Of the other probe-based systems, Scorpions (Whitcombe *et al.*, 1999) are the most interesting, although their use is fairly specific. Scorpions convert sequence-specific priming and probing into a unimolecular event, making the signalling reaction extremely fast by having the fluorophore attached to the 5′ end of an oligonucleotide that is made up of two separate sequences: (i) a probe sequence complementary to the extension product of the primer, (ii) a target specific primer. Both are linked via a non-amplifiable monomer to prevent copying of the probe sequences during the polymerization step of the PCR. The quencher is linked to the 3′-end of a second oligonucleotide that is fully complementary to the probe sequence. Following extension of the Scorpion primer, and after denaturation and re-annealing, the specific probe sequence is able to hybridize to its complement within the same strand of DNA. Fluorescence is no longer quenched and a fluorescent signal is observed. This unimolecular hybridization is kinetically more favourable than the quencher oligonucleotide reannealing with the Scorpion and allows the introduction of rapid cycling conditions together with a significantly stronger signal strength when compared with both hydrolysis probes and Molecular beacons (Thelwell *et al.*, 2000).

There are several other probe chemistries, all with their own advantages and disadvantages. These include Hybeacons, which require only a single fluorophore and make use of the quenching properties of DNA. This makes them easy to design and synthesize (French *et al.*, 2002). Light-up probes are composed of thiazole orange conjugated to peptide nucleic acid (PNA) and combine the excellent hybridization properties of PNA, which allows the use of shorter probes, with the extraordinary fluorescence enhancement of asymmetrical cyanine dyes upon binding to nucleic acids (Timcheva *et al.*, 1997). Eclipse probes are linear probes that have a minor groove binder (MGB) and quencher on the 5′-end and the fluorophore on the 3′-end (Afonina *et al.*, 2002).

Real-time PCR assays generally use symmetric

primers. However, this results in the reactions typically slowing down and entering the plateau phase in a stochastic manner, because reannealing of the template strands gradually outcompetes primer and probe binding to the template strands. This can be a particular problem when the aim is to detect very low copy number targets, for example in single cell analysis. Asymmetric PCR potentially circumvents the problem of amplicon strand reannealing by using unequal primer concentrations. However, asymmetric amplification is much less efficient and requires extensive optimization to identify the optimal primer ratios, the amounts of starting material, and the number of amplification cycles that can generate reasonable amounts of product for individual template/target combinations. A qPCR variant termed linear-after-the-exponential PCR (LATE-PCR) uses unequal primer concentrations but takes into account the effect of the actual primer concentrations on primer T_m (Sanchez *et al.*, 2004). LATE-PCR features a correction for the fact that the T_m of the limiting primer is often several degrees below the T_m of excess primer and allows the asymmetric PCR to proceed as efficiently as symmetric PCR. Furthermore, single strand amplicons are generated with predictable kinetics for many cycles beyond the exponential phase of the PCR. This permits uncoupling of primer annealing from product detection. As a result, the T_m of the probe no longer needs to be higher than the T_m of either primer. This permits the use of low-T_m probes, which are inherently more allele-discriminating, generate lower background, and can be used at saturating concentrations without interfering with the efficiency of amplification, making them particularly suited for allelic discrimination and the detection of nucleic acids from single cells (Pierce and Wangh, 2011). At the same time, the generation of single-stranded amplification products allows rapid and accurate sequencing of the PCR amplicons (Jia *et al.*, 2010).

Finally, the introduction of DNA analogues such as locked nucleic acids (LNA) (Kumar *et al.*, 1998) has simplified probe design and improved assay quality. Structural studies have shown LNA to be an RNA mimic, fitting seamlessly into an A-type duplex geometry. In LNA, the furanose ring conformation is restricted by a methylene bridge that connects the 2′-oxygen position of ribose to the 4′-carbon. This bridge reduces the conformational flexibility of the ribose, which provides superior affinity and specificity in binding complementary sequences of DNA or RNA. At the same time, this binding affinity is decreased substantially when the hybrids contain even a single mismatched base pair; the change in T_m caused by a mismatch is significantly greater with a LNA/DNA duplex than with a DNA/DNA duplex, resulting in enhanced specificity for SNP/mutation analyses (Johnson *et al.*, 2004). Using appropriate dyes, it is possible to use probes as short as seven nucleotides long for mismatch discrimination, making it possible to generate universal genotyping reagents. It is also possible to substitute DNA with LNA in primers, with the potential to improve primer specificity and performance (Latorra *et al.*, 2003).

qPCR evolution

qPCR technology continues to develop in response to the need for ever-more specific, convenient and cost-effective applications, with several recent improvements refining qPCR assay performance and accuracy as well as simplifying data analysis:

1. The introduction of more reliable and labour-saving, yet less expensive reagents, coupled with the launch of affordable, but more robust qPCR thermal cyclers make reaction assembly simpler and improve flexibility, performance and consistency of qPCR assays. GMP-compliant manufacturing and formulation facilities permit the production of oligonucleotides with rigorous quality control and the documentation required for clinical trials and diagnostic expression analysis.
2. Automated workflows (Grisold and Kessler, 2006; Marshall *et al.*, 2007), more reliable multiplexing assays (Westh *et al.*, 2009; Beck *et al.*, 2010), the launch of instruments using 384- and 1536-well formats and of miniaturized qPCR systems (Morrison *et al.*, 2006), including microfluidic digital PCR (Zhang and Xing, 2007), has resulted in the

expansion of qPCR technology into high throughput applications. Miniaturization offers several potential advantages such as short assay time, low reagent usage and rapid heating/cooling rates, as well as integration of multiple processing modules to further reduce size and power consumption (Neuzil *et al.*, 2006; Marcus *et al.*, 2006). High-density array-based formats and support for faster assay conditions enables shorter assay times, allowing the parallel screening of hundreds of targets at vastly reduced reagent cost where qPCR can provide an alternative to DNA microarrays for accurate expression analysis with high sensitivity (Dixon *et al.*, 2009; Wagner and Arany, 2009) or even be used for routine diagnostic services (Gunson *et al.*, 2008). They also allow powerful multiway profiling, where quantification of reporter genes permits the characterization of samples as functions of multiple factors such as time and genetic make-up (Stahlberg *et al.*, 2008).

3 The increased emphasis on high throughput profiling ties in with the increased understanding for the need for constant and consistent quality qPCR data control. This has resulted in the development of software programs that aim to minimize subjective human input and maximize objective multilevel analysis. The publication of numerous advanced algorithms (Burns *et al.*, 2005; Larionov *et al.*, 2005; Jin *et al.*, 2006; Nordgard *et al.*, 2006; Smith *et al.*, 2007; Batsch *et al.*, 2008; O'Connor and Runquist, 2008; Platts *et al.*, 2008; Rutledge and Stewart, 2008; Spiess *et al.*, 2008; Verderio *et al.*, 2008) and their emergent inclusion in the development of more intuitive analysis software helps with assay set-up (Gallup and Ackermann, 2008), project management and data analysis (Hellemans *et al.*, 2007) to allow robust, accurate and biologically relevant quantification (see below).

Assay considerations

The relative simplicity of a qPCR assay means that it is not difficult to obtain numerous data points from many samples in a short amount of time. The difficulty is deciding whether a Cq reported by the qPCR instrument is real, or whether it is the result of a problem with the assay (Fig. 1.7).

Sample handling and processing

Sample acquisition, handling and processing routines constitute obvious sources of experimental variability and have a significant effect on the power of the assay to deliver biologically relevant qPCR results. Unsurprisingly, the literature is full of different protocols describing methods for sample collection, storage and nucleic acid extraction, with little attempt to standardize any of the procedures. Since such samples are used for ever-increasingly sophisticated analysis, it is important to verify the variability introduced during sample handling. This is particularly important for experiments targeting RNA. While DNA is remarkably stable, RNA molecules are highly unstable and so mRNA profiles may be altered during sample collection, storage or processing. As a general guide, the faster RNA is stabilized the better; therefore sample preparation and speed of extraction are key to successful RNA extraction (Bustin, 2000). Conventionally, successful RNA extraction requires that samples are processed immediately after collection from the biological source (e.g. fresh tissue) or are snap frozen in liquid nitrogen (fresh-frozen). However, there are practical issues surrounding either of these approaches: for example, in a clinical setting, biological material may be left at room temperature for several minutes or hours prior to stabilization and sample storage and transport to the laboratory, which are major concerns.

Assay design

Another critical limitation of qPCR assays is associated with assay design in general and primer design in particular. One obvious question addresses the specificity of the target to be amplified, since a search of any sequence repository for a specific nucleic acid target often reveals a number of variants, making it challenging to deduce which particular microorganism, DNA or RNA has been targeted. Providing an accession number is the most basic, yet frequently omitted piece of information in publications. Common starting points for qPCR target searches are the National

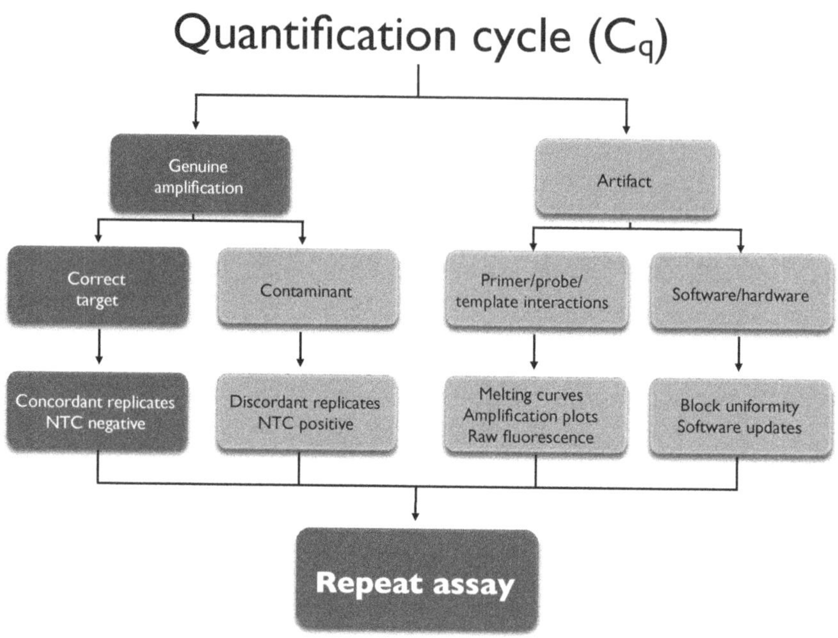

Figure 1.7 Considerations following the reporting of a Cq. The Cq may be a result of genuine amplification or due to some artefact. If genuine, it may still be caused by contamination, which is an ever-present danger for any nucleic acid amplification procedure. It is essential to take note of discordant replicates and positive NTCs, since these frequently allow an immediate assessment of the validity of a result. Alternatively, positive Cq values can be the result of primer dimer amplification or other oligonucleotide/template interactions or some software or hardware problem. Analysis of amplification plots, melt curves and raw fluorescence data are often sufficient to allow an informed decision and there are testing procedures available that check instrument function. In any case, it is essential to repeat an assay, ideally using biological replicates, especially when very small changes in target copy numbers are being quantified.

Center for Biotechnology Information (NCBI) nucleotide search site (http://www.ncbi.nlm.nih.gov/sites/nuccore) or gene search (http://www.ncbi.nlm.nih.gov/gene) websites.

It is now well established that amplification efficiency, and hence assay sensitivity, is directly associated with good primer design (Nolan *et al.*, 2006a; Bustin and Nolan, 2009). In theory, it should be easy to utilize the tens of thousands of published qPCR assays to obtain suitable primer and probe sequences. The qPrimerDepot (http://primerdepot.nci.nih.gov) provides RT-qPCR primers for 99.96% of human RefSeq sequences, most of which cross an exon-exon border which overlaps one of the largest introns. An alternative choice is RTPrimerDB (http://medgen.ugent.be/rtprimerdb/), which contains more than 8000 qPCR assays for over 5500 genes. When an assay is submitted there is a requirement for appropriate validation and optimization data, with the option for further users to leave feedback (Pattyn *et al.*, 2003, 2006; Lefever *et al.*, 2009b). The database provides a freely accessible data retrieval system and an in silico assay evaluation pipeline for all listed qPCR assays. Specifically, assay reports contain gene information, oligonucleotide sequences, reaction conditions, publication information, users' experimental evaluation, feedback and submitter's contact details. Additional information is provided on the alignment of primer and probe sequences on known transcript variants of a gene, along with SNP positions and peptide domain information as well as the secondary structure of the amplicon sequence.

The use of such properly validated assays has numerous advantages, obviating the requirement for time-consuming primer design and empirical optimization and critically, introducing more uniformity and standardization among different laboratories. RTPrimer DB is particularly flexible and can be queried using the official gene name or symbol, Entrez or Ensembl Gene identifier, SNP identifier, or oligonucleotide sequence. Queries can be restricted to a particular application, e.g. RNA quantification, genomic DNA copy number quantification, SNP detection, organism or detection chemistry.

However, no database can accommodate every gene or variant; new variants may have been identified, or suitable assays have never been designed for the organism of interest. Consequently, there are a large number of conditions where good assay design know-how is essential. In practice this means designing several primer combinations for each target and testing the various permutations to obtain the optimal combination. Although this has been well documented and is easily and cheaply implemented (Nolan *et al.*, 2006a), it is still an aspect of assay design that is all too frequently neglected and often results in primers being used that have not been optimized, both in terms of design as well as reaction conditions.

In many ways, the requirements for design of a PCR or qPCR primer are similar, in that it is undesirable to include repeats of the same base or palindromic motifs within the primer sequence and GC content should be around 50%. The degree of clamping at the 3′ should be sufficient to secure hybridization but not to promote non-specific priming. It is a good general guide to aim for two of five bases at the 3′ to be G or C and the remaining three to be A or T. Primer dimers are avoided by rejecting primers with three or more bases matches at the 3′ end and six bases or more matching within the rest of the primer. A measure of the Gibbs free energy (ΔG) can be helpful to estimate primer secondary structure stability. A highly negative value indicates a stable structure; hence primers with negative self dimerization scores are likely to form primer dimers. Similarly, individual primers adopting a hairpin structure with negative ΔG values should be rejected. At the very least, the stem must not remain closed at the annealing temperature of the primer. Fig. 1.8 shows examples of two primers targeting two different mRNAs designed by the authors using Beacon Designer (Premier Biosoft; see below). Both generate very robust assays and both assays result in single melting curve peaks characteristic of unique amplification products.

In contrast to conventional PCRs, the qPCR amplicon length is ideally kept to within 60–150 bp. Amplification of a shorter fragment is much more efficient than of a larger fragment and so keeping the amplicon short drives the reaction towards doubling of the target at each cycle and therefore ensures a greater accuracy in quantification. In addition, the target sequence should be free of secondary structure; this can be predicted by entering the sequence into a folding prediction program such as mfold (Zuker, 2003). Amplification of folded regions is less efficient that of open structures because folding inhibits priming and elongation of the sequence by DNA polymerase.

Probe design and location differs slightly depending on the application. Hydrolysis probes are designed to be located close to the 3′ of the primer on the same strand when being used for quantification; for SNP detection they are positioned centrally in the amplicon and such that the SNP is in the centre of the probe. This is also the case when designing Molecular beacons or LNA-modified probes. All probes have an annealing temperature of 7–10°C higher than the primers; as with primers it is important to ensure that the probe does not contain repetitive regions, and will not adopt secondary structure. In addition, runs of four repeats of the same base should be avoided and the fluorophore should not be attached to a G because this naturally suppresses fluorescence.

Inhibition of the PCR assay by factors co-purified during nucleic acid extraction can be an important factor contributing to poor assay performance (Nolan *et al.*, 2006b). It has recently been shown that the general assumption that inhibitors affect all PCRs to the same extent, i.e. that the reaction of interest and the control reaction are equally susceptible to inhibition, is wrong (Huggett *et al.*, 2008). Instead it appears that susceptibility to inhibition is highly variable between reactions, inhibitors and targets. These findings have serious implications for all PCR-based gene expression studies, including the relatively new PCR array method, and for both qualitative and quantitative PCR-based molecular diagnostic assays, suggesting that careful consideration

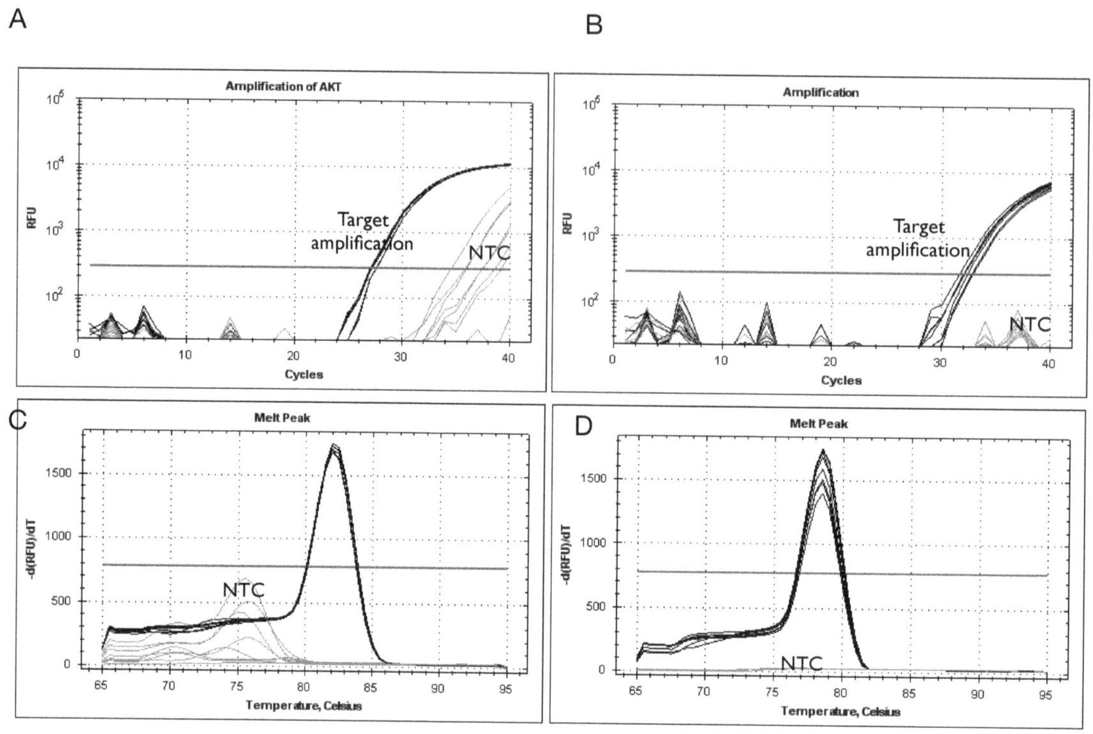

Figure 1.8 Robust qPCR assay design. (A) The primers initiate similarly efficient amplification reactions with SYBR Green I dye as the reporter over a wide temperature annealing range (55–65°C). Positive signals are seen in the no template controls (NTC), but can be minimized by choosing the appropriate annealing temperature. (B) The melting profile clearly distinguishes correctly amplified amplicon from the NTC. (C) Again the assay is characterized by its tolerance to a wide temperature gradient, although clearly there is a little more variability than seen in A. However, at none of the temperatures do the NTC samples generate positive fluorescence signals. (D) A single melt peak demonstrates that the assay generates unique PCR amplicons. The lower G/C content and shorter length of the amplicon is reflected in the lower temperature of the melt peak.

should be given to inhibition compatibility when conducting PCR analyses.

It is becoming routine practice to use specific software packages for qPCR assay design, especially when these automatically screen for structure within the amplicon, oligos and repetitive sequences. Beacon Designer (www.premierbiosoft.com), which is used to design all qPCR assays, not just Molecular beacons, can be used for RNA expression or genotyping assays and the associated package Allele ID also incorporates a function to select sequences from homology line-up plots. Alternatively, there are free design services available from oligonucleotide synthesis companies, such as Sigma who will provide a custom design to the specification of the experiment. There are also commercially available assays, but it is important to ensure that the assay sequences are available in accordance with new guidelines on the requirements for reporting qPCR assays (MIQE guidelines; Bustin *et al.*, 2009).

RT-qPCR considerations

qPCR is the ideal application for the detection and discrimination of microorganisms, mutations or SNP analyses. This is because sample handling and DNA template preparation, assay protocols and data interpretation are all relatively straightforward. In contrast, although RT-qPCR is widely used for the detection of cellular mRNA (Pignot *et al.*, 2008), micro RNAs (Schmittgen *et al.*, 2008), as well as microorganisms (Lu *et al.*, 2008), there are significant biological as well as technical limitations that make its use much more problematic (Bustin, 2008b). A significant contribution to variability in gene expression studies, and one that constitutes 'baseline' variability, arises from the fact that *in vivo* RNA is subject to constant degradation by complex interactions of deadenylation and decapping enzyme complexes as well as exonucleases and endonucleases (Coller and Parker, 2004). There are numerous reasons for technical variability although, unlike biological variability, these can be addressed by the use of appropriate standing operating procedures. The most obvious causes are inconsistent sample selection, handling, and RNA isolation. RNA extraction is a laborious process, often involving solid samples that must be ground to a fine powder while ensuring the sample remains frozen to minimize the activity of nucleases after cell lysis. In many cases, tissue disruption is accompanied by simultaneous addition of a strong protein denaturation agent containing a chaotropic compound such as guanidinium salt. This serves to inactivate nucleases and solubilize cell membranes. Particularly challenging sample types include formalin-fixed and paraffin-fixed archival sections. These are typically treated with proteinase K to degrade the tissue structure prior to guanadinium-based extractions. The nucleic acid purified from these samples is usually very fragmented, around 200 bases, and RNA is difficult to work with due to cross-linking between bases and also to proteins (Lewis *et al.*, 2001; Lewis and Maughan, 2004).

Crucially, the accuracy of gene expression profiling is highly dependent on RNA quality, both in terms of its integrity as well as in terms of the lack of inhibitors co-purified during the extraction procedures (Gallup and Ackermann, 2006; Nolan *et al.*, 2006a,b). Unfortunately, insufficient attention is paid to the analysis of RNA quality, and a survey of the working practices of 100 experienced qPCR users revealed that around half did not adequately quality assess their RNA (Bustin, 2005). A recent survey of papers published in 2007/08 revealed that >60% of papers did not even mention mRNA quality (Bustin, 2008a) and an analysis of papers published in 2010 showed that the higher the impact factor of the journal, the less likely it is that any relevant technical information is included (Huggett and Bustin, 2011).

Technical problems are clearly illustrated by the incorrect RT-qPCR data analysis that resulted in the retraction (Bohlenius *et al.*, 2007) of a paper with Science magazine 'Breakthrough of the Year' status (Huang *et al.*, 2005). One of the most egregious examples of the enormous implications for the health and lives of individuals that result from inappropriate use of this technology is provided by the use of RT-qPCR data that purported to demonstrate the presence of measles virus (MV) RNA in children with developmental disorders (Uhlmann *et al.*, 2002). It provided sustenance to the controversy surrounding the triple measles, mumps and rubella (MMR) virus vaccine, as the data were interpreted as providing hard evidence for a link between MMR, gut pathology and autism. However, a detailed analysis of the raw data underlying that report carried out by one of the authors, acting as an expert witness to the UK High Court and the US Vaccine Court, revealed that these data were obtained amongst a catalogue of mistakes, inaccuracies and inappropriate analysis methods as well as contamination and poor assay performance (Bustin, 2008d). A reanalysis of the data concluded that the assay had been detecting DNA and since MV is an RNA-only virus, the RT-qPCR data had been erroneously interpreted. A recent paper that included two of the authors from the initial report was unable to reproduce the original findings and concluded that there was no link between autism and enteropathy (Hornig *et al.*, 2008). Sadly, despite this and other evidence (Afzal *et al.*, 2006; D'Souza *et al.*, 2007), the authors of the 2002 report have not retracted their original findings, in spite of the immense repercussions their original report has had upon public health.

Clearly, it is essential that the quality of RNA should be determined as a matter of routine whenever possible. Since it is advisable to use approximately the same amount of RNA for cDNA synthesis when comparing different samples, quantification of RNA in extracted samples is important. There are several procedures for the measure of nucleic acid concentration including spectrophotometric or Nanodrop analysis, the Lab901/Agilent/BioRad/Qiagen microfluidics-based systems, or Ribogreen fluorescence detection. Unfortunately, each system produces a different result and so it is imperative that a single system is selected and data obtained using the different methods are not directly compared (Doma and Parker, 2007), especially when performing RT reactions. The purity of the sample includes a consideration of the extent of genomic DNA contamination that can interfere with RT reactions and also cause an increased estimate of copy number. As discussed above, it is important to establish that the RNA sample does not contain factors that may inhibit downstream RT reactions or the qPCR assay.

Since all currently available thermostable polymerases are DNA dependent, RNA must be converted ('reverse transcribed') into cDNA prior to its amplification reaction. Unfortunately, some of the assumptions carried through from legacy RT-PCR are now known to be unsound: primarily that the process of RT is reproducibly quantitative, linear and maintains proportionality between genes when the cDNA for each is produced. Furthermore, cDNA synthesis protocols are highly diverse and utilize different types of RNA-dependent DNA polymerase enzymes that are characterized by low fidelity. Commercial suppliers often provide blends of multiple enzymes to maximize the benefit from the best features of each. Nonetheless, different enzyme, buffer and RT temperature combinations have dramatic effects on cDNA synthesis, resulting in huge differences in cDNA yield of different targets (Stahlberg *et al.*, 2004a,b). There are several choices for cDNA priming strategies that involve reverse transcription from target-specific primers, oligo-dT and anchored variants thereof, various random priming procedures as well as combinations of the above. A specific primer targeting the RNA sequence can be used to direct RT towards the specific region of interest. In many cases, this will be one of the PCR primers, and conventional 'one-step' protocols include a RT step with the reaction containing both primers. Priming with a target specific primer is particularly vulnerable to template secondary structure and so the assay design must include steps to ensure that the primer targets a region that is predicted to be open under the RT conditions determined for the primer. Alternatively, it may desirable to produce cDNA that is representative of the entire RNA population. This is achieved by priming all transcripts using either a primer consisting of a string of thymidine residues (oligo-dT) (for eukaryotic organisms only) or collection of primers of random sequence (random primers). Reliable cDNA priming using oligo-dT is critically dependent on RNA integrity and targets both the 3′ polyA tract of mRNA as well as A-rich regions of mRNA and other RNA molecules. In order to increase specificity, many oligo-dT primers have two additional bases at the 3′ end ('anchored' primers) and RT proceeds from these towards the 5′-end of the RNA. Random priming will reverse transcribe mostly rRNA and may lose sensitivity, but is the most efficient way of cDNA priming an RNA sample for the analysis of multiple targets. Unfortunately, the choice of priming strategy results in large differences in cDNA yield for given targets (Stahlberg *et al.*, 2004a). Specifically, targets present at lower copy number (e.g. those typically of great interest) appear to be more vulnerable to cDNA yield concentration fluctuations, while those at higher copy number (e.g. often adopted as reference genes) are less variable. Similar phenomena have also been observed when using oligo-dT priming. For this reason it is important to validate the reproducibility of RT protocols, ensuring that protocols are designed such that all samples can be processed together or that an experimental calibrator is included so that batch-to-batch variations in the relative yield of each target can be corrected (Lacey *et al.*, 2005). Fig. 1.9 demonstrates that it is possible to obtain highly reproducible RT-qPCR results, if consistent concentrations of a high quality RNA sample and a robust assay are used by a careful operator.

Normalization

From the above, it is clear that the problems associated with inherent variability of RNA, variability of extraction protocols that may co-purify RT or PCR inhibitors, as well as different reverse transcription and PCR efficiencies, mean that it is important that an accurate method of normalization is chosen to control for experimental variation introduced during the multistage process of carrying out a qPCR assay. Unfortunately, normalization remains one of qPCRs most difficult problems to address when designing a qPCR

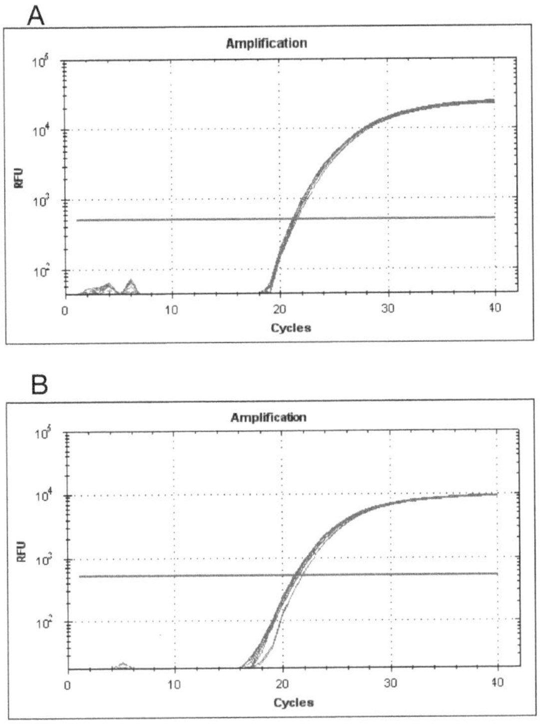

Figure 1.9 Reliability and repeatability of RT-qPCR. (A) Five separate RT reactions were set up simultaneously and subjected to separate PCR assays. The resulting C_qs are virtually identical (mean Cq=21.24±0.07). (B) Single RT reactions were set up on five consecutive days, before being subjected to separate PCR assays. Although there is a little more variation, the Cq value are essentially the same (21.27±0.08).

experiment. Several strategies have been proposed that range from ensuring that a similar sample size is chosen to using multiple internal reference genes (Huggett *et al.*, 2005). These approaches are not mutually exclusive and can be incorporated into a protocol at many stages.

Normalization by relating the quantity of a target gene to total sample mass or volume is a legacy approach remaining from northern blotting techniques. In this case gel loading and sample transfer to filter paper were validated by probing for rRNA or a housekeeping gene such as GAPDH. It was assumed that expression of this housekeeping gene was stable between individuals, experimental conditions or physiological states. While still an integral part of RT-qPCR measurements, there are several disadvantages to measuring mRNA levels relative to sample mass or volume. For example, in a comparison of samples of different origin, it is incorrect to assume that the same tissue mass contains similar cell numbers, or that the relative distribution of cells is equal. Normalization of the specific target quantity to total RNA may underestimate the expression of target genes in some samples. This approach may be more suitable when a precise number and similar cells are targeted; even then this approach is not ideal. A related technique, again reminiscent of northern blotting, is to normalize target quantity to DNA or RNA concentration. While measuring gene copy number relative to input DNA concentration is a perfectly valid approach, the situation is more complicated for transcript quantification. When RNA concentration is determined, the vast majority of the RNA component is ribosomal RNA (rRNA). Transcription of rRNA is independent of the transcription of mRNA since it is transcribed by different enzymes. Furthermore, as rRNA makes up ~80% of the RNA fraction, quantification of rRNA would mask subtle changes in the mRNA component, which typically constitutes 2–5%. In addition, this approach does not take into account variations in the template RNA quality. Nonetheless, while not ideal, there are situations where there are no other alternatives but to measure relative to total RNA and in many cases the data are reliable. Some analysis packages such as GenEx software from MultiD (www.multid.se) allow the total RNA concentration to be evaluated as a normalization technique alongside other approaches such as those below. One theoretical solution would be to purify mRNA and quantify this. Unfortunately, the purification process introduces inaccuracies and an extra processing step that is undesirable and in many cases there is insufficient RNA sample to purify mRNA.

A very common approach to correct for sample differences is to express the target concentration relative to that of a single internal reference gene usually with little or no validation. Although still used, this approach was largely discredited a long time ago (Bustin, 2000, 2002). In order for single reference gene normalization to be valid, the expression of the reference gene must remain constant between all experimental samples. It is rare to find such a convenient target and all too common and misguided is to select this gene at random, without validation; GAPDH, β-actin and 18S or 16S being particular favourites. When the reference gene is not stably expressed between all samples, a ratio of the target gene of interest to the reference gene will reflect the expression changes of both targets. This is unhelpful when the expression behaviour of neither target has been defined and can lead to inaccuracies and false results. An alternative approach, which is gaining momentum, is to express the data relative to the behaviour of multiple reference genes and use geometric averaging to measure and control for error induced trends in the relative reference gene expression. There are various approaches such as those offered by geNorm (Vandesompele *et al.*, 2002) that allow the quantity of potential reference genes to be compared in selected groups of experimental and control samples. These use the combined variation of several genes to produce a stable reference factor for normalization and can be very accurate, potentially providing 0.5-fold resolution (Hellemans *et al.*, 2004). In an analogous manner, gene expression analysis of miRNA molecules in eukaryotic cells requires accurate and reliable normalization to relevant targets and the mean expression value of all expressed microRNAs in a given sample has been shown to provide optimal normalization for miRNA qPCR data (Mestdagh *et al.*, 2009). Clearly, such extensive reference gene validation is taxing on time, sample and resource. In future, alternative approaches such as normalization to expressed ALU repeats (EARs) or other repeats may provide less labour-intensive normalization strategies (Marullo *et al.*, 2010).

Data analysis

The successful generation of high-quality qPCR experimental runs and acquisition of Cq values is followed by the critical data analysis step. The increasing penetration of 384-well and higher throughput qPCR instruments, together with PCR array methods such as the Biotrove and Fluidigm systems has resulted in data analysis becoming one of the biggest bottlenecks in qPCR experiments. Coupled with this is the fact that many biologists may not be able to analyse and interpret appropriately the quantitative data from their qPCR assays. Hence it has become increasingly obvious that, even for research applications, data analysis, as an essential component of the complete assay, should be carried out using dedicated software. A wide range of analysis packages can be accessed from http://www.gene-quantification.de/download.html.

One of the first tools was REST (Relative Expression Software Tool) (Pfaffl *et al.*, 1998), which was developed to address the problem of comparing target gene expression levels relative to reference genes indifferent of samples. Importantly, these calculations also accommodated differences in the efficiency of each of the PCR assays. The use of ratios for gene expression measurements makes it complex to perform traditional statistical analysis, as ratio distributions do not have a standard deviation. The latest REST (http://www.gene-quantification.de/rest-2009.html) is a standalone software package and has been extended with advanced algorithms using randomization and bootstrapping techniques to take into account issues of multiple reference gene normalization and quantification of the uncertainty inherent in any measurement of expression ratios. The program also allows measurement of not only the statistical significance of deviations but also their likely magnitude, even in the presence of outliers. A visually attractive graphing ability provides a visual representation of variation for each gene, highlighting potential issues such as distribution skew.

In terms of power and sophistication, two commercial software tools stand out: GenEx and qBase plus.

GenEx (http://www.multid.se/) is a powerful package for processing and analysing qPCR data. It offers a vast range of tools, ranging from basic data editing and management to advanced, cutting-edge data analysis including selection and validation of reference genes, classification of samples, gene grouping, monitoring of time dependent processes and basic and sophisticated statistical and graphic capacity. Statistical features include parametric and non-parametric tests, clustering methods, principal component analysis and artificial neural networks. Most qPCR experiment- processes have a nested design with technical replicates at each step, where each step introduces more variance to the data. The noise contribution from each step can be analysed and used to optimize the experimental design for future experiments. One interesting feature is the experimental design optimization module, which calculates how to minimize the total variation for the experimental design with consideration of the experimental budget. GenEx can be used for the analysis of both relative quantification (ΔΔCq) and 'absolute' quantification, i.e. relative to a dilution curve, with powerful presentation tools generating visually attractive illustrations. For accurate and robust normalization of qPCR data, GenEx has the advantage of incorporating both NormFinder (Andersen *et al.*, 2004) and geNorm (Vandesompele *et al.*, 2002) in the software. geNorm (http://medgen.ugent.be/~jvdesomp/genorm/) is used to select an optimal set of reference genes from a larger set of candidate genes. It is critical that the tested candidate genes are not co-regulated. The geNorm process performs a reiterative calculation of the M-value, which describes the variation of a potential reference gene compared with all other candidate genes. The gene with highest M-value is eliminated until there are only two reference genes left. Usually the last two or three remaining genes are recommended as the optimum candidate reference genes. NormFinder (http://www.mdl.dk/publicationsnormfinder.htm) is also used to find the optimum reference genes but, in contrast to geNorm, takes information of expression in groupings of samples into account. The result is an optimum pair of reference genes that might have compensating expression, so that one gene is slightly overexpressed in one group, but the other gene is correspondingly underexpressed in the same group.

qBase plus (http://www.biogazelle.com/products/qBase plus) is a relatively new software tool developed by members of the group that invented the geNorm algorithms (Vandesompele *et al.*, 2002; Hellemans *et al.*, 2007). It has the capacity to determine PCR replicate variation, assessment of positive and negative control samples, determination of reference gene expression stability and evaluation of deviating normalization factors.

Technical replicates (repeated measurements of the same sample in the same run) are useful as they allow for quality control of the technical reproducibility of the qPCR data, provide better accuracy, and allow the generation of results when individual qPCR reactions failed. Low reproducibility of the qPCR reaction is often the first indication of an unstable assay that requires further optimization. qBase plus and GenEx automatically deal with technical replicates or repeated measurements, outliers can be excluded so they do not contribute to a significant bias and large errors. However, it is important to note that biological variability is often much larger than RT technical variability, which is also much larger than PCR technical variability.

The second type of quality control allows an evaluation of the positive and negative sample controls. An amplification signal in the NTC sample indicates a potential contamination issue or formation of primer dimers; hence both packages flag suspicious no-template control (NTC) samples based on a user defined or default thresholds.

Reference gene or normalization factor stability is the third type of quality control. The user is able to choose a minimal acceptable normalization factor stability by defining a threshold value for two indicators of expression stability of the used reference genes; the geNorm expression stability value of the reference gene (M) (Vandesompele *et al.*, 2002) and the coefficient of variation (CV) of the relative quantities (Hellemans *et al.*, 2007). qBase plus has extended and improved the functionality of geNorm by allowing ranking of candidate reference genes up to the single most stable gene, by combining the calculation of relative quantities and geNorm analysis and by allowing the processing of experiments with missing data in a way that has the lowest impact on the overall analysis through intelligent retention of as many data points as possible.

Inspection of normalization factors is a very useful feature and constitutes a fourth type of quality control. qBase plus displays the geometric mean of selected reference genes (the normalization factor) for each sample. These factors should be similar for all samples if approximately equal amounts of equal quality input material were used. High variability of the normalization factors indicates large differences in starting material quantity or quality, or a problem with one of the reference genes, which may be either not stably expressed or not adequately measured.

There are alternatives, usually web-based analysis tools available: QPCR (https://esus.genome.tugraz.at/rtpcr/) allows storage, management, and analysis of qPCR data (Pabinger *et al.*, 2009). It comprises a parser to import data from qPCR instruments and includes a variety of analysis methods to calculate cycle-threshold and amplification efficiency values. The analysis pipeline includes technical and biological replicate handling, incorporation of sample or gene specific efficiency, normalization using single or multiple reference genes, inter-run calibration, and fold change calculation. Moreover, the application supports an assessment of error propagation throughout all analysis steps and allows conducting of statistical tests on biological replicates. Results can be visualized in customizable charts and exported for further investigation. Calculation of amplification efficiencies for RT-PCR experiments (CAmpER (http://camper.cebitec.uni-bielefeld.de) is a more limited web-based tool for the basic analyses of qPCR assays run on the Roche Lightcycler, the Biorad Opticon and the Qiagen Rotorgene.

MIQE

As discussed above, technological advances mean that there is an ever-increasing choice of platforms, chemistries, protocols as well as applications and targets for qPCR analysis. Whilst this serves to underline the pre-eminence of qPCR as a technique, is exciting and generating a vast amount of data, it has also become clear that a high percentage of publications utilizing qPCR technology, and especially those aiming to profile cellular RNA levels, report poorly designed, executed and interpreted experiments and results (Bustin, 2010). Considerations of mRNA transcription, *in vivo* stability, regulation by miRNAs, tissue-specificity of splice variants, allele-specific differences in expression, the lack of concordance between most mRNAs and their specified proteins, the critical importance of post-translational modifications and questions of tissue heterogeneity all describe serious issues that are not being addressed in an adequate manner.

Trust in the accuracy and integrity of the scientific literature is an essential prerequisite for maintaining scientific excellence and advancing knowledge. This requires urgent consideration by researchers, reviewers and editors who need to agree a basic set of quality criteria and adhere to elementary procedures that are prerequisites for the publication of reliable and reproducible data. Such a list must include delineating minimum quality standards for template preparation, validation and consistent use of cDNA priming methods, enzymes, protocols and, equally critically, appropriate analysis of data. Furthermore, it is entirely unacceptable that most publications do not address the critical issue of RNA quality assessment. It is equally intolerable that data are not normalized in a proper way. In addition, it is vital that the conventions underlying data acquisition and analysis become more transparent. Consequently, it is essential that the editors of scientific publications consider issuing agreed checklists specifying the key information to be included when reporting experimental results. There are significant efforts under way to organize such 'minimum information' checklists, with the 'Minimum information for biological and biomedical investigations' (MIBBI) project (www.mibbi.org) offering a common portal aimed at promoting gradual data integration (http://mibbi.sourceforge.net).

The problem of inadequate information being provided with publications is being addressed by the minimum information for publication of quantitative real-time PCR experiments (MIQE) guidelines (Bustin *et al.*, 2009). MIQE constitutes a set of publication guidelines for researchers, journal reviewers and editors that list the minimum information required to allow potential reproduction, as well as unambiguous quality assessment, of a qPCR-based experiment. MIQE is modelled on analogous rules drawn up for DNA microarray analysis (Brazma *et al.*, 2001), proteomics experiments (Taylor *et al.*, 2007), genome sequence specification (Field *et al.*, 2008) and those under discussion for RNAi work (Haney, 2007) and metabolomics (Sansone *et al.*, 2007), initiatives coordinated under the auspices of MIBBI (Taylor *et al.*, 2008).

The four key areas of standardization that

define any qPCR experiment are study design, technical detail, analysis methods and statistics. MIQE addresses these under a set of captions that describe a large number of individual elements: 'Experimental design, sample, nucleic acids, reverse transcription, target, primers and probes, assay details, PCR cycling and data analysis'. At first sight, these look daunting, arduous and overexacting. In practice, it is clear that most, if not all of these parameters, describe information that would be obtained as a matter of course during the experimental design, optimization and validation stages. Importantly, there is a clear hierarchy with some parameters, labelled 'E' (essential) in the published guidelines, indispensable for an adequate description of the qPCR assay, whereas other components, labelled 'D' (desirable) more peripheral, yet constituting an effective foundation for the realization of best practice protocols. Of course, these parameters are not set in stone and are open for discussion (Bustin *et al.*, 2010), but the guidelines encompass a core of indispensable information that should be accessible to reviewers and readers.

Finally, another development complementary to MIQE concerns the problems associated with attempting to share qPCR data between different laboratories and users. Currently, easy data interchange between instrument software and third-party data analysis packages, colleagues and collaborators, and between authors, peer reviewers, journals and readers, is difficult. One solution to this problem is the development of the Real-time PCR Data Markup Language (RDML), a structured and universal data standard for exchanging quantitative PCR (qPCR) data (Lefever *et al.*, 2009a). The aim is to enable transparent exchange of annotated qPCR data by providing sufficient information for a third party to be able to understand the experimental set-up, re-analyse the data and interpret the results. RDML files contain all measured data as well as information about the samples and targets being analysed; thus this data standard provides sufficient information to allow the reader to understand the experimental set-up, re-analyse the data and interpret the results. A useful tool for converting instrument-specific data to the RDML format for convenient data interchange is found on the RDML website (http://medgen.ugent.be/rdml/chooseTool.php).

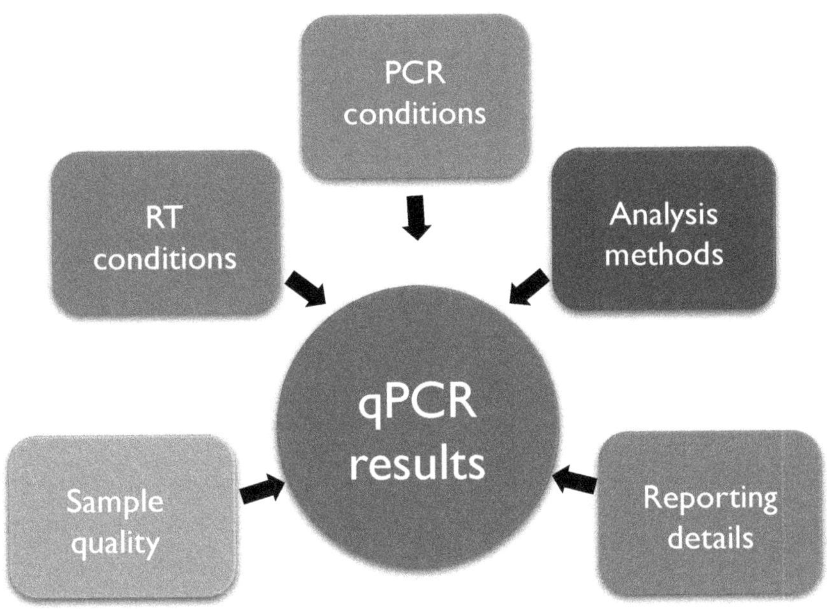

Figure 1.10 Successful qPCR results depend on samples being acquired, handled and processed appropriately, on suitable RT and PCR conditions being established, fit-for-purpose analysis methods being applied and, finally, experimental protocols being reported in a transparent and complete manner.

Conclusions

Some technologies arrive at a time that is 'just right' and drive the development of new and unexpected applications. Advances in enzymes, buffers, chemistries and instrumentation have allowed qPCR to progress from a cutting edge technology used as a research tool in specialist laboratories to becoming a routine technique across the life sciences, making it a technology whose time has come (Walker, 2002). Currently, it provides the best compromise between ease of use, sensitivity, reliability and specificity of any molecular diagnostic assay.

Biologically relevant qPCR results depend on several parameters that must all come together in an appropriate, controlled and transparent manner (Fig. 1.10). The main issues remaining are those concerning the potential for observing false positive or negative results, nucleic acid quality control issues, the requirement for good assay design and optimization and transparency of protocols and analysis methodologies. Therefore, it cannot be over emphasized that there remains a constant need for assay validation to ascertain the reliability and reproducibility of established, modified or newly developed technologies for routine analysis. Finally, the development of lab-on-a-chip devices combining miniaturized reactors, advanced microfluidics and sophisticated detection technologies will open up a whole new field for qPCR applications, bringing new opportunities and new challenges.

References

Afonina, I.A., Reed, M.W., Lusby, E., Shishkina, I.G., and Belousov, Y.S. (2002). Minor groove binder-conjugated DNA probes for quantitative DNA detection by hybridization-triggered fluorescence. Biotechniques *32*, 940–949.

Afzal, M.A., Ozoemena, L.C., O'Hare, A., Kidger, K.A., Bentley, M.L., and Minor, P.D. (2006). Absence of detectable measles virus genome sequence in blood of autistic children who have had their MMR vaccination during the routine childhood immunization schedule of UK. J. Med. Virol. *78*, 623–630.

Andersen, C.L., Jensen, J.L., and Orntoft, T.F. (2004). Normalization of real-time quantitative reverse transcription-PCR data: a model-based variance estimation approach to identify genes suited for normalization, applied to bladder and colon cancer data sets. Cancer Res. *64*, 5245–5250.

Batsch, A., Noetel, A., Fork, C., Urban, A., Lazic, D., Lucas, T., Pietsch, J., Lazar, A., Schomig, E., and Grundemann, D. (2008). Simultaneous fitting of real-time PCR data with efficiency of amplification modeled as Gaussian function of target fluorescence. BMC Bioinformatics *9*, 95.

Beck, E.T., Jurgens, L.A., Kehl, S.C., Bose, M.E., Patitucci, T., LaGue, E., Darga, P., Wilkinson, K., Witt, L.M., Fan, J., *et al.* (2010). Development of a rapid automated influenza A, influenza B, and respiratory syncytial virus A/B multiplex real-time RT-PCR assay and its use during the 2009 H1N1 swine-origin influenza virus epidemic in Milwaukee, Wisconsin. J. Mol. Diagn. *12*, 74–81.

Bohlenius, H., Eriksson, S., Parcy, F., and Nilsson, O. (2007). Retraction. Science *316*, 367.

Bonnet, G., Tyagi, S., Libchaber, A., and Kramer, F.R. (1999). Thermodynamic basis of the enhanced specificity of structured DNA probes. Proc. Natl. Acad. Sci. U.S.A. *96*, 6171–6176.

Brazma, A., Hingamp, P., Quackenbush, J., Sherlock, G., Spellman, P., Stoeckert, C., Aach, J., Ansorge, W., Ball, C.A., Causton, H.C., *et al.* (2001). Minimum information about a microarray experiment (MIAME)-toward standards for microarray data. Nat. Genet. *29*, 365–371.

Burns, M.J., Nixon, G.J., Foy, C.A., and Harris, N. (2005). Standardisation of data from real-time quantitative PCR methods – evaluation of outliers and comparison of calibration curves. BMC Biotechnol. *5*, 31.

Bustin, S.A. (2000). Absolute quantification of mRNA using real-time reverse transcription polymerase chain reaction assays. J. Mol. Endocrinol. *25*, 169–193.

Bustin, S.A. (2002). Quantification of mRNA using real-time reverse transcription PCR (RT-PCR): trends and problems. J. Mol. Endocrinol. *29*, 23–39.

Bustin, S.A. (2004) A-Z of Quantitative PCR (IUL Press, La Jolla, CA).

Bustin, S.A. (2005). Real-time, fluorescence-based quantitative PCR: a snapshot of current procedures and preferences. Expert Rev. Mol. Diagn. *5*, 493–498.

Bustin, S.A. (2008a). Molecular medicine, gene-expression profiling and molecular diagnostics: putting the cart before the horse. Biomarkers Med. *2*, 201–207.

Bustin, S.A. (2008b). Real-time polymerase chain reaction – towards a more reliable, accurate and relevant assay. Eur. Pharm. Rev. *6*, 19–27.

Bustin, S.A. (2008c). Real-time quantitative PCR-opportunities and pitfalls. Eur. Pharm. Rev. *4*, 18–23.

Bustin, S.A. (2008d). RT-qPCR and molecular diagnostics: no evidence for measles virus in the GI tract of autistic children. Eur. Pharm. Rev. Dig. *1*, 11–16.

Bustin, S.A. (2010). Why the need for qPCR publication guidelines? The case for MIQE. Methods *50*, 217–226.

Bustin, S.A., and Nolan, T. (2009). qPCR Assay design. Eur. Pharm. Rev. 3, 26–32.

Bustin, S.A., Benes, V., Garson, J.A., Hellemans, J., Huggett, J., Kubista, M., Mueller, R., Nolan, T., Pfaffl, M.W., Shipley, G.L., *et al.* (2009). The MIQE guidelines: minimum information for publication of quantitative real-time PCR experiments. Clin. Chem. *55*, 611–622.

Bustin, S.A., Beaulieu, J.F., Huggett, J., Jaggi, R., Kibenge, F.S., Olsvik, P.A., Penning, L.C., and Toegel, S. (2010). MIQE precis: Practical implementation of minimum standard guidelines for fluorescence-based quantitative real-time PCR experiments. BMC Mol. Biol. *11*, 74.

Chakravorty, S., Aladegbami, B., Burday, M., Levi, M., Marras, S.A., Shah, D., El-Hajj, H.H., Kramer, F.R., and Alland, D. (2010). Rapid universal identification of bacterial pathogens from clinical cultures by using a novel sloppy Molecular beacon melting temperature signature technique. J. Clin. Microbiol. *48*, 258–267.

Coller, J., and Parker, R. (2004). Eukaryotic mRNA decapping. Annu. Rev. Biochem. 73, 861–890.

D'Souza, Y., Dionne, S., Seidman, E.G., Bitton, A., and Ward, B.J. (2007). No evidence of persisting measles virus in the intestinal tissues of patients with inflammatory bowel disease. Gut *56*, 886–888.

Dixon, J.M., Lubomirski, M., Amaratunga, D., Morrison, T.B., Brenan, C.J., and Ilyin, S.E. (2009). Nanoliter high-throughput RT-qPCR: a statistical analysis and assessment. Biotechniques 46, ii–viii.

Doma, M.K., and Parker, R. (2007). RNA quality control in eukaryotes. Cell *131*, 660–668.

El-Hajj, H.H., Marras, S.A., Tyagi, S., Shashkina, E., Kamboj, M., Kiehn, T.E., Glickman, M.S., Kramer, F.R., and Alland, D. (2009). Use of sloppy Molecular beacon probes for identification of mycobacterial species. J. Clin. Microbiol. 47, 1190–1198.

Field, D., Garrity, G., Gray, T., Morrison, N., Selengut, J., Sterk, P., Tatusova, T., Thomson, N., Allen, M.J., Angiuoli, S.V., *et al.* (2008). The minimum information about a genome sequence (MIGS) specification. Nat. Biotechnol. *26*, 541–547.

French, D.J., Archard, C.L., Andersen, M.T., and McDowell, D.G. (2002). Ultra-rapid DNA analysis using HyBeacon probes and direct PCR amplification from saliva. Mol. Cell Probes *16*, 319–326.

Gallup, J.M., and Ackermann, M.R. (2006). Addressing fluorogenic real-time qPCR inhibition using the novel custom Excel file system 'FocusField2-6GallupqPCRSet-upTool-001' to attain consistently high fidelity qPCR reactions. Biol. Proced. Online *8*, 87–152.

Gallup, J.M., and Ackermann, M.R. (2008). The 'PREXCEL-Q Method' for qPCR. Int. J. Biomed. Sci. *4*, 100–120.

Gibson, U.E., Heid, C.A., and Williams, P.M. (1996). A novel method for real time quantitative RT-PCR. Genome Res. *6*, 995–1001.

Grisold, A.J., and Kessler, H.H. (2006). Use of hybridization probes in a real-time PCR assay on the LightCycler for the detection of methicillin-resistant *Staphylococcus aureus*. Methods Mol. Biol. *345*, 79–89.

Gunson, R.N., Bennett, S., Maclean, A., and Carman, W.F. (2008). Using multiplex real time PCR in order to streamline a routine diagnostic service. J. Clin. Virol. *43*, 372–375.

Haney, S.A. (2007). Increasing the robustness and validity of RNAi screens. Pharmacogenomics *8*, 1037–1049.

Heid, C.A., Stevens, J., Livak, K.J., and Williams, P.M. (1996). Real time quantitative PCR. Genome Res. *6*, 986–994.

Hellemans, J., Preobrazhenska, O., Willaert, A., Debeer, P., Verdonk, P.C., Costa, T., Janssens, K., Menten, B., Van Roy, N., Vermeulen, S.J., *et al.* (2004). Loss-of-function mutations in LEMD3 result in osteopoikilosis, Buschke-Ollendorff syndrome and melorheostosis. Nat. Genet. *36*, 1213–1218.

Hellemans, J., Mortier, G., De Paepe, A., Speleman, F., and Vandesompele, J. (2007). qBase relative quantification framework and software for management and automated analysis of real-time quantitative PCR data. Genome Biol. *8*, R19.

Higuchi, R., Dollinger, G., Walsh, P.S., and Griffith, R. (1992). Simultaneous amplification and detection of specific DNA sequences. Biotechnology *10*, 413–417.

Higuchi, R., Fockler, C., Dollinger, G., and Watson, R. (1993). Kinetic PCR analysis: real-time monitoring of DNA amplification reactions. Biotechnology *11*, 1026–1030.

Holland, P.M., Abramson, R.D., Watson, R., and Gelfand, D.H. (1991). Detection of specific polymerase chain reaction product by utilizing the 5′–3′ exonuclease activity of Thermus aquaticus DNA polymerase. Proc. Natl. Acad. Sci. U.S.A. *88*, 7276–7280.

Hornig, M., Briese, T., Buie, T., Bauman, M.L., Lauwers, G., Siemetzki, U., Hummel, K., Rota, P.A., Bellini, W.J., O'Leary, J.J., *et al.* (2008). Lack of association between measles virus vaccine and autism with enteropathy: a case–control study. PLoS ONE 3, e3140.

Huang, T., Bohlenius, H., Eriksson, S., Parcy, F., and Nilsson, O. (2005). The mRNA of the *Arabidopsis* gene FT moves from leaf to shoot apex and induces flowering. Science *309*, 1694–1696.

Huggett, J., and Bustin, S.A. (2011). Standardisation and reporting for nucleic acid quantification. Accred. Qual. Assur. *16*, 399–405.

Huggett, J., Dheda, K., Bustin, S., and Zumla, A. (2005). Real-time RT-PCR normalisation; strategies and considerations. Genes Immun. *6*, 279–284.

Huggett, J.F., Novak, T., Garson, J.A., Green, C., Morris-Jones, S.D., Miller, R.F., and Zumla, A. (2008). Differential susceptibility of PCR reactions to inhibitors: an important and unrecognised phenomenon. BMC Res. Notes *1*, 70.

Jia, Y., Osborne, A., Rice, J.E., and Wangh, L.J. (2010). Dilute-'N'-Go dideoxy sequencing of all DNA strands generated in multiplex LATE-PCR assays. Nucleic Acids Res. *38*, e119.

Jin, N., He, K., and Liu, L. (2006). qPCR-DAMS: a database tool to analyze, manage, and store both relative and absolute quantitative real-time PCR data. Physiol. Genomics *25*, 525–527.

Johnson, M.P., Haupt, L.M., and Griffiths, L.R. (2004). Locked nucleic acid (LNA) single nucleotide polymorphism (SNP) genotype analysis and validation using real-time PCR. Nucleic Acids Res. *32*, e55.

Kreuzer, K.A., Bohn, A., Lass, U., Peters, U.R., and Schmidt, C.A. (2000). Influence of DNA polymerases on quantitative PCR results using TaqMan(TM)probe format in the LightCycler(TM) instrument. Mol. Cell Probes *14*, 57–60.

Kumar, R., Singh, S.K., Koshkin, A.A., Rajwanshi, V.K., Meldgaard, M., and Wengel, J. (1998). The first analogues of LNA (locked nucleic acids): phosphorothioate-LNA and 2′-thio-LNA. Bioorg. Med. Chem. Lett. *8*, 2219–2222.

Lacey, H.A., Nolan, T., Greenwood, S.L., Glazier, J.D., and Sibley, C.P. (2005). Gestational profile of Na^+/H^+ exchanger and Cl–/HCO3– anion exchanger mRNA expression in placenta using real-time QPCR. Placenta *26*, 93–98.

Larionov, A., Krause, A., and Miller, W. (2005). A standard curve based method for relative real time PCR data processing. BMC Bioinformatics *6*, 62.

Latorra, D., Arar, K., and Hurley, J.M. (2003). Design considerations and effects of LNA in PCR primers. Mol. Cell Probes *17*, 253–259.

Le Guillou-Guillemette, H., and Lunel-Fabiani, F. (2009). Detection and quantification of serum or plasma HCV RNA: mini review of commercially available assays. Meth. Mol. Biol. *510*, 3–14.

Lee, L.G., Connell, C.R., and Bloch, W. (1993). Allelic discrimination by nick-translation PCR with fluorogenic probes. Nucleic Acids Res. *21*, 3761–3766.

Lefever, S., Hellemans, J., Pattyn, F., Przybylski, D.R., Taylor, C., Geurts, R., Untergasser, A., and Vandesompele, J. (2009a). RDML: structured language and reporting guidelines for real-time quantitative PCR data. Nucleic Acids Res. *37*, 2066–2069.

Lefever, S., Vandesompele, J., Speleman, F., and Pattyn, F. (2009b). RTPrimerDB: the portal for real-time PCR primers and probes. Nucleic Acids Res. *37*, D942–5.

Lewis, F., and Maughan, N.J. (2004). Extraction of Total RNA from Formalin-Fixed Paraffin-Embedded Tissue. In A-Z of Quantitative PCR, Bustin, S.A., ed (IUL, La Jolla, CA), pp. 591–603.

Lewis, F., Maughan, N.J., Smith, V., Hillan, K., and Quirke, P. (2001). Unlocking the archive – gene expression in paraffin-embedded tissue. J. Pathol. *195*, 66–71.

Lopez, M.M., Llop, P., Olmos, A., Marco-Noales, E., Cambra, M., and Bertolini, E. (2009). Are molecular tools solving the challenges posed by detection of plant pathogenic bacteria and viruses? Curr. Issues Mol. Biol. *11*, 13–46.

Lu, X., Holloway, B., Dare, R.K., Kuypers, J., Yagi, S., Williams, J.V., Hall, C.B., and Erdman, D.D. (2008). Real-time reverse transcription-PCR assay for comprehensive detection of human rhinoviruses. J. Clin. Microbiol. *46*, 533–539.

Marcus, J.S., Anderson, W.F., and Quake, S.R. (2006). Parallel picoliter rt-PCR assays using microfluidics. Anal. Chem. *78*, 956–958.

Marshall, R., Chernesky, M., Jang, D., Hook, E.W., Cartwright, C.P., Howell-Adams, B., Ho, S., Welk, J., Lai-Zhang, J., Brashear, J., *et al.* (2007). Characteristics of the m2000 automated sample preparation and multiplex real-time PCR system for detection of Chlamydia trachomatis and Neisseria gonorrhoeae. J. Clin. Microbiol. *45*, 747–751.

Marullo, M., Zuccato, C., Mariotti, C., Lahiri, N., Tabrizi, S.J., Di Donato, S., and Cattaneo, E. (2010). Expressed Alu repeats as a novel, reliable tool for normalization of real-time quantitative RT-PCR data. Genome Biol. *11*, R9.

Mestdagh, P., Van Vlierberghe, P., De Weer, A., Muth, D., Westermann, F., Speleman, F., and Vandesompele, J. (2009). A novel and universal method for microRNA RT-qPCR data normalization. Genome Biol. *10*, R64.

Morrison, T., Hurley, J., Garcia, J., Yoder, K., Katz, A., Roberts, D., Cho, J., Kanigan, T., Ilyin, S.E., Horowitz, D., *et al.* (2006). Nanoliter high throughput quantitative PCR. Nucleic Acids Res. *34*, e123.

Morrison, T.B., Weis, J.J., and Wittwer, C.T. (1998). Quantification of low-copy transcripts by continuous SYBR Green I monitoring during amplification. Biotechniques *24*, 954–962.

Neuzil, P., Zhang, C., Pipper, J., Oh, S., and Zhuo, L. (2006). Ultra fast miniaturized real-time PCR: 40 cycles in less than six minutes. Nucleic Acids Res. *34*, e77.

Nolan, T., Hands, R.E., and Bustin, S.A. (2006a). Quantification of mRNA using real-time RT-PCR. Nature Protocols *1*, 1559–1582.

Nolan, T., Hands, R.E., Ogunkolade, B.W., and Bustin, S.A. (2006b). SPUD: a qPCR assay for the detection of inhibitors in nucleic acid preparations. Anal. Biochem. *351*, 308–310.

Nordgard, O., Kvaloy, J.T., Farmen, R.K., and Heikkila, R. (2006). Error propagation in relative real-time reverse transcription polymerase chain reaction quantification models: the balance between accuracy and precision. Anal. Biochem. *356*, 182–193.

O'Connor, W., and Runquist, E.A. (2008). Error minimization algorithm for comparative quantitative PCR analysis: Q-Anal. Anal. Biochem. *378*, 96–98.

Pabinger, S., Thallinger, G.G., Snajder, R., Eichhorn, H., Rader, R., and Trajanoski, Z. (2009). QPCR: Application for real-time PCR data management and analysis. BMC Bioinformatics *10*, 268.

Pfaffl, M., Schwarz, F., and Sauerwein, H. (1998). Quantification of insulin-like growth factor-1 (IGF-1) mRNA: modulation of growth intensity by feeding results in inter- and intra-tissue-specific differences of IGF-1 mRNA expression in steers. *106*, 514–521.

Pierce, K.E., and Wangh, L.J. (2011). LATE-PCR and allied technologies: real-time detection strategies for rapid, reliable diagnosis from single cells. Meth. Mol. Biol. *688*, 47–66.

Pignot, G., Bieche, I., Vacher, S., Guet, C., Vieillefond, A., Debre, B., Lidereau, R., and Amsellem-Ouazana, D. (2008). Large-scale Real-time Reverse Transcription-PCR approach of angiogenic pathways in human transitional cell carcinoma of the bladder: identification of VEGFA as a major independent prognostic marker. Eur. Urol. *56*, 678–688.

Platts, A.E., Johnson, G.D., Linnemann, A.K., and Krawetz, S.A. (2008). Real-time PCR quantification using a variable reaction efficiency model. Anal. Biochem. *380*, 315–322.

Reuter, T., Gilroyed, B.H., Alexander, T.W., Mitchell, G., Balachandran, A., Czub, S., and McAllister, T.A. (2009). Prion protein detection via direct immuno-quantitative real-time PCR. J. Microbiol. Methods *78*, 307–311.

Ririe, K.M., Rasmussen, R.P., and Wittwer, C.T. (1997). Product differentiation by analysis of DNA melting curves during the polymerase chain reaction. Anal. Biochem. *245*, 154–160.

Rutledge, R.G., and Stewart, D. (2008). A kinetic-based sigmoidal model for the polymerase chain reaction and its application to high-capacity absolute quantitative real-time PCR. BMC Biotechnol. *8*, 47.

Saiki, R.K., Scharf, S., Faloona, F., Mullis, K.B., Horn, G.T., Erlich, H.A., and Arnheim, N. (1985). Enzymatic amplification of beta-globin genomic sequences and restriction site analysis for diagnosis of sickle cell anemia. Science *230*, 1350–1354.

Sanchez, J.A., Pierce, K.E., Rice, J.E., and Wangh, L.J. (2004). Linear-after-the-exponential (LATE)-PCR: an advanced method of asymmetric PCR and its uses in quantitative real-time analysis. Proc. Natl. Acad. Sci. U.S.A. *101*, 1933–1938.

Sansone, S.A., Fan, T., Goodacre, R., Griffin, J.L., Hardy, N.W., Kaddurah-Daouk, R., Kristal, B.S., Lindon, J., Mendes, P., Morrison, N., *et al.* (2007). The metabolomics standards initiative. Nature Biotechnol. *25*, 846–848.

Schmittgen, T.D., Lee, E.J., Jiang, J., Sarkar, A., Yang, L., Elton, T.S., and Chen, C. (2008). Real-time PCR quantification of precursor and mature microRNA. Methods *44*, 31–38.

Smith, C.J., and Osborn, A.M. (2009). Advantages and limitations of quantitative PCR (Q-PCR)-based approaches in microbial ecology. FEMS Microbiol. Ecol. *67*, 6–20.

Smith, M.V., Miller, C.R., Kohn, M., Walker, N.J., and Portier, C.J. (2007). Absolute estimation of initial concentrations of amplicon in a real-time RT-PCR process. BMC Bioinformatics *8*, 409.

Spiess, A.N., Feig, C., and Ritz, C. (2008). Highly accurate sigmoidal fitting of real-time PCR data by introducing a parameter for asymmetry. BMC Bioinformatics *9*, 221.

Stahlberg, A., Elbing, K., Andrade-Garda, J.M., Sjogreen, B., Forootan, A., and Kubista, M. (2008). Multiway real-time PCR gene expression profiling in yeast *Saccharomyces cerevisiae* reveals altered transcriptional response of ADH-genes to glucose stimuli. BMC Genomics *9*, 170.

Stahlberg, A., Hakansson, J., Xian, X., Semb, H., and Kubista, M. (2004a). Properties of the reverse transcription reaction in mRNA quantification. Clin. Chem. *50*, 509–515.

Stahlberg, A., Kubista, M., and Pfaffl, M. (2004b). Comparison of reverse transcriptases in gene expression analysis. Clin. Chem. *50*, 1678–1680.

Taylor, C.F., Field, D., Sansone, S.A., Aerts, J., Apweiler, R., Ashburner, M., Ball, C.A., Binz, P.A., Bogue, M., Booth, T., *et al.* (2008). Promoting coherent minimum reporting guidelines for biological and biomedical investigations: the MIBBI project. Nat. Biotechnol. *26*, 889–896.

Taylor, C.F., Paton, N.W., Lilley, K.S., Binz, P.A., Julian, R.K., Jr., Jones, A.R., Zhu, W., Apweiler, R., Aebersold, R., Deutsch, E.W., *et al.* (2007). The minimum information about a proteomics experiment (MIAPE). Nat. Biotechnol. *25*, 887–893.

Thelwell, N., Millington, S., Solinas, A., Booth, J., and Brown, T. (2000). Mode of action and application of Scorpion primers to mutation detection. Nucleic Acids Res. *28*, 3752–3761.

Timcheva, I.I., Maximova, V.A., Deligeorgiev, T.G., Gadjev, N.I., Sabnis, R.W., and Ivanov, I.G. (1997). Fluorescence spectral characteristics of novel asymmetric monomethine cyanine dyes in nucleic acid solutions. FEBS Lett. *405*, 141–144.

Tyagi, S., and Kramer, F.R. (1996). Molecular beacons: probes that fluoresce upon hybridization. Nat. Biotechnol. *14*, 303–308.

Uhlmann, V., Martin, C.M., Sheils, O., Pilkington, L., Silva, I., Killalea, A., Murch, S.B., Walker-Smith, J., Thomson, M., Wakefield, A.J., *et al.* (2002). Potential viral pathogenic mechanism for new variant inflammatory bowel disease. Mol. Pathol. *55*, 84–90.

Vandesompele, J., De Preter, K., Pattyn, F., Poppe, B., Van Roy, N., De Paepe, A., and Speleman, F. (2002). Accurate normalization of real-time quantitative RT-PCR data by geometric averaging of multiple internal control genes. Genome Biol. *3*, 0034.1–0034.11.

Verderio, P., Pizzamiglio, S., Gallo, F., and Ramsden, S.C. (2008). FCI: an R-based algorithm for evaluating uncertainty of absolute real-time PCR quantification. BMC Bioinformatics *9*, 13.

Wagner, B.K., and Arany, Z. (2009). High-throughput real-time PCR for detection of gene-expression levels. Meth. Mol. Biol. *486*, 167–175.

Walker, N.J. (2002). A technique whose time has come. Science *296*, 557–559.

Westh, H., Lisby, G., Breysse, F., Boddinghaus, B., Chomarat, M., Gant, V., Goglio, A., Raglio, A., Schuster, H., Stuber, F., *et al.* (2009). Multiplex real-time PCR and blood culture for identification of bloodstream pathogens in patients with suspected sepsis. Clin. Microbiol. Infect. *15*, 544–551.

Whitcombe, D., Theaker, J., Guy, S.P., Brown, T., and Little, S. (1999). Detection of PCR products using self-probing amplicons and fluorescence. Nat. Biotechnol. *17*, 804–807.

Wittwer, C.T., Herrmann, M.G., Moss, A.A., and Rasmussen, R.P. (1997). Continuous fluorescence monitoring of rapid cycle DNA amplification. Biotechniques *22*, 130–138.

Zhang, C., and Xing, D. (2007). Miniaturized PCR chips for nucleic acid amplification and analysis: latest advances and future trends. Nucleic Acids Res. *35*, 4223–4237.

Zhang, X.X., Zhang, T., and Fang, H.H. (2009). Antibiotic resistance genes in water environment. Appl. Microbiol. Biotechnol. *82*, 397–414.

Zuker, M. (2003). Mfold web server for nucleic acid folding and hybridization prediction. Nucleic Acids Res. *31*, 3406–3415.

DOI: https://doi.org/10.21775/9781912530243.03

Homogeneous Fluorescent Chemistries for Real-time PCR

3

Martin A. Lee, David J. Squirrell, Dario L. Leslie and Tom Brown

Abstract

The development of fluorescent methods for the closed tube polymerase chain reaction has greatly simplified the process of quantification. Current approaches use fluorescent probes that interact with the amplification products during the PCR to allow kinetic measurements of product accumulation. These probe methods include generic approaches to DNA quantification such as fluorescent DNA binding dyes. There are also a number of strand-specific probes that use the phenomenon of fluorescent energy transfer. In this chapter we describe these methods in detail, outline the principles of each process, and describe published examples. This text has been written to provide an impartial overview of the utility of different assays and to show how they may be used on various commercially available thermal cyclers.

Introduction

The fluorescent real-time polymerase chain reaction (PCR) (Saiki *et al.*, 1985; Mullis *et al.*, 1987) can provide both qualitative and quantitative analysis in various applications. Real-time PCR differs from earlier methods of analysis in that additional components are required to carry out the process. These components include an optical system integrated into the thermal cycler, and a probe that reports amplification during the course of the PCR process. The real-time PCR thermal cycler is discussed in Logan and Edwards (2009a) in detail. In this chapter we discuss the probe technologies or 'reporting chemistries.' Since the first edition of this text new methods have been reported in the literature and have become commercial products. This edition expands to include these developments.

In order to give the reader a complete understanding of current technology, the principles of real-time analysis will initially be presented outside the context of any specific commercial platform. The number of these is increasing and it is certain that the capabilities on offer will undergo continued improvement and allow new fluorescent approaches to be realized. At this point it is important to highlight to new users the relationship between the choice of probe system and the instrument. These are inextricably linked: the optical specification and the analysis tools on any given platform greatly influence the applicability and utility of different probe systems. Whilst the main factors for the choice of instrument are often driven by throughput requirements and the initial purchase cost, careful consideration of the reporting chemistry for the required application should be made before purchase since the operation of a particular chemistry may be greatly compromised on some platforms.

Equally important is the technical support provided by suppliers. This may be limited for non-supported chemistries and so called 'open platforms' from manufacturers that do not support any specific chemistry, or who may not be able to provide technical advice for their chemistry of choice. In this chapter we break down probe technology into a number of components to enable the reader to understand how the different assay systems may be used on current and future fluorimetric thermal cyclers. For new users this will facilitate the implementation of various assays on commercial platforms.

We first present a background to the limitations of PCR without real-time detection and introduce the two main classes of real-time chemistries. A second section on the basics of fluorescent resonance energy transfer (FRET) will allow the reader to understand how dye technologies may be applied in strand-specific applications. An outline of the function and utility of real-time PCR probe methods will provide the reader with an understanding of how such technology may be best applied. Finally, experimental considerations and specific examples will be provided although the reader should refer to the relevant chapters in this text for detailed information.

Background to real-time PCR

Prior to the introduction of the first commercial real-time PCR instruments in 1996 the utility of the PCR was limited. It could only be easily applied as a qualitative method. Analysis of amplification products was carried out at the end of thermal cycling using techniques such as gel electrophoresis or PCR ELISA. The dynamic range for quantitative PCR using these approaches was limited because the accumulation of specific product tends to plateau when the reaction is allowed to progress to completion. At high cycle numbers the amount of product is often unrelated to the initial amount of target nucleic acid. Factors that contribute to the plateau effect include, amongst others, substrate depletion (nucleotides, amplimers, etc.), specific inhibition (competitive binding of products-to-products rather than amplimer-to-product) and non-specific product inhibition (amplimer artefact accumulation, mis-priming and the accumulation of pyrophosphate). Quantitative analysis, with a limited dynamic range, could only be achieved by stopping thermal cycling before this plateau was achieved. Competitive PCR (utilizing molecular mimics that compete for amplimers) could be used to achieve a wider dynamic range. However, this approach is less accurate and more time consuming than the real-time methods described here.

The fluorescent real-time approach requires a thermal cycler that can interrogate the sample throughout the course of amplification. These instruments are discussed in Logan and Edwards (2009a) and the basic kinetic theory of real-time PCR is in Logan and Edwards (2009b). The signal collected during the exponential stage of amplification is of most use for the determination of the number of initial target nucleic acid molecules. At this stage the reaction efficiency is so high that the number of amplicon molecules effectively doubles every cycle. Amplification is observed as a two-fold increase in signal as it rises above that of the background noise. The amount of signal noise in a real-time assay is a function of the type of chemistry utilized, the optical and thermal (since fluorescence has a reciprocal relationship with temperature) performance of the instrument as well as other experimental considerations such as volume effects and the degree of mixing of reaction cocktails.

There are two main classes of real-time fluorescent chemistries: 'generic' and 'strand-specific' methods. Generic methods use probes that bind non-specifically to DNA and include the intercalators and other DNA binding dyes. Strand-specific methods use nucleic acid probes that target the amplicon (product) between the amplimer binding regions. It should be noted that most of the systems described utilize DNA for the probe. The use of modified bases was proposed in 1991 (Holland *et al.*, 1991). Methods for peptide nucleic acid (PNA) probes utilizing FRET have been reported (Ortiz *et al.*, 1998) and described for application in real-time PCR (Svanik *et al.*, 2000a,b), and for those probes that use hybridization alone (rather than hydrolysis) it should be possible to utilize PNA and other analogues such as locked nucleic acid (LNA) (Koshkin *et al.*, 1998) in the chemistries. Nucleotide analogues increase the melting temperature (TM) of the probe improving signalling in most methods, in particular where the target is AT rich. The strand-specific methods have the advantage that amplification of reaction artefacts such as amplimer-dimers do not contribute to the observed signal. Generally they provide higher specificity and better signal–noise ratios than the generic methods that are described next. The type and mode of action of different probe systems is illustrated in Table 1.1.

Generic detection using DNA binding dyes and melting point analysis

DNA-binding dyes have been used extensively in molecular biology research for the direct analysis and quantification of nucleic acids. These dyes bind to dsDNA with enhanced fluorescence. Ethidium bromide, which is used conventionally for staining agarose gels, has been used in real-time PCR (Higuchi *et al.*, 1992, 1993). Other dyes, such as those from Molecular Probes (YoPro® and YoYo®) (Ogura *et al.*, 1994), have also been reported. The fluorescent signal is usually monitored towards the end of the extension step in a 3step PCR.

Of these probes, by far the most reported are the minor groove-binding dyes SYBR®Green-1 (Becker *et al.*, 1996) and SYBR®Gold. These dyes typically exhibit a 20- to 100-fold fluorescence enhancement on binding and are commonly used because their emission maxima closely match that of fluorescein and the optics in most commercial real-time instruments are set to detect in this (circa 520 nm) wavelength range. The SYBR® dyes are popular for real-time detection because they are readily available from PCR reagent suppliers and their use requires little additional experimental design.

The optimum concentration for these dyes for a number of instrument platforms is published in the open literature. For SYBR® dyes this is typically 1:30,000 to 1:100,000 dilution of the reference solution supplied by the manufacturer, although the concentration is dependent on the tube format (e.g. composite glass or native polypropylene) and the optical efficiency of the instrument. Ready-to-go cocktails are available from suppliers with the dye already in the reagent mix.

Binding dye chemistries are excellent for assay optimization. The signal is proportional to the total nucleic acid concentration and therefore directly related to the PCR process. In an optimized assay, they may thus be used to determine accurately the reaction efficiency. It should be noted that this is not always the case when using labelled nucleic acid probes, where the signal generated is related to both the assay and the 'probe efficiency'. In some assay systems the probe efficiency can have a major effect on the signal obtained and this is discussed later for each assay type. In both strand specific and generic RT-PCR, the signal obtained is additionally dependent on the efficiency of the reverse transcriptase step.

Melting point analysis may be carried out on most commercial real-time instruments, usually at the end of the PCR amplification (Fig. 1.1). Temperature-dependent fluorescence measurements are made whilst slowly increasing the temperature of the reaction products

Table 1.1 Summary of fluorogenic reporting chemistries for real-time PCR. The schematics show the probe structure and mode of action

Method	Principle	Self-probing embodiment	**Principle** Probe structure	Nucleic acid interaction
DNA binding agent	Enhancement of fluorescence DNA binding into DNA duplex	N/A	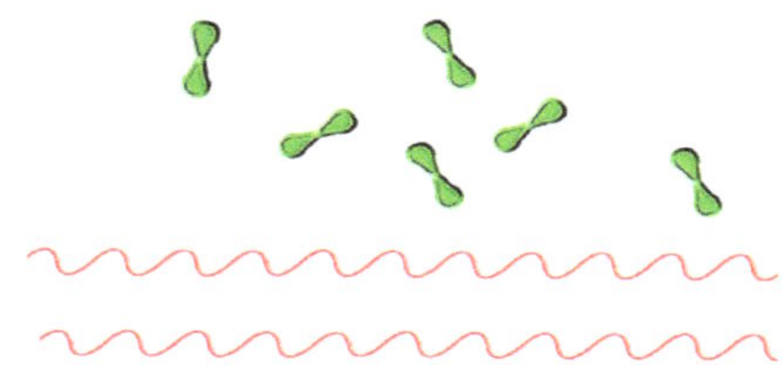	
5′ nuclease assay	Hydrolysis of probe by concomitant hydrolysis of probe during primer extension	Intrataq		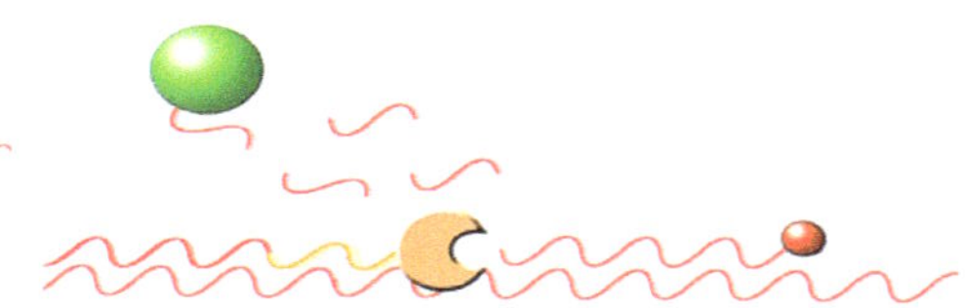
Dual hybridization probe (*cis*)	Hybridization of two linear probes to nascent PCR strand	Embodiment of Scorpions®	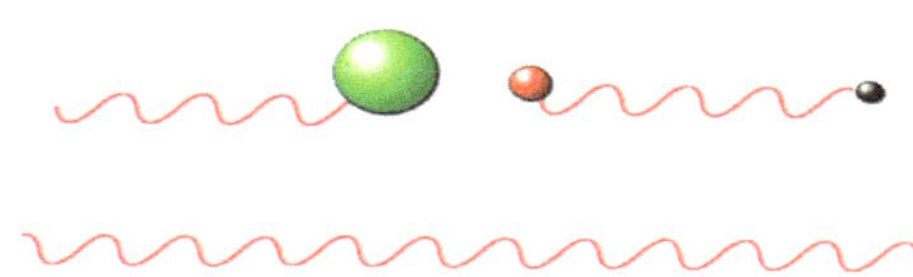	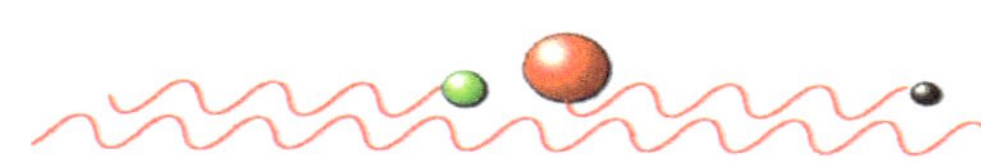
Dual hybridization probe (*trans*)	Hybridization of linear probe to nascent PCR strand generated by a labelled primer	Not reported	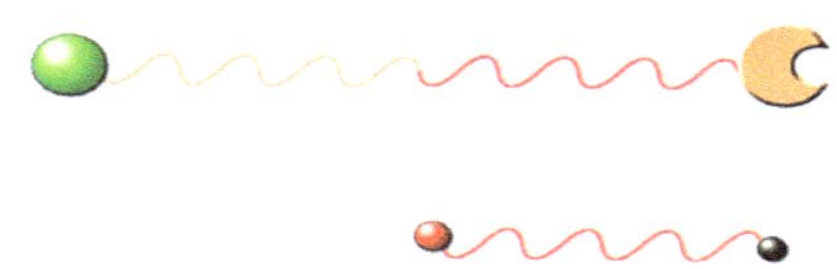	

Table 1.1 (Continued)

Method	Principle	Self-probing embodiment	Principle: Probe structure	Principle: Nucleic acid interaction
ResonSense®	Hybridization of linear probe to nascent PCR strand. Energy is transferred through a DNA binding agent	Angler®	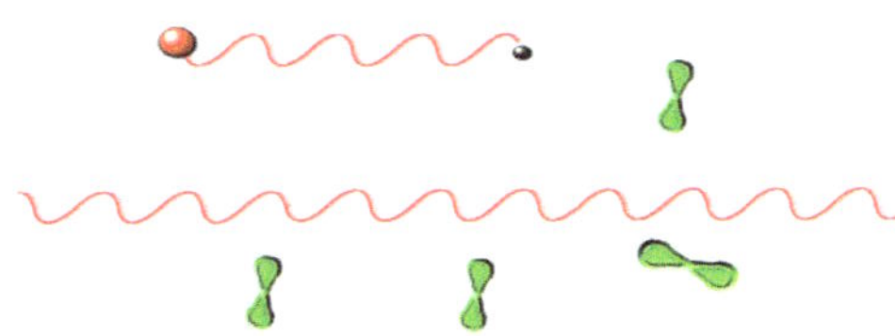	
Hybeacons®	Hybridization of linear probes.	Not reported	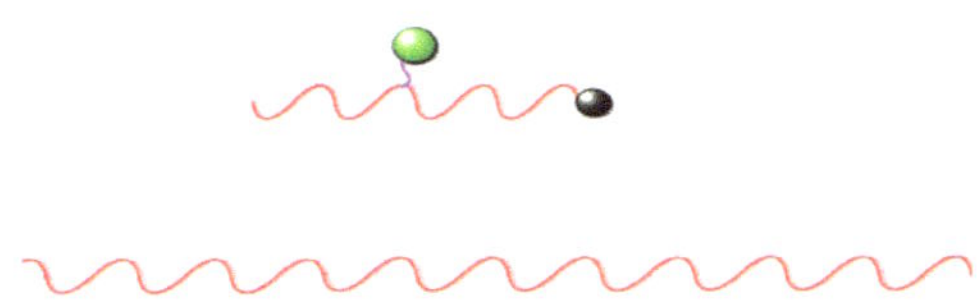	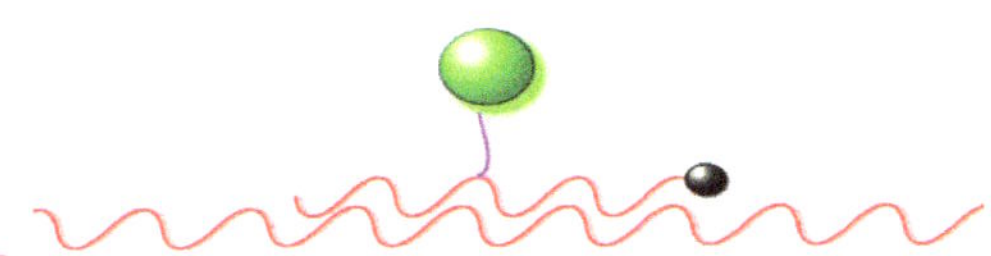
Eclipse®	Linear conformational change. Short probe is stabilized to target using a minor groove binding agent	Not reported	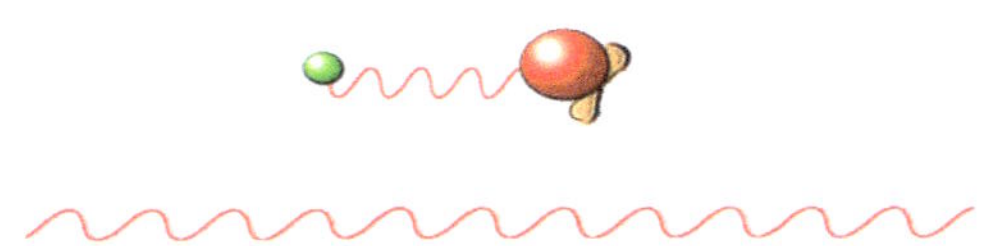	
Light-Up®	Fluorescent enhancement upon hybridization with DNA-binding agent covalently linked to probe	Not reported	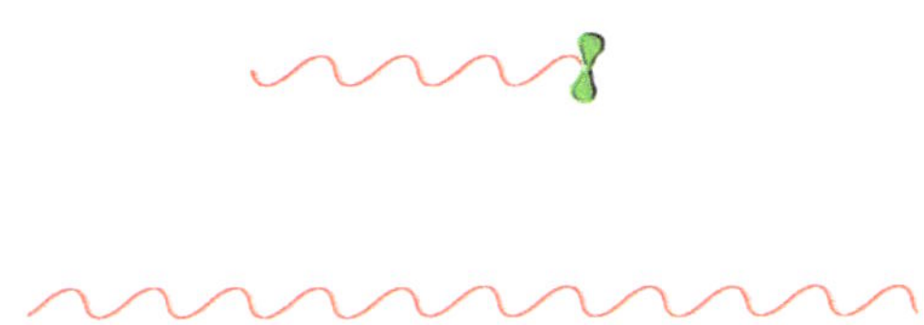	

Method	Principle	Self-probing embodiment	**Principle** Probe structure	Nucleic acid interaction
Molecular Beacons	Conformational change. Native probe stabilized with hairpin structure	Scorpions®	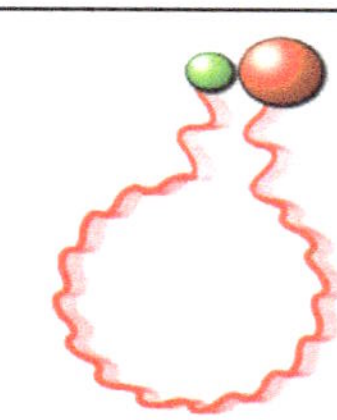	
Ying–Yang®	Competitive binding of linear probe	Duplex Scorpions®	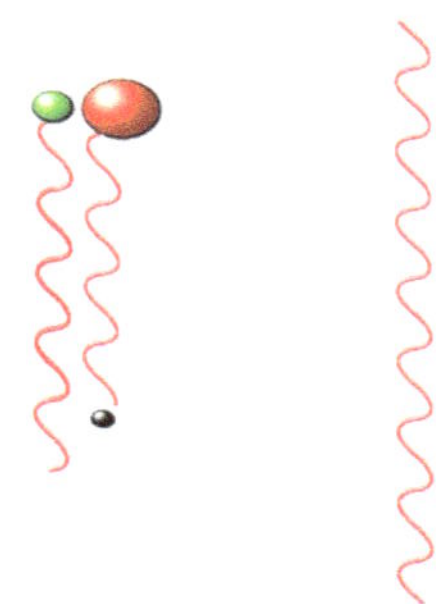	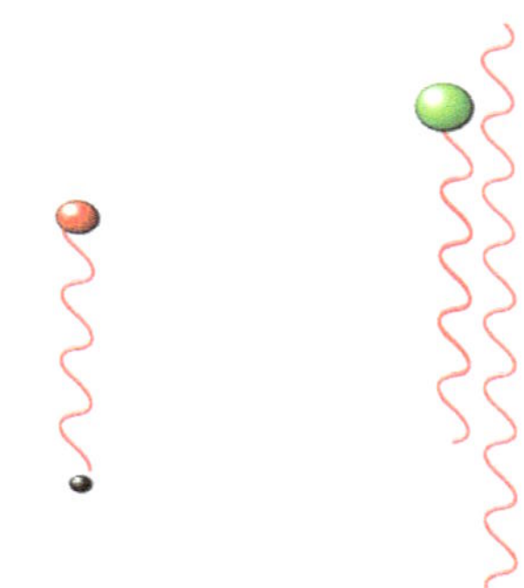

Key: DNA, donor dye, acceptor dye, DNA binding agent, minor groove binder, polymerase, 3' blocker.

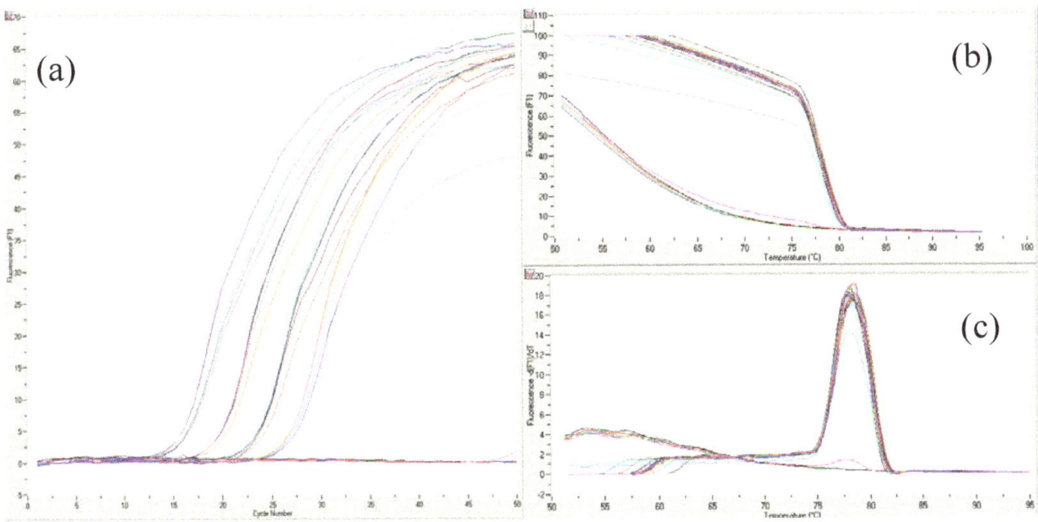

Figure 1.1 Amplification using generic binding dyes (minor-groove binding dye SYBR®Gold) on the LightCycler®. (a) The amplification plot shows 50 cycles (to completion) of amplification of four dilutions (four replicates) of a 10-fold dilution series and four no-template controls, one of which produces a low amplification signal. The reaction utilised anti-Taq antibody hot start and UNG carry-over protection and the data utilise the background subtraction algorithm. (b) Melting point analysis of the amplified products, and (c) the first negative differential of the fluorescence, with respect to Temperature, plotted against Temperature. From this plot it can observed that the signal in the positive no-template control has the same melting point as the specific amplification in positive samples, and therefore amplification is most likely a results of cross-contamination.

from around 50°C to around 95°C. As DNA duplexes melt the fluorescence decreases as the bound dye is released. Most instruments provide an analysis of this data by plotting the first negative differential of the fluorescence signal with respect to temperature, against temperature. This plot appears as one or more peaks representing the point(s) at which the maximum rate(s) of change in fluorescence occurs corresponding to a particular dsDNA product. Specific reaction products generally melt at a higher temperature than artefacts such as amplimer dimers and misprimed products. Melting temperature is a function of GC content and to a lesser extent product length, which allows the artefact to be observed to melt at a higher temperature than specific products in some reactions, although this is a rare observation.

The melting peak method is analogous to that of agarose gel electrophoresis in that it does not unequivocally determine the presence of the correct sequence. Products with similar molecular masses or melting peaks may not be resolved by either technique. Gel electrophoresis has a higher resolution than that of melting point analysis and the quality of the melting point data collected varies greatly between hardware platforms. Those instruments where the temperature is accurately controlled and recorded, and where the correlation between temperature and fluorescence is good, will produce the highest resolution melting point data.

The major limitation with binding dye chemistries is that they bind to total nucleic acid, and in a PCR of suboptimal efficiency a significant amount of signal will be derived from reaction artefacts. This leads to a dramatic loss in the dynamic range of quantification of samples and standards with low (< 1000) initial target copies. However, a melting point analysis allows non-specific artefacts to be discriminated, to some degree, from specific amplicons (Ririe *et al.*, 1997). Many dyes become limiting at the concentrations they are used in the reaction. At higher concentrations they may inhibit the PCR. When dyes become limiting, melting analysis loses resolution because of competition between products for free dye. New dyes such as LC-green can be used at higher concentration and they have facilitated high resolution melting on specialist instruments; the analysis of polymorphisms within products without the use of probes.

The melting point analysis adds significantly to the utility of fluorescent DNA-binding dyes since the data can be used to determine a number of important variables. Amplimer complements can be synthesized and used in melting point experiments to empirically determine annealing temperatures. The melting temperature of the product when determined may be used to improve the reaction in two respects on some instruments. Firstly, sequentially decreasing the denaturing step to a few degrees above the product melting temperature ensures that high molecular mass sample DNA is not denatured in later cycles which reduces the risks of mis-priming events in later cycles. Secondly, this lowering of the denaturing temperature allows a shortening of the time taken for thermal cycling. If the melting temperature of reaction artefacts is significantly lower than that of the amplicon then it is possible to monitor fluorescence at a temperature just below the melting point of the amplicon, but higher than that of artefacts (Morrison *et al.*, 1998). This serves to reduce the artefact contribution to the specific amplification signal and may improve the dynamic range of quantification for a given assay.

An important consideration when using such dyes is the incorporation of a 'hot-start' to reduce the formation of artefacts and so improve the dynamic range of quantification (Morrison *et al.*, 1998; Lee *et al.*, 1999). It should be noted that most of these dyes are both toxic and mutagenic to varying degrees, due to their ability to bind DNA. Storage conditions for these reagents are critical since they may be photo-bleached by both ambient and artificial light. The use of light-tight containers such as darkened microfuge tubes will reduce this, but most are labile on long-term storage.

When optimized, a SYBR®Green-1 or SYBR®Gold assay can be used for most applications (such as expression studies). The ability to multiplex these probes is limited because the binding, which generates the fluorescence, is non-specific. However, it is possible to multiplex by melting point using binding dyes, providing that the melting peak of each species can be resolved. Whilst allelic variation and mutation detection can be carried out, for example, using the amplification refractory mutation system (ARMS) approach (Newton *et al.*, 1989), this requires assays for both variants to be carried out in separate tubes. Because of these limitations assays that require high-confidence results are best addressed using the strand specific approaches that are discussed next.

Strand-specific detection using labelled nucleic acid probes

The strand-specific methods employ fluorophore-coupled nucleic acids that interact with reaction products, probing accumulating PCR products for the presence of the target sequence. Such PCRs exhibit greater specificity than the respective assay without the probe component. However, the probe targets only a portion of the amplicon sequence and therefore the signal for the product only reports the presence of sequences that are complementary to the probe rather than the entire amplicon. The basis of detection of the product can be classed by the mode of action of the probe.

The efficiency of the probe is important if it is to produce a signal that correlates closely with the concentration of amplicon in the exponential stage of amplification, and thus is used to determine the number of initial target copies. With an inefficient probe the signal may not increase exponentially and not correlate with initial target copy number. Inefficient probes will produce an inaccurate value for the PCR efficiency. The signal generated by a probe is dependent on, for example, sufficient hybridization and/or cleavage as is later described. The target sequence is the main factor affecting probe efficiency because of different sequence-dependent factors for a given probe type. Therefore the benefits of selecting a probe type should be carefully considered when designing an experiment and also when purchasing an instrument, since the preferred assay system may not easily be applied on every instrument.

The optimum composition of the core chemistry may vary with different platforms and this is discussed in Logan and Edwards (2009b). The optical specifications of platforms differ such that they will be optimized for the specific fluorophores. The concentration of probes may also vary between platforms but typically probes are used at a final concentration of around 0.1–1 μM.

In this text probe types are grouped for discussion by mode of action. The evolution of real-time methods is not presented chronologically, although reference is made to how and when they were implemented. Most of these methods use the phenomenon of fluorescence energy transfer (FET). More specifically, many of these methods use fluorescence resonance energy transfer (FRET), in which energy is transferred between two fluorophores, rather than, for example, between a fluorophore and a non-fluorescent quencher. This phenomenon will be discussed first and then the application and experimental considerations of the various methods will be outlined.

Fluorescent resonance energy transfer probes (FRET) and the passive reference

FRET or Förster energy transfer involves the non-radiative (no photon emitted) transfer of energy between fluorophores and has been used for a number of biological assays. There are excellent reviews of principles and applications (Cardullo *et al.*, 1988; Selvin, 1995; Clegg, 1992), so there is no need to include an in depth description of the process in this text. Here we present an overview and the two main approaches to detection and multiplexing using FRET as used in real-time PCR experiments. Reference will be made to specific instruments to illustrate the process.

For FRET to occur there must be at least two fluorophores in close proximity. One fluorophore, called the 'donor,' is excited by an incident light source and emits light at a longer wavelength that falls within the excitation spectrum of the second fluorophore, called the 'acceptor.' When the donor and acceptor are on the same or neighbouring molecule(s) then the emission energy of the excited donor may be transferred through a number of distance and orientation dependent effects including dipole–dipole interactions between the donor and acceptor. Variations in the intensity of emissions of fluorophores on nucleic acids resulting from spatial changes during the course of amplification provide the basis for detecting amplicon accumulation (Fig. 1.2). The choice of donor-acceptor dye pairs

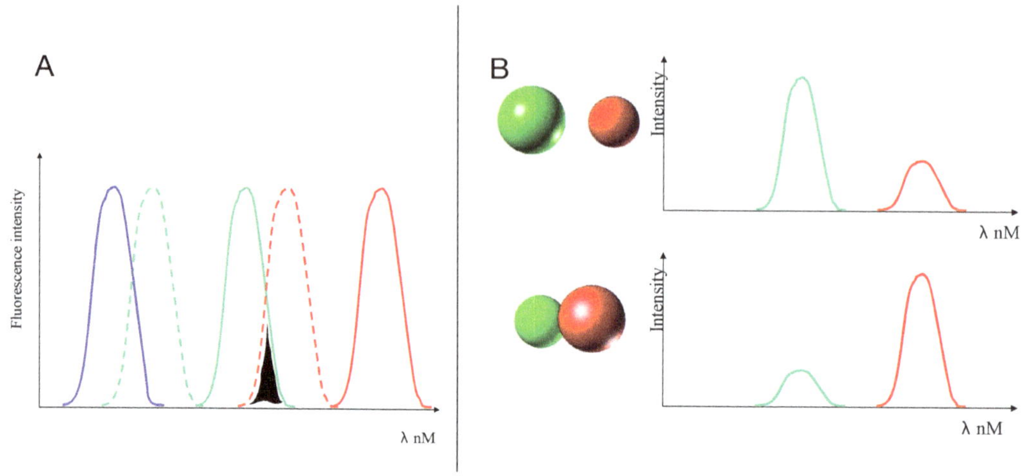

Figure 1.2 Fluorescence energy resonance transfer. (A) The emissions of a light source (blue line) falls within the excitation spectrum (green line dashed) of a fluorophore and cause this dye to fluoresce. The emissions (green solid) of this fluorophore fall within the excitation spectrum (red dashed line) of a second fluorophore that emits fluorescence at longer wavelengths (red solid). When the second fluorophore is close to the first on the same or neighbouring molecule, the energy that would be emitted by the first fluorophore (top of B) can be transferred to the second through a number of energy transfer mechanisms such that the second fluorophore emits this energy at its own emission wavelengths (bottom of B).

is important in achieving a good signal: noise ratio. The Förster distance (the distance between the dye pair at which the efficiency of energy transfer is 50%) varies for different dye combinations.

Analysis utilizing FRET involves monitoring either a rise or drop in fluorescence of either fluorophore depending on the assay type and instrument. On some platforms it is possible to present this data as a change in the quotient of the emissions of the two dyes. For example, if the basis of detection is the donor fluorophore being liberated spatially from the acceptor then one could express the data as donor/acceptor. If the converse is true the basis of detection is the donor being brought in close proximity to the acceptor one could express the data as acceptor/donor. This simple type of analysis is useful in a non-multiplexed real-time PCR context since subtle changes in sample volume due to pipetting errors and/or changes in sample volume due to evaporation can be normalized using this approach.

Of importance to the signal–noise ratio of assay systems utilizing FRET is the ratio of dyes in a FRET pair. In particular this is important where two dyes may be used on the same probe (dual labelled) as described later. These probes are more complicated in terms of manufacture since they require incorporation of both dyes into the probe molecule. These types of probes are therefore more expensive to synthesize. These oligonucleotide probes are usually made by one of two common methods. The first dye, if this is at the 3′-end of the probe, is part of the support matrix on which the probe is synthesized by sequential addition of each base in the sequence. The second dye is either directly incorporated by the use of fluorophore-labelled monomer, or by the inclusion of a monomer with a reactive group, and the dye is subsequently covalently cross-linked after synthesis. In the latter case, any free dye needs to be removed by a separation step, and this is usually carried out using high-performance liquid chromatography (HPLC). In cases where neither dye is at the

3′-end of the probe (e.g. Scorpions), both dyes are incorporated as fluorophore labelled monomers. In dual-labelled probe types the donor is liberated from the acceptor. If the stoichiometry of the donor and acceptor fluorophore incorporation is not balanced (i.e. the probe is impure) then a low signal and/or a high background may be obtained. The quality of probes can vary greatly between suppliers and even between scales of synthesis. Quality control of such probes is important to avoid wasting time trying to optimize assays with poor quality dual-labelled probes. The authors would recommend that probes should be checked following manufacture by any one or more of a number of techniques to ensure the integrity of each probe. These include mass spectrometry, HPLC separation using analysis of the fractions at the relevant wavelengths. Degradation and subsequent analysis utilizing a spectrofluorimeter is an ideal method of determining dye stoichiometry. Some suppliers offer one or more methods as a service or as part of their quality control system.

Another experimental consideration that is important in improving the signal–noise ratio of assays is the sequence dependent quenching of fluorescence. In particular, a fluorescein dye may be partially quenched by a neighbouring guanine base (G) in the sequence, which should be avoided. Whilst both dyes in a FRET pair may be fluorescent it is becoming increasingly common to find the use of non-fluorescent acceptors (quenchers). These are often termed 'dark quenchers' and can provide collisional or Dexter quenching. No light is emitted from these molecules, rather the energy is dissipated to solvent. Their use is now more common because a non-fluorescent moiety on a dual-labelled probe frees the instrument's bandwidth, increasing both the utility of instrument for multiplexing, and the dynamic range of a given detector where fluorescent emissions may have otherwise overlapped. Such quenching moieties are now commonly employed on dual-labelled probes. One limitation of this approach we have observed is that it is difficult for the end user to determine the efficiency of the incorporation of these quenchers into dual-labelled probes. In practice this may lead to a higher noise for the reasons described above (Zahng *et al.*, 2001). The use of gold nanoparticles as efficient quenching moieties has also been reported (Dubertret *et al.*, 2001); however, this has yet to be demonstrated in a homogenous PCR.

Since most instruments can excite and detect the emissions of fluorescein, a combination of fluorescein and any dye that can act as an acceptor may be used to report amplification. On most instruments the rise or drop in either fluorescein or the acceptor would be suitable to report amplification in various assay types. However, when a multiplexed assay is required, the instrument specification narrows the choice of dyes and assay types that can be used. Two principal approaches are used by different manufacturers to allow multiplexing. Here we describe these as the universal donor method and the universal acceptor method (see Fig. 1.3):

- *Universal donor.* In this arrangement the system is set up to multiplex assays by monitoring the rise in acceptor emissions. Different acceptors are used to report amplification from two or more amplifications in the same tube. Typically the machine has one excitation source and detectors at longer wavelengths that are used when multiplexing, e.g. the LightCycler®.
- *Universal acceptor.* In this arrangement the system is set up to multiplex by monitoring the rise in donor emissions that are being quenched by a longer wavelength or non-fluorescent acceptor. Different donors are used to report amplification from two or more amplifications in the same tube. Typically the systems detect donors that have short

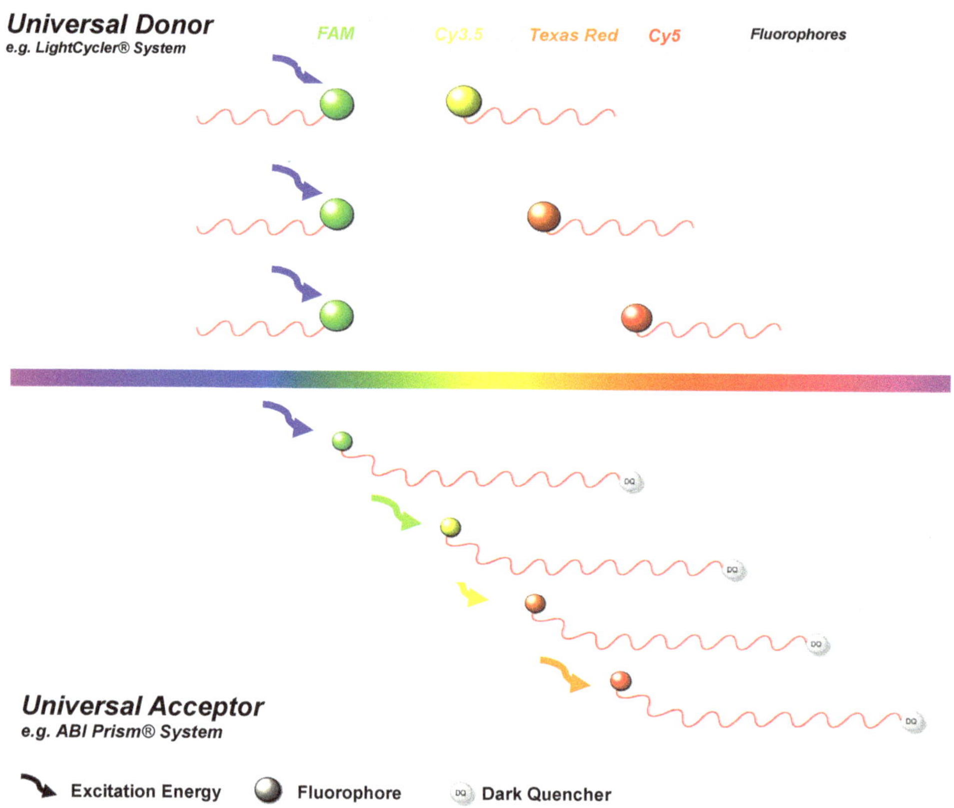

Figure 1.3 Schematic showing the two principal methods for multiplexing using FRET: (top) the universal donor on the LightCycler® and (bottom) the universal acceptor on the ABI 7700.

wavelength emission and are all excited by the one excitation source, e.g. The Applied Biosystems 7700, 7000. Alternatively, dyes that are spectrally well separated with longer wavelengths may be used on some instruments that have the ability to both excite at more than one wavelength, and measure emission at more than one wavelength, e.g. BioRad iCycler® and Cepheid SmartCycler®.

In theory, most fluorimetric instruments can be used to detect either universal donor or universal acceptor dye combinations, but the software on most platforms is not flexible enough to allow practical application and analysis using both approaches without first exporting the data. The universal acceptor arrangement is the most common because of its effectiveness with quenched dual-labelled probes (5′-nuclease probes, molecular beacons etc.) that are widely used. However, instruments such as the Corbett RotaGene® are flexible enough in optical design and analysis to allow the use of most, if not all fluorescent methods. It should be noted that generally the signal obtained when measuring acceptor fluorescence in universal donor methods, is of a lower value (because of energy transfer inefficiencies) and may inherently generate more noise than the universal acceptor methods, where donor fluorescence is measured. That is not to say that a comparative signal–noise is not achievable with universal donor methods. In particular this is possible where the probe and FRET efficiencies are good and the dyes are spectrally well separated, such that the donor moieties do not significantly contribute to the high background when detecting the acceptor emissions, as is achieved with the LightCycler® dye combinations (fluorescein, LC640 and LC705).

One approach to allow the universal acceptor method to be used on a system optimized for the universal donor method uses the 'big dye' approach to transfer the energy of the excitation source of the detector to the reporting dye (Solinas *et al.*, 2001). This approach has been previously applied in DNA sequencing chemistries and instruments (Ju *et al.*, 1995). For example, on the LightCycler® the dyes for the longer wavelength detectors are only minimally excited by the blue LED. A fluorescein molecule that is chemically coupled to such a dye may now be detected in the longer wavelength detectors since the energy will be transferred by the fluorescein, in turn this energy may be transferred to a longer wavelength or dark acceptor dye. Although less efficient, the method does work and these probes are available from a few suppliers. However, the trend with new instruments is to move away from supporting universal donor chemistries alone, for example the LightCycler® 480 instrument now has multiple excitation wavelengths facilitating multiplex universal acceptor methods and negating the necessity for dye deconvolution algorithms for some applications. Table 1.2 provides a summary of commonly used dye pairs.

A passive reference is used in some systems. This dye is detected and its signal is used principally to normalize for optical variations in the instrument (such as edge effects), but also normalizes for differences in volume that occur between wells and subtle differences in the optical properties of tubes and wells. These differences may be due to either volumetric errors in pipetting during preparation, and/or changes that occur during the amplification as evaporation occurs, causing a drift in the background signal. The reference dye takes no part in the reaction itself. On Applied Biosystems and similar instruments ROX (carboxy-X-rhodamine) dye is commonly used for this purpose. On instruments where there is little surface area for evaporation to occur, such as capillary systems like the Idaho & Roche LightCycler®, and in centrifugal systems where the volume and concentration is maintained in the reaction well, such as the Corbett RotaGene®, a passive reference is less important.

Colour correction algorithms

A prerequisite of multiplexing using fluorescence is discriminating the emission signals of different fluorescent dyes at different wavelengths. The requirement for an algorithm is dependent largely upon the fluorescent principles for which the instrument platform is optimized. For example, on platforms which use the universal donor system a colour compensation algorithm is required to analyse all dyes in a given multiplex assay. This is because there is usually only a single excitation source. When, for example, these systems are applied to universal acceptor chemistries they still require a colour compensation algorithm because the resulting measured energies will be a composite of all fluorophore emissions within a given multiplex. Those systems designed for using the universal acceptor system which have excitation sources that excite only specific fluorophores in a multiplex do not require a colour compensation algorithm. The authors would note that this is technically difficult to achieve. In practice 'cross talk' is usually addressed algorithmically on platforms intended for high-confidence applications such as diagnostics. Therefore, the provision of colour correction processes artificially segments the platform market into instruments intended for applications other than research and 'research-only' instruments.

Table 1.2 Some of the common fluorophore combinations that can be used effectively in energy transfer pairs

Fluorophore	Excitation maximum (nm)	Emission maximum (nm)	As a DONOR when used with	As an acceptor when used with
Cascade Blue	396	410	DABCYL	Fluorescein, SYBR®Gold, SYBR®Green-1
Cy3	552	570	DABCYL, BHQ-2, ElleQuencher, Eclipse	Fluorescein, SYBR®Gold, SYBR®Green-1
Cy3.5	581	596	DABCYL, ElleQuencher	Fluorescein, SYBR®Gold, SYBR®Green-1
Cy5	643	667	DABCYL, BHQ-2, ElleQuencher, Eclipse	Fluorescein, SYBR®Gold, SYBR®Green-1
Cy5.5	675	694	DABCYL, BHQ-2, ElleQuencher	N/A
EDANS(5-((2-aminoethyl) amino)naphthalene-1-sulfonic acid)	335	493	DABCYL	N/A
FAM (carboxyfluorescein)	494	518	TAMRA, ROX, Cy3, Cy3.5, Cy5, Cy5.5, DABCYL, BHQ-1/2, Eclipse	
Fluorescein	492	520	TAMRA, ROX, Cy3, Cy3.5, Cy5, Cy5.5, DABCYL, BHQ-1/2, Eclipse	
HEX (carboxy-2′,4,4′,5′,7,7′-hexachlorofluorescein)	535	556	TAMRA, DABCL, BHQ-1/2, Eclipse	N/A
JOE (carboxy-4′, 5′-dichloro-2′,7′-dimethoxyfluorescein	520	548	TAMRA, DABCL, BHQ-1, ElleQuencher	N/A
LC640	625	640	N/A	Fluorescein
LC705	685	705	N/A	Fluorescein
ROX (carboxy-X-rhodamine)	585	605	DABCYL, BHQ-2, ElleQuencher	Fluorescein
SYBR®Gold	495	537	Cy5, Cy5.5, LC 640, LC705	N/A
SYBR®Green-1	494	521	Cy5, Cy5.5, LC 640, LC705	N/A
TAMRA (carboxytetramethylrhodamine)	564	605	DABCYL, BHQ-2, ElleQuencher	Fluorescein
TET (carboxy-2′,4,7,7′-tetrachlorofluorescein)	521	544	TAMRA, DABCL, BHQ-1/2, Eclipse	N/A
Texas Red	583	603	BHQ-2	N/A
VIC	538	554	TAMRA	N/A
Yakima Yellow	526	549	Eclipse	N/A

DABCYL (((4-(dimethylamino)phenyl)azo)benzoic acid), Black Hole Quencher 1 and 2 (BHQ), ElleQuencher(TM) and Eclipse ™ are non-fluorescent 'dark' quenching moieties.

The authors find that those new to the field who require flexibility in their PCR platform, for example an instrument that may be required as a departmental resource, often make a poorly informed first purchase. The trend in new platforms is to increase the choice of excitation and emission filters which may be used together in different combinations. On the basis of the optical specification alone one might expect that such systems would have excellent flexibility. However, whilst they almost certainly will be able to analyse singleplex assays in all fluorescent configurations (universal donor and acceptor) they may be limited in their flexibility without a colour compensation algorithm. Systems without a colour compensation algorithm will almost certainly not be able to analyse multiplex universal donor configurations. To some degree this makes the flexibility of these 'open' systems rather troublesome rather than offering real utility. However, some third-generation instruments (such as the LightCycler® 480 family) are truly flexible in their ability to multiplex in all fluorophore configurations; having multiple excitation and emission filters and a colour compensation algorithm.

A second important aspect of a colour compensation algorithm is its ability to account for the temperature dependent effects of fluorophore emissions. It will be perhaps familiar to most readers that the spectral nature of fluorophores means there is a requirement to numerically deconvolute a spectrum when carrying out multiplex analyses. What is perhaps less apparent is that most fluorophore emissions are reciprocal with respect to temperature. This is largely to do with the ionic environment of the PCR buffer. For example, Tris-based buffers drop in pH as temperature increases and therefore the resulting fluorophore emission decreases. It is noted here that the ionic composition of the calibrator is an important consideration in the calibration experiment as discussed below. Some colour compensation algorithms also numerically account for these variations. The key benefit is that one may multiplex using melting point analysis combined with strand specific probes as well as and/or multiplex by fluorophore wavelength.

Different approaches are available to achieve this colour compensation:

1 A fluorospectrometer that can measure a given area of the visible spectrum and use a calibration algorithm and software to deconvolute the different dye components. For example Applied Biosystems 7700 and 7900.
2 A fluorimeter with discrete optical detectors using a calibration algorithm and software to deconvolute the different dye components. For example Applied Biosystems 7000, Roche LightCycler®.
3 A fluorimeter with discrete optical detectors and utilizing a 'factory set' calibration to deconvolute the different dye components. For example Cepheid SmartCycler®.
4 A fluorimeter with discrete optical detectors with sufficient bandwidth between dyes and narrow cut off filter to minimize cross-talk. For example Corbett/Qiagen RotorGene®.

All approaches are equally applicable but we would recommend, for high-confidence applications, the use of an algorithm to minimize the risk of false-positives due to erroneous data interpretation.

A good algorithm is one that completely compensates for the contribution of other dyes at the wavelength of the dye being monitored. This is usually achieved using a calibration with pure dyes. For systems that are designed to provide multiplexing using probes and melting point analysis, this includes heating each pure dye through the PCR operating range. The colour compensation process could be broken down and described as:

1 *Optical calibration.* Using a defined preparation of fluorophore that is optimum for the assay to be analysed, 'a calibrator.' This may include several replicate wells of the dye to produce an arithmetic mean of the resulting spectra. Empty or background noise samples also normally form part of the reagents required.
2 *Numerical calibration.* The software of the machine produces a numerical transformation of the fluorophore and background (calibrator) sample data to produce a set of coefficients that characterize each fluorophore's emission energy across all channels of the instrument.
3 *Dye deconvolution.* The coefficients are applied to real data from PCR experiments in a transformation which is an inversion of the calibration process. These data may then be used for secondary analysis such as amplification and melting point analysis.

The flexibility of many commercial systems means that users may deviate from those fluorophore combinations recommended by the system provider. The key benefit in doing so is a greater choice of fluorophores, lower oligonucleotide synthesis costs from a greater range of suppliers, and the application of a growing number of commercial kits often with bespoke dye combinations. As systems have become more complex in terms of their optical configurations the authors find that the successful application of these bespoke colour compensations are too difficult for many users

There are some important practical aspects that one should adhere to in order to ensure that the colour compensation processes works correctly. To a great extent this may be largely negated by the use of system provider calibration kits. A poor calibration will result in either under or over correction which manifests itself as a false-positive signal in an intended negative channel. This will be as a result of 'spill' or 'bleed' over of optical signal from a neighbouring positive channel. These are sometimes evident as inverted amplification and/or melt curves when colour correction is poor. If these signals are above the threshold levels of one's secondary analysis (amplification or melt analysis) then the colour compensation process requires refinement. The important experimental considerations are:

1 *Fluorophore concentration.* Some fluorophores are not efficiently excited by the excitation source of the chosen platform. This is implicit in their intended use within a universal donor configuration. All dyes must be applied in a calibration at a concentration to which they are sufficiently excited as to be within the working dynamic range of the detector (linear range) in the channel to be used for analysis. The same applies to those additional channels to which they are known to significantly overlap. In practice this means that one must manually adjust dye concentrations until they satisfy these criteria. Trial and error using the analysis of raw data is the best method to ensure success. Too high (detector saturation) or too low (below background noise) a signal will lead to an incorrect numerical process and poor secondary data analysis (amplification or melting point). For those systems that provide for melting point analysis with probes, this is more complex since the dye concentration must also satisfy this criterion across the temperature range of the analysis. It should be noted that the solubility of dyes (due to increasing molecular mass) generally decreases with increasing wavelength and

therefore there may be a practical limit to achieving the ideal fluorophore signal on some platforms.

2 *Fluorophore and the ionic environment.* The fluorophore emission spectra may change with respect to wavelength and/or temperature depending upon the ionic environment. The pure dye calibrator should ideally be made using the same solutes as the PCR for which its calibration coefficients are to be applied. The buffer match is crucial since formulations change frequently with different suppliers and assay requirements. For example, *Taq*, *Pfu* and similar DNA-dependent DNA polymerases function in Tris-based buffer systems. RNA/DNA-dependent DNA polymerases such as *Tth* and r*Tth* will be supplied with bicine or tricine buffers and manganese rather than magnesium as their preferred bivalent ion source. Other solutes that are commonly used as PCR adjuncts, for example co-solvents such as DMF and DMSO may also produce significant changes in spectral and thermal emission responses. To a lesser extent other co-solutes and adjuncts such as salts (potassium chloride, ammonium sulphate), sugars (sucrose, glycerol) can also have some effect. There is an increasing trend towards stabilized PCR mixes which may have their own functional components such as freeze drying excipients, including sugars and structural polymers (oligosaccharides and polyphenols). Suffice to say that PCR formulations from different suppliers (and even different batches) may differ and therefore may perform differently with respect to the colour compensation. Therefore, it is important to have an appreciation of the content of such mixes if they are to be applied in experiments that require colour compensation, one may then prepare suitable calibrators.

3 *Oligonucleotides.* Some fluorophore labels are influenced by the primary sequence to which they are attached. This is often termed 'proximal quenching' and is exploited in some fluorogenic chemistries. In some experiments this proximal affect is so pronounced that it will perturb the colour compensation process. It may require that the pure dye calibration be carried out using the oligonucleotide utilized in the desired PCR experiment. If the donor probe in a multiplex experiment is affected then this can cause significant problems. The authors would recommend redesign of the oligonucleotide and/or assay in question. Pooling of different donor oligonucleotides is often recommended but will produce a poor result. It should also be noted that multiplexing of proximal probes using a single excitation source and a colour compensation process is extremely difficult since the calibration can never satisfy the two spectral states of each fluorophore that may change as a result of amplification.

A number of factors can affect the requirement for re-calibration; these include the drift in detector sensitivities and excitation sources. Solid state devices such as LEDs and photodiodes are more robust with time than lamps and photon multiplier tubes. The calibration is also highly dependent on the optical geometry, so any servicing of the instrument may initiate the requirement for recalibration.

The 5′-nuclease assay

The principle of a DNA synthesis-dependent 5′–3′ exonuclease process using the enzyme *Taq* polymerase was first described in 1989 (Gelfand, 1989). The application of this process for analysis of PCR amplification using isotopic labelling and a separation analysis was later reported (Holland *et al.*, 1991). A fluorogenic approach for monitoring the hydrolysis of HIV-1 peptides in real time (Matayoshi *et al.*, 1990) had been described and this principle was subsequently applied to the 5′-nuclease assay. A fluorogenic 5′-nuclease PCR assay was first reported in 1993 (Lee *et al.*, 1993).

The 5′-nuclease assay is now by far the most commonly used strand specific probe method. This is partly because this assay was the principal detection chemistry for the first commercial real-time PCR instrument, the Applied Biosystems 7700 launched in 1996 (Bassam *et al.*, 1996). Applied Biosystems market the 5′-nuclease assay for research applications under the registered TaqMan® trademark of Roche Molecular Systems Inc.

The probe is dual labelled. Typically, one dye of a FRET pair is located at the 5′-end of the probe, usually the donor, and the second is labelled at or near the 3′-end, usually the acceptor. However, the reverse is acceptable and more efficient quenching may be obtained by locating the acceptor closer to the donor. When dyes are coupled in close proximity the signal–noise ratio of the assay improves compared to terminally labelled probes. The optimum distance of the two dyes on the probe is dependent upon the properties of the dyes used in the FRET pair. Typically a spacing of six nucleotides is optimal. If the dyes are too close then there may be inefficient transfer of energy (Heller *et al.*, 1987) and they risk being cleaved together on the same fragment. The probe is protected at the 3′-end to prevent it being extended as an amplimer by the *Taq* polymerase. A phosphate group is normally employed for this but other blocker-groups, e.g. propandiol, work equally well.

The 5′-nuclease assay uses a probe that is designed to be cleaved by the concomitant 5′–3′ exonuclease activity of the *Taq* polymerase (Holland *et al.*, 1991). This occurs whilst the *Taq* enzyme is extending the nascent 3′-end of a neighbouring amplimer bound to the complementary strand. The products of hydrolysis have been shown to include one or two nucleotide products and it is proposed that the polymerase displaces the first one or two base pairs it encounters before beginning cleavage (Holland *et al.*, 1991). The 5′-nuclease assay is fundamentally different to the other strand specific probe assays because the signal is generated by the 5′-exonuclease process. Probe hybridization is necessary, but not sufficient, for this to occur. The probe is consumed within the reaction and therefore the method is unsuitable for melting point analysis.

The Applied Biosystems 7700 instrument utilizes dyes that were previously optimized for Applied Biosystems Sequencers, so the first commercially available 5′-nuclease probes used FAM (carboxyfluorescein), JOE (carboxy-4′, 5′-dichloro-2′,7′-dimethoxyfluorescein), TET (carboxy-2′,4,7,7′-tetrachlorofluorescein), and HEX (carboxy-2′,4,4′,5′,7,7′-hexachlorofluorescein) as the donors, and TAMRA (carboxytetramethylrhodamine) as the universal acceptor. The ROX (carboxy-X-rhodamine) sequencing dye was used as a passive reference. In practice a large number of dye combinations are possible and a multiplex of up to six reporter dyes in one reaction has been used for SNP end-point scoring (Lee *et al.*, 1999).

A number of software tools can be used to design 5′-nuclease probes. Similar instruments that followed the 7700 were developed with optics that supported dual probe systems using the a universal acceptor arrangement, and a large number of oligonucleotide synthesis companies supply the dual-labelled probes with the correct dyes to support the 7700 and later instruments. Applied Biosystems, who market TaqMan® probes, recommend that the design of such probes should follow these recommendations for optimum signal generation: the GC content of the probe should be in the 30–80% range, the strand with the greater number of Cs rather than Gs should be selected; runs of identical nucleotides should be avoided; there should be no Gs at the 5′-end (which would otherwise partially quench the fluorophore); the probe should have a melting temperature that is at least 5°C greater than the amplimers; and, finally, the probe sequence should not overlap with the amplimer regions (Gelmini *et al.*, 1997).

The 5′-nuclease assay was the first probe system reported which used dye pairs that are covalently linked prior to molecular cleavage. The spacing of the dyes may be optimized with respect to Förster distance. Following cleavage, separation of donor from acceptor is maximized, although some spatial effects are observable on binding to the complementary strand. Providing the experimental parameters are optimized, the 5′-nuclease system produces a signal–noise ratio that is usually much better than that achievable with other assay methods (Fig. 1.4). The method is very versatile and can be used on a large number of platforms. However, there are limitations and these will be outlined briefly here.

Applicability of the 5′-nuclease assay is limited by the target and probe sequence requirements for efficient hydrolysis. A PCR may be shown to be highly efficient, but the signal from a hydrolysis probe may be low. Therefore design of an efficient probe is the key to an efficient PCR signal. As for most of the described systems the assay is designed around the probe and not the amplimers *per se*. In particular, the use of this assay system for AT-rich DNA can be problematic. To obtain a TM sufficient to allow efficient cleavage of the probe it should be ~ 5–10°C greater than that of the amplimers. For some sequences this would require the synthesis of very long probes that could only be produced in low yields. The development of the minor groove binding (MBG) probe technology that increases the

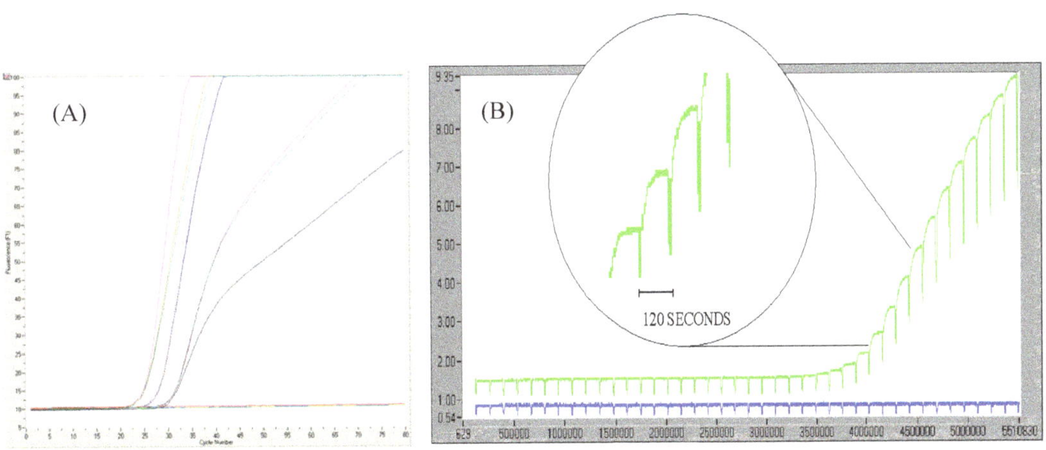

Figure 1.4 Amplification using the 5′ nuclease assay on the LightCycler®: (A) the amplification plot shows 80 cycles (in excess of completion) of thermal cycling of three dilutions (three replicates) of a 10-fold dilution series and three no-template controls on the Roche LightCycler®. The reaction utilised anti-Taq antibody hot start and UNG carry-over protection. The reaction used the primers and probes from the Applied Biosystems Human β-Actin kit which utilises a FAM-TAMRA internally quenched TaqMan® probe. The amplification plot shows raw data (no analysis/ background subtraction) and illustrates the excellent signal–noise ratio achievable with this chemistry. (B) Shows the same reaction amplified on the Idaho LightCycler® carried out by continuous monitoring of a positive and negative sample throughout all stages of amplification. The data shows that ~120 seconds are required each cycle to take the hydrolysis of bound probe to completion (plateau) each cycle.

TM of such probes (Kutyavin *et al.*, 2000; Salmon *et al.*, 2002) has reduced this problem somewhat but adds to the complexity and cost of the probe. Similarly locked nucleic acids may be used to improve the probe TM (Letertre *et al.*, 2003). The difficulties associated with the synthesis of dual label probes and the signal: noise ratios obtained was discussed in 'Fluorescent resonance energy transfer probes (FRET) and the passive reference', above. We have observed this from almost all suppliers although this is less problematic in small-scale syntheses from reputable companies. The use of an additional complementary probe labelled with a quenching moiety may be useful in improving the signal–noise ratio of problematic probes (Nurmi *et al.*, 2002).

On many platforms the temperature transitions make up a large proportion of total assay time. On instruments with faster temperature transitions such as the Roche/Idaho LightCyler®, Corbett RotorGene® and Cepheid SmartCycler®, the overall PCR assay time may be significantly reduced. Whilst the PCR process has been shown to proceed at considerable speed on fast systems, the signal–noise ratio of the 5′-nuclease assay may be dramatically improved with an increase in hold time at the annealing step. The 5′-nuclease process has been shown to generate signal much slower than the PCR process itself (Whitcombe *et al.*, 1999a; Lee *et al.*, 2002) and this is illustrated in Fig. 1.4.

The 5′-nuclease assay has been reported for a large number of applications and is supported by the majority of oligonucleotide synthesis companies and suppliers of instruments, such that for many new users, the trademark TaqMan® is used as a synonym for real-time PCR. These applications include expression studies, detection of pathogens, allelic discrimination, and genomic applications to name but a few.

Probe library

One of the most recent innovative developments in probe technology is 'Probe Library' (Mouritzen *et al.*, 2005). This is a commercial off-the-shelf array of probes. The probes are not specific; rather each probe is pre-designed, being a 10- to 15-mer sequence that is reproduced frequently throughout a specific genome. The PCR assay design is carried out by the end-user with a web-based design package. A sequence is entered and the software then designs primers around probes which exist within the library. There are probe libraries for different genomes; if a genome is not catered for then a closest match may provide a solution.

The probes themselves are dual-labelled probes. The length of each probe is maintained short such that the sequence occurs with enough frequency to be useful. The TM of the probe is raised using locked nucleic acids such that it hybridizes to the amplicon. The signalling is produced from the 5–3′ hydrolysis of the probe concomitant with primer extension, i.e. the biochemical process is the same as the 5′ to 3′ nuclease assay except that the probe is not designed specifically for the amplicon. The specificity of the reaction is derived only from the amplimers. When the assay is optimized the end-user is able to carry out a large scale synthesis of the sequence or purchase it from the supplier. This should allow the assay to be applied further on a wide variety of instrument platforms. Additional benefits will be realized from the kit-scale synthesis and the associated quality control that is required for such a product. Problems with dye stoichiometry observed in research-scale synthesis of dual-labelled probes will be minimized.

Probe library is useful for the rapid development of quantitative real-time PCR assays.

Market literature suggests an assay can be optimized in several days. The main useful application will probably be in the rapid development of assays for gene expression studies. A useful feature for this application is that the probes are designed to have similar efficiencies. This significantly simplifies the optimization process and eliminates some aspects of experimental error. The authors would not consider this probe system to be suitable for high-confidence applications such as diagnostics. The high background in some samples may lead to a false-positive signal. It may be possible to carry out an allele specific multiplex assay using two or more probes over a polymorphism but this has not been reported as-of-yet.

Direct hybridization methods

Direct hybridization methods use one or more probes that hybridize to the product strands during the low-temperature stages of the PCR each cycle and are released from the product strand during the high temperature stages. Unlike the 5′ nuclease assay the signal is not cumulative, but does increase each cycle as the amount of probe binding rises with product accumulation. Since the signal is derived from a hybridization event it is proportional to the product concentration, but is also a function of the probe efficiency. The signal from hybridization assays may be based on direct binding of product, or on the competitive formation of different hybrids. Generally, where the hybridization event is simple the effect of probe efficiency will decrease. Here we describe the different types of hybridization probe in detail.

Linear hybridization probes

A number of different probe systems fall into this class. These use the direct hybridization of linear oligonucleotides; these are oligonucleotides that are designed to have minimal secondary structure when not bound to target. Here we will describe the various types and use of these probes in more detail.

The most often reported probes of this class are dual hybridization probes that were the strand-specific chemistry developed for PCR applications on the LightCycler® instrument (Cardullo *et al.*, 1988; Wittwer *et al.*, 1997). The term hybridization probe is often used as a synonym for this probe method. This probe system uses two oligonucleotide probes each labelled with one dye of a FRET pair. Typically one dye is located at the 3′-end of one oligonucleotide and the other dye is located at the 5′-end of the second oligonucleotide. The second oligonucleotide is also protected to stop primer extension using a 3′-phosphate or other blocker molecule.

These oligonucleotides are designed to hybridize adjacently on one of the nascent PCR product strands such that the dyes are in a *cis*-configuration with respect to the nucleic acid. Both probes are designed to bind to the non-amplimer region of one of the amplicon strands. As product accumulates both probes bind to the strand and the basis of detection is to monitor the drop in donor fluorescence, or the rise in acceptor fluorescence, or to monitor the quotient of acceptor/donor in non-multiplexed reactions. Typically fluorescence is measured at the end of the annealing step and requires 10–30 seconds to generate sufficient signal for this application.

Dual hybridization probes were first described for use in PCR on the Idaho LightCycler®. The two channel optical systems of these machines used fluorescein as the donor moiety and the Amersham dye Cy5 as the acceptor. The later three channel Roche LightCycler®s use three dyes, two of which are optimized for the LightCycler®, LC640 and LC705, although it is possible to multiplex using other dyes with emission wavelengths that correspond to the discreet optical detectors of the LightCycler™. The assay is applicable to other platforms such as the Corbett RotorGene® in either a universal donor arrangement or a universal acceptor arrangement by using a number of dye combinations and a non-fluorescent quenching moiety.

For applications requiring accurate quantification of initial target copy numbers a requirement is that the probes be designed to bind at approximately the same temperature, but above that of the amplimers so that efficient signal is generated during the annealing step, and melt at a temperature that will allow efficient extension by the *Taq* polymerase. This typically requires the probes to have a melting temperature of 65–75°C. If the probe TM is too low, insufficient signal will be generated each cycle. If the probe TM is too high, the probes may clamp the PCR and reduce amplification efficiency. The probe efficiency (the amount of signal generated by hybridization) can be significantly reduced by probe–probe interactions if there exists any sequence complementary between probes. This is important in multiplexed reactions where the concentration and number of probes increases. Ideally the probes should be designed to bind to one strand, close to, but not overlapping the reverse amplimer region. The TM of probes can be calculated using software tools such as the Idaho Technology applet found on their website, or using other commercial packages such as that supplied by Roche for use on the LightCycler®. If the design of probes is carried out as described then the probe efficiency will be high enough to provide accurate quantification of initial target molecules. A commonly observed effect in this and other hybridization assays is a decrease in probe signal in later cycles. This 'hook' effect is a result of increased specific product accumulation and competitive displacement of probes as the product strands anneal upon cooling (Fig. 1.5). This effect does not reduce the ability to accurately quantify initial target numbers. A hook is a good indicator that the amplification is preceding with a high efficiency.

One of the useful features of hybridization probes generally is the ability to carry out a multiplex qualitative analysis utilizing melting point analysis. For example, the same fluorescent dye on two different probes may be used to detect two amplification species in the same reaction providing their melting motifs can be resolved by the instrument and software. If the software can (a) determine more than one peak and (b) calculate the relative areas under these peaks using integral analysis, then a relative quantitative value for each species can be determined within a single reaction. Detection of internal controls using this approach is discussed in Lee *et al.* (2009).

Likewise, one probe may be used to detect strand variation by designing the probe to bind over a site where strand variation may occur. For example, allelic variation is readily detected using this approach. In the case of dual hybridization probes one probe is designed to bind to a conserved (non-variant) sequence that is adjacent to the sequence variation. This first probe is known as the 'anchor' probe as it is designed to hybridize to the target over a temperature range at which the TM of the second 'mutation' probe will change depending on the specific sequence. The mutation probe can be observed to melt at different temperatures depending on the presence of either variant. For example, the genotype for a given sample may be determined using this approach. A single peak would indicate a homozygous genotype, and if both peaks occurred then this would be a heterozygous genotype. This is a useful method

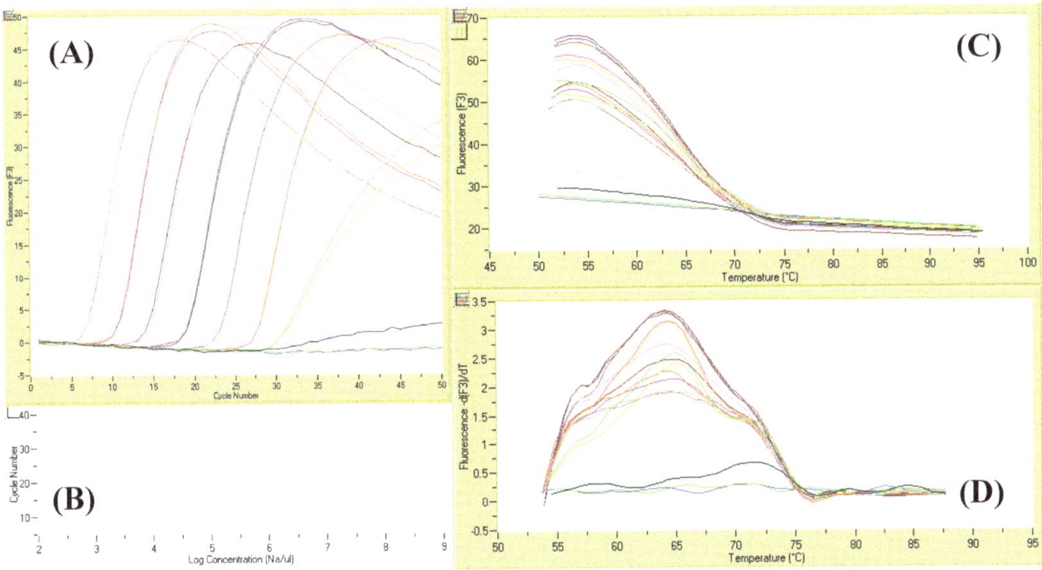

Figure 1.5 Amplification using dual hybridization probes on the LightCycler®: (A) the amplification plot shows 50 cycles of thermal cycling of replicates of eight dilutions of a 10-fold dilution series of a reference for the target, and no-template controls on the Roche LightCycler®. The amplification shows the data analysed using the background subtraction algorithm. The reaction utilised anti-Taq antibody hot-start and UNG carry-over protection. The reaction used the primers and probes for the Human β-globin gene using a 3′ labelled fluorescein donor probe and a 5′ Cy5.5 acceptor probe. The acceptor probe was blocked to prevent 3′ extension using an octendiol moiety. A 5′-3′ exonuclease Taq polymerase was used to illustrate that the 'hook' effect that is observed in linear hybridization probes as a result of probe displacement by product in later cycles. (B) The derived standard curve from the amplification plot. (C) The melt curve showing probe disassociation from product. (D) The first negative differential of fluorescence with respect to temperature showing the peak at which the maximum rate of probe dissociation occurs.

because a single probe can detect two variants, and this may be multiplexed by fluorescent emission to increase throughput if required. This approach can also be used to study the allele frequency in a mixed pool for a given population. For example, using the integral analysis of melting peaks described above allows a relative value for genetic variation, e.g. single nucleotide polymorphisms (SNPs), to be determined for a given population.

A less widely reported variation of the dual hybridization probe is one that uses two probes with the dyes in a *trans* configuration with respect to the probed strand (Bernard *et al.*, 1998). In this embodiment one 'free' probe is used in combination with a fluorescence-labelled amplimer. The energy from the dye on the amplimer may be transferred to the specific probe during the low temperature steps and the probe system may be similarly applied. Whilst this method is contemporary with the *cis* method it has been reported in fewer articles. The main drawback of the embodiment as reported in the original seminal paper was that it required internal labelling of the primer using an acceptor dye (Cy5). Internal labelling with dyes such as Cy5 is still not a general offering of many oligonucleotide synthesis houses. In addition the original method was reported for application on the LightCycler® system and the proprietary Roche dyes are also not generally offered in similar configurations. However, the authors have carried out a study of the labelling requirements and successfully produced

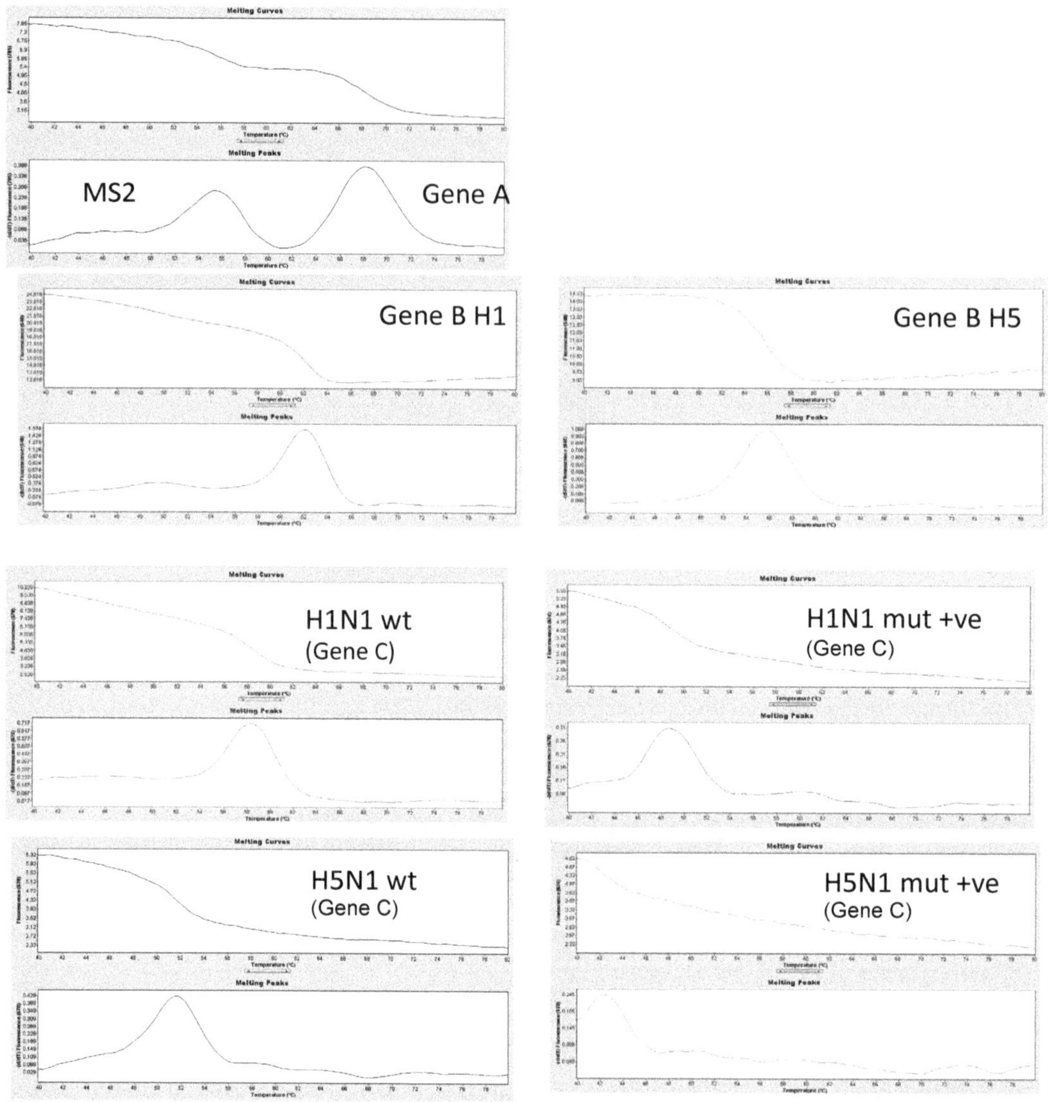

Figure 1.6 This shows experimental data for a multiplex assay using the *trans* dual Hybridization probe assay for characterization of pandemic Influenza A: Amplification was configured in the Universal Donor arrangement where a FAM label was present at the 3′-end of each specific probe. Reporter dyes (IDT dyes TEX615, TYE665 and TYE 705) were used at the 5′-end of specific primers. Signalling was achieved through energy transfer between the FAM label on the probe and the respective reporter on the primer. The amplification was carried out using the multiplex analysis of the LightCycler 2.0 which required the application of a colour correction process to the data generated using these fluorophores. The assay was designed to detect the following targets using the following reporters and melting point motifs: Pan Influenza A using TYE705 (~68°C). The assay was designed such that the primers and probes are directed against highly conserved regions of an Influenza A target (GENE A) (a). An MS2 process control (~56°C). MS2 assay component was designed such that the virus is introduced into the sample extraction process to control for all assay processes. Influenza HA subtyping component using Tex615 (b): this assay component was designed such that two assays (each two primers and one probe) are directed against highly conserved regions of a conserved target (Gene B) within Pandemic serotypes of Influenza A; HA1 (~62°C), HA5 (~56°C). Influenza NA subtyping and selection for oseltamivir resistance marker (polymorphism in H1N1-2009 and putative H5N1 mutation) using TYE665 in a conserved target GENE C. This component used a single assay (two primers and one probe) to amplify and type using highly conserved regions of

a multivalent assay for eight amplification species using only three reporter fluorophores which is described in Fig. 1.6. The system clearly has greater potential as this method does not use the full optical and analysis methods of the system utilized.

The impetus to construct an assay using this system was driven by the following factors. The method uses short probes based upon hybridization. Therefore, the kinetics of amplification and signalling are fast. The region of conserved sequence available to the probes may be very small. The primers and probes which are labelled only require a single donor fluorophore on the probe(s) and this may be on the 3′-end. This negates the additional requirement for protection from polymerase extension (steric blocking). This makes their implementation extremely low cost, in particular when compared to dual-labelled methods. Dyes may be shared between assays because the multiplex signals may be analysed by melting point. As the system uses FRET as the energy transfer system, rather than proximal quenching, colour compensation may be easily applied providing the universal donor is used on the probes (3′-fluoroscein) and the reporters are on the primers (5′-labels, for example). This allows for multiplexing by melting point analysis and fluorophore emission. The method is easily applied to parallel optic systems that use a single excitation source combined with multiple emission filters and an applied colour compensation algorithm. Such systems, e.g. the LightCycler® 2.0 and Enigma FL® and ML® platforms provide a rapid analysis since the emission energy at all wavelengths may be collected and analysed together. Channels that have only a single reporter may be analysed by their amplification signals and/or melt analysis. Channels that have two or more reporters may be analysed by end-point melt analysis. This combination of attributes is unique across the real-time methods. The main drawback of the method (in the described embodiment) is that primers contain a fluorophore label. Non-specific reaction artefacts may be generated that produce an amplification signal. However, the authors have found that careful design and empirical optimization can eliminate this occurrence. In addition, melt analysis may be used to discriminated them should this happen.

Yin–Yang probes are dual linear probes that work on the basis of competitive hybridization between themselves and the PCR product (Li *et al.*, 2002). This principle was first demonstrated for the end-point detection of PCR products in 1989 (Morrison *et al.*, 1989). Two labelled oligonucleotides are used. A fluorophore on the reporter probe is quenched by

---

the H1N1 and H5N1 genomes (c). The probe is designed to span this polymorphism, a missense mutation *TAT* to *CAT*; H1N1 (2009) wt (~58°C), H1N1-2009 mutation (~49°C), H5N1 wt (~51°C), H5N1 putative mutation (~42°C). All assay components (MS2, Gene A, B and C assays) were freeze dried into a environmentally stable cake which could be re-suspended with extracted viral material before amplification in a single reaction. Amplification is designed to be multivalent rather than a true multiplex where the relevant species amplify in the presence of a given viral isolate. Mixed samples may produce more complex amplification signals than these shown here. However, the assay illustrates how a small number of fluorophores (three) may be used to generate signal within a high level multiplex (eight possible targets). The data shown here are laboratory data from a LightCycler® 2.0 instrument but the final build is applied on the Enigma ML® point of use instrument. This instrument includes an automated data interpretation algorithm such that the operator receives clinically relevant information rather than experimental data. The assay may be applied to either screen, test, or confirm individuals for the pandemic strains (in 2009) of influenza A in combination with the oseltamivir resistance polymorphisms in H1N1-2009 and putative H5N1. This may be used directly by the clinician to administer the appropriate chemotherapy. The authors would like to thank Enigma Diagnostics Ltd (Dr Martin Lee, Dr Mark Laverick, Dr Paul Martin, Mr David Styles and Dr Samanth Spiers) and GSK (Dr Heather Madsen and Mr Tom White) for the kind permission to publish these data.

an acceptor on a second shorter complementary probe. Upon the increase in PCR product accumulation the acceptor probe is displaced by the hybridization of the longer target. The same principle is applied to the 'molecular zipper' probe, which is in fact a generic priming method utilized in rolling circle amplification (RCA). Displacement of the complement is facilitated by the amplification process which creates a stable hybrid complement that binds preferentially over the quenching complement (Yi *et al.*, 2006).

A limitation of these approaches is that two oligonucleotide probes are required. The application of dual hybridization probes may be difficult for some applications where there is little or no conserved region for one or either probe. The empirical optimization of two probes may increase the assay optimization time. ResonSense®, Hybeacons® and Light-up® probes are examples of single linear probe systems that have been described for a limited number of applications. The experimental considerations for the design and implementation of these probes will be similar to dual hybridization probes because of their similar mode of operation. These probes will be briefly described.

ResonSense® uses a single linear probe and a DNA-binding dye as a FRET pair. Signal is generated when the probe binds to the accumulating product and energy transfer occurs between the dye on the probe and the DNA binding agent in the DNA duplex (Cardullo *et al.*, 1988; Lee *et al.*, 1999; Taylor *et al.*, 2001). ResonSense® has been reported for use in real-time quantitative PCR (Lee *et al.*, 2002). The ResonSense® probe may be labelled with the fluorophore at either end or internally. A 3′ probe label will prevent primer extension. If the label is at the 5′-end or is internal then an additional 3′-blocker molecule will be required. The probe system may be used in either a universal donor arrangement (Lee *et al.*, 2002), or in a universal acceptor arrangement using a non-fluorescent quenching moiety. For example, the minor groove binding dye SYBR®Gold may be used to transfer energy to probes labelled with Cy3, Cy5, Cy5.5, to name a few. The broad emission spectra of SYBR®Gold allows for a variety of acceptor dyes to be utilized. SYBR®Gold and related dyes have two excitation peaks, one of which is in the UV. The use of a UV excitation source would allow for efficient excitation of the SYBR®Gold whilst minimizing the excitation of much longer wavelength dyes as described. Alternatively, the drop in donor dye emission using dyes such as fluorescein can be monitored by using a longer wavelength binding dye such as SYTO 63. This Universal Acceptor embodiment provides a method of applying ResonSense® on instruments other than the LightCycler®. The utility of ResonSense® for genotyping applications using LATE-PCR as a PCR end-point has also been reported (Sanchez *et al.*, 2006).

Hybeacons™ are linear probes that use a fluorescent moiety attached to an internal nucleotide that exhibits enhanced fluorescence with hybridization to a specific target (French *et al.*, 2001). The probe is 3′-protected to prevent primer extension during the course of the PCR. The increase in signal allows for direct quantification of initial target copy numbers and qualitative analysis of closely related sequences using melting point methodology. The simple structure of the probe and mechanism of signal generation make it ideal for multiplexing in strand variation analysis. Indeed, a six colour multiplex has been reported by applying a novel labelling strategy (Richardson *et al.*, 2010).

Light Up™ probes (Svanik *et al.*, 2000a,b; Isacsson *et al.*, 2000) are peptide nucleic acid (PNA) probes that use a DNA binding agent to report the hybridization of the PNA to specific DNA sequence. The application of Light Up™ probes for real-time PCR has been described (Wolffs *et al.*, 2001). In this probe method the probe binds to the target and brings the binding agent spatially close to the probe allowing it to bind with enhanced fluorescence. The dye reported in Light-Up™ probes is Thiazole orange (Lee *et al.*, 1986). PNA has been shown previously to be a useful molecule in some PCRs, being used to clamp amplification (Kyger *et al.*, 1998). The PNA has been shown not to clamp the PCR in this application due to the probe target sequence not being located near the amplimer regions.

The molecular beacon and other conformational probes

The use of a conformational change in fluorescent labelled probes to detect nucleic acid hybridization was first reported in 1987 (Heller *et al.*, 1987), and was later proposed for a homogeneous assay, in the context of nucleic acid amplification, in 1995 (Livak *et al.*, 1995; Young, 1996). The temperature dependent change in FRET occurring as a DNA probe, with a donor and acceptor label at either end, hybridizes to its specific complement is sufficient to provide a signal (of the event) in a homogeneous assay format. The Eclipse® probe system works on this principle but utilizes a minor-groove binder moiety to stabilize the structure (Afonia *et al.*, 2002). It is chemically similar to the MGB probes that may be used in the 5′-nuclease assay but the probe is typically labelled with the MGB and dark acceptor on the 5′-end to prevent the probe being hydrolysed. This type of probe can be used for a number applications and melting point analysis can be used to identify sequence variation as previously described.

The molecular beacon was described first in 1995 (Tyagi *et al.*, 1996, 1998; Salvatore *et al.*, 2006) and has been reported for a number of applications including direct 16SRNA detection (Schofield *et al.*, 1997), RNA detection in nucleic acid sequence-based amplification (NASBA) (Leone *et al.*, 1998), and recently the detection of DNA binding proteins (Heyduk *et al.*, 2002).

The chemical structure of the molecular beacon may be identical to that of the probe used in the 5′-nuclease assay in that it may be labelled at the 5′ and 3′-ends with a donor and an acceptor moiety respectively, and may also be protected from primer extension at the 3′-end. An alternative embodiment uses a single fluorophore label in the loop rather than two labels in the stem. The signalling is generated because the fluorescence is quenched when hybridized to target as opposed to fluorescing in the ssDNA loop (Hwang *et al.*, 2004).

The process of detection is different from TaqMan in that the molecular beacon is designed not to be consumed during the course of the reaction. The 5′ and 3′-sequences of the beacon are designed to be complementary, using a sequence not present in the target, such that the two fluorescent moieties are brought close enough to allow effective energy transfer in a 'hairpin' structure. The amplification is detected by the sequence dependent opening of a 'loop' sequence of the probe that is complementary to the target. Therefore these probes may be monitored for certain applications using melting point methodology (Robinson *et al.*, 200). Signalling may be poor in some molecular beacon assays since the stem-loop may preferentially fold, being an intramolecular process as opposed to the probe target duplex binding being a biomolecular process (Tsourkas *et al.*, 2003). Nucleic acid analogues have been used to improve the template probe hybridization (Ortiz *et al.*, 1998). One approach to increase the signal–noise ratio in a molecular beacon is the inclusion of non-nucleoside linkages below the stem structure which effectively reduce the chances of base mispairing (Häner *et al.*, 2010). These reduce unfavourable structures in the energy transfer pairing scheme that could produce a high background.

The difference in mode of operation of a hydrolysis probe and a probe such as a molecular beacon merits some discussion. Although the probe is chemically similar in structure, the process of detection is different. However, when the process is carried out using similar reaction conditions the signal in both assays will contain a component derived from both the conformational change in hybridization and a component derived from any strand specific hydrolysis by the 5′–3′ exonuclease activity of the enzyme contributing to the increase in the Förster distance of the fluorophores. The contribution of either component will depend on the assay design and the enzyme used, as some enzymes have different activities and may entirely lack 5′–3′-exonuclease activity, e.g. Stoffel fragment of *Taq* polymerase. The advantages and application of a molecular beacon to utilize both signalling processes has been reported (Kong. *et al.*, 2002) and the utility of either process for SNP scoring has been discussed (Täpp *et al.*, 2000).

Typically a universal acceptor arrangement is used since it can be carried out on all instruments that can analyse the 5′-nuclease assay. The original work used EDANS(5-((2-aminoethyl)amino)naphthalene-1-sulfonic acid) that is excited in the UV and emits in the blue region of the visible spectrum, and DABCYL (((4-(dimethylamino)phenyl)azo) benzoic acid) a dark acceptor. The use of a dark quencher for molecular beacons is still commonplace; however, the donor moiety is more commonly one that can be used on current instruments, although any FRET dye pair will work and some have been reported (Zhang *et al.*, 2001). The use of different fluorophores for quantitative multiplexing is possible (Vet *et al.*, 1999). A review of the use of such structures for a number of applications is given in Broude *et al.* (2002).

One application of the hairpin structure has been the development of a generic detection chemistry. This has been available in the past as Sunrise® and is now available as Amplifluor®. This chemistry uses a molecular beacon structure that is present as a tail on one amplimer (Nazarenko *et al.*, 1997). During PCR the *Taq* polymerase synthesizes the complement sequence so that upon hybridization the hairpin structure does not form, and the donor and acceptor are liberated as the basis of detection. A further simplified embodiment uses a unique sequence that is incorporated into the tail of one amplimer called a 'z' sequence. A generic Amplifluor® probe called a UniPrimer® can initiate primer extension since it has specificity for the z-sequence. When the complement is synthesized in the next cycle of PCR then the two dyes are liberated as previously described. The authors have seen many probe methods applied in a similar manner to the UniPrimer® including strand-specific approaches such as the 5′-nuclease assay (Whitcombe *et al.*, 1998; Zahng *et al.*, 2003).

Recent improvements include the development of the Lux® primer method (Nazarenko *et al.*, 2002). This uses the same principal as Amplifluor®, but only one fluorophore is required. The fluorophore must be two or three nucleotides from the 3′-end attached at the C5 position of thymidine. The primer has a hairpin structure such that the hairpin loop must terminate with CG or GC. This structure causes intrinsic *or* 'proximal' quenching of the dye. Upon extension of the complementary sequence a dsDNA structure promotes fluorescence of the dye. A generic system based upon energy transfer and the displacement of a complementary probe (rather than a hairpin) has been reported for application in rolling circle amplification (RCA) (Yi *et al.*, 2006). Whilst the 'molecular zipper' has been applied in RCA the principle could equally be applied to PCR.

Whilst Amplifluor® is a generic detection approach, the probe having no specificity for the internal sequence of the amplicon; it does offer some advantages over DNA binding approaches for detection. Firstly the signal–noise ratio of a FRET-based assays such as this is typically better than that obtained using a DNA binding dye, although non-specific priming may still generate signal. Secondly the probe may be a common probe sequence. A similar approach using a primer-incorporated sequence in a generic context utilizing common hydrolysis probes has also been reported for single tube genotyping (Whitcombe *et al.*, 1998). Finally, this approach allows for a generic approach to multiplexing using different fluorophore emissions.

The conformational principle has been applied to other 'me-too' approaches. Whilst each approach may offer slightly different utility, their main advantage is possibly with respect to freedom to practice the process. Some approaches, such as that reported by Xiao *et al.* (2009) have merit for applications where the PCR process itself is not essential.

Self-probing amplicons

A self-probing amplicon is an amplicon that has a probe chemically incorporated into the PCR product. The principle was first reported in 1999 and used a molecular beacon that was connected to the 5′-end of an amplimer via a linker molecule ((Newton *et al.*, 1993; Whitcombe *et al.*, 1999a,b). This probe is an oligonucleotide containing a sequence that is complementary to a region synthesized by the extension of the amplimer to which it is attached. The probe sequence also contains stem sequence elements at either end that hybridize and bring two fluorophores in a FRET pair spatially close together. The linker (PCR stopper) prevents the enzyme generating the complement of the probe during the extension of the reverse strand. Following denaturation of the duplex at the high temperature step of PCR, the probe binds to its complementary sequence that now forms part of the same molecule. The energetics favour this structure rather than that of the complementary stem structures within the probe sequence. The binding of the probe sequence to target increases the Förster distance allowing signal generation. The self-probing unimolecular mechanism has been proven to be the dominant mechanism of signal generation, rather than the bimolecular mechanism of the probe binding to other amplicons (Thelwell *et al.*, 2000). This is a key advantage over molecular beacons. It has also been shown that the cleavage of such probes does not occur due to the 5′–3′ exonuclease activity of the enzyme (Thelwell *et al.*, 2000).

In the original work the probe used fluorescein as the donor moiety and methyl red as a non-fluorescent acceptor. These probes are known commercially as Scorpion® probes.

One advantage of this method is that the probe is made in the same synthesis as the amplimer and therefore a separate probe is not required. The speed of signal generation has been reported to be much faster than that of the respective native molecular beacon which utilizes bimolecular probing (Whitcombe *et al.*, 1999a; Thelwell *et al.*, 2000), and actually more efficient when using short annealing times (Thelwell *et al.*, 2000). Since the probe becomes linked to the sequence it is probing, the event is in effect a zero order reaction (Saha *et al.*, 2001). The equal (1:1) stoichiometry of the probe to target means that the amplification signal is directly proportional to the amount of initial target species. This has also been illustrated utilizing a ResonSense® probe attached to an amplimer (Lee *et al.*, 2002). This type of probe has been termed Angler®. These types of assays will be of most use in improving throughput on faster thermal cyclers.

Other embodiments of the Scorpions® assay include the duplex Scorpion primer method (Solinos *et al.*, 2001). In this method the amplimer is labelled with a linear probe that has specificity for the strand the amplimer will generate. This probe contains one fluorophore of a FRET pair. A second probe that is the complement of the probe is free in solution. This probe is labelled with the second fluorophore in the FRET pair. Prior to the priming by the amplimer the probes interact and FRET occurs. When the amplimer extends the first probe is able to bind to the newly generated complement in an intramolecular fashion. The first probe is then spatially separated from the second and the change in FRET indicates amplification.

IntraTaq™ is the self-probing embodiment of the 5′-nuclease assay that has been shown to be more efficient in terms of the signal–noise ratio and the speed of the reaction (Solinas, *et al.*, 2002). In theory it should be possible to link any probe system to the amplimer in a similar fashion, and one would expect a commensurate improvement in signal with respect to speed for assays based on hybridization. One advantage of the self-probing amplicon that is not immediately apparent is that the probe is better controlled for, compared to a separate probe, since it is synthesized on the same molecule (Lee *et al.*, 2002). The main disadvantage of such probes is that the synthesis is more complex and therefore more costly, and the availability is limited to those synthesis companies that can reproduce these chemistries.

Intellectual property

In this section the authors do not provide definitive advice on what methodologies an individual may or may not have legal rights to use with respect to the available technologies. Regional variations in patent law mean that what intellectual property is protected varies greatly. In some territories patent protection may not have been sought so there will be no barriers to the use of the technology. Additionally, where patents are granted, they may not provide the same protection since the filing may have been examined and opposed on the basis of different prior art. At times patent status may also be the subject of appeals such that the protection of the intellectual property may change. To summarize, the patent landscape, and therefore the freedom to use the PCR process and various probe technologies, will vary depending where and when the process is to be used.

The history of the development of the PCR process has influenced the licensing programme structure of the core technologies. Cetus Corporation worked with Applied Biosystems on the development of early thermal cycling instrumentation and they protected the associated IP. The process became such a valuable tool to the life science researcher that when Roche acquired the process from Cetus the process was made available through an open licensing programme. Researchers that purchase licensed enzymes, reagents and thermal cycling instruments have the rights to carry out the process for their research activities. The authors have had contact with many groups who have misunderstood the licensing programme and have assumed that rights of use are automatically obtained for all areas under this programme. This has been observed mostly in academic, government, or non-profit making organizations that may have less commercial interest than pharmaceutical and diagnostics companies. Other activities such as service diagnostics require the end user to either obtain an end ‘field’ licence or purchase licensed diagnostic platforms and reagents. However, the key patents covering the core process are reaching the point where they will lapse and the requirement for licensing will diminish over the next few years. After this period licensing will only be a requirement if one uses technology covered by more recent filings such as those patents covering newer technologies such as new enzymes, some real-time processes and instruments. There is an 'extended licensing' programme from the key stakeholders covering their real-time processes and modified enzymes many of which are only now being given a granted patent status. This should in effect continue the research open licensing programme with additional provision for some real-time methods. By far the largest application area for real-time methods is in measurement of gene expression. However, many researchers will be developing a process that they would like to freely use. These researchers should be aware of the licensing programmes that make the technology available for use and be aware of third party IP and how that may be accessed. Given the cost and development times of validated assays, a naive consideration of IP could prove costly. Therefore these users should consider and further research the information provided here.

PCR technology has been a commercial success for many players. The patent landscape is complex and the exploitation of new technology has created a competitive litigious market for nucleic acid amplification technology. Some long court cases have proved costly. The requirement for a licence can be broken down into the key technology groups required. These are summarized below. The authors recommend that a detailed appraisal of the key patents status should be carried out by end users who plan to use a PCR process for application in fields other than purely research.

Enzymes

Taq polymerase and other enzymes that fall within the scope of the *Taq* patents will require the user to have a licence for use until the associated patents lapse. However, there is at least one commercial enzyme that is claimed to fall outside of this patent. Over the last 10 years the chemical hotstart methods have become the choice for many users. The processes to make and use these enzymes constitute new IP and therefore researchers may require licensing to use them.

Instruments

The automation patent covering the process of amplification in an automated process under computer control varies geographically. Therefore the requirement for an instrument to be licensed will depend upon the heating technology it uses, and specific embodiments covered by other key patents. Optical instruments are protected in some territories and therefore freedom to use these machines will vary such that the choice of instrument may change the requirement for licensing in some geographical regions. Some instruments purchased prior to the grant of the associated patents may not be licensed for some real-time methods. Indeed some licensed instruments with this capability were sold with a licence that did not cover their use for optical analysis because of the early status of these patent families.

Probe methods

Many probe methods have been developed. The rights to use probes for real-time PCR will depend upon which probes you choose since some probe technologies, albeit functioning by a different process, may well be covered by the main real-time patents. For example, in some patent territories probes with two or more fluorogenic labels and 3′ protection will fall within granted patents. Therefore the requirement to research your application and choose a suitable probe method may be invaluable in obtaining the rights of use.

Dye technology

The dye labels on purchased oligonucleotides may have been synthesized with a limited licence. Fortunately there are many dyes, some of which are generic for popular labels and may be more generally available for application.

Core technology

Many improvements to the PCR portfolio have come from third parties. These have been in the areas of enzyme 'hot start', carryover prevention and allele-specific priming methods to name a few.

Sequences

Many primer and probe sequences and genomes are the subject of granted patents. The ability to use these depends upon the local patent law. The right to those sequences for certain applications may require a licence.
To summarize, the choice of technologies that you choose to carry out you real-time method may reduce the cost of applying your process. There are some useful websites that can help researchers understand the patent landscape.

Summary and future improvements

There are many chemistries that may be used effectively to carry out the real-time detection of PCR amplification. Since the first edition of this text was published in 2004 there has been little improvement in the technology offering over that provided by existing fluorescent methods. Chemistries may differ in their utility for both quantitative and qualitative applications. For example, the 5′-nuclease assay has achieved broad appeal across all technical and market sectors using fluorogenic PCR. This commercial success is largely the result of early adoption, dominant marketing and support from many instrument suppliers. Few methodologies, with perhaps with the exception of the dual hybridization probe approach, offer a sensible solution across most quantitative and qualitative applications. However, the costs associated with the intellectual property rights to practise these established methodologies, in particular within the diagnostics sector, have been a driving force in the development of the many 'me-too' approaches presented here. Such approaches often have little technical merit over the established core real-time methodologies. As these methods enter their third decade, patent protection will cease, and the commercial value of these newer methods will diminish. The commercial freedom for any user to exploit their work beyond the research sector is now ensured by careful planning and commercial research.

However, the expansion of other enabling technology such as novel bases, new fluorophores and better instrumentation has improved the possibilities for those who use both the established and new methodologies. The user should look beyond the specific probe embodiment of these augmented technologies when considering their own requirements.

The authors would direct new users to the use of established processes to ensure a high rate of experimental success. In particular those methods that use linear probes are simple to design using known principles and generic molecular software tools. Therefore they have a have a high rate of success. Methods that do not utilize two labels within a single oligonucleotide will generally be of a lower synthesis cost for an equivalent yield of probe, even if the method requires more than one oligonucleotide.

The suitability of all assays will depend upon the application and the instrument platform used, the suitability of the software for correct data analysis, and the target sequence. The specifications of new instruments are improving, generally becoming faster and more flexible in operation. New PCR real-time detection methods such as fluorescence polarization, time resolved fluorescence and application of quantum dot technology are possible but are yet to be fully realized within mainstream laboratories. The authors would expect that advantages of bioluminescent methods, with respect to the relative simplicity of luminometry compared with fluorimetry, will be realized in the near future with emerging commercial systems. One can appreciate the possibilities by looking at the development of the parallelism achieved in bioluminescent sequencing.

Such methods could improve the speed, robustness, and costs associated with real-time PCR.

Acknowledgements

The authors would like to thank the editors for the invitation to submit a revised version for the second edition.

References

Afonia, I.A., Reed, M.W., Lusby, E., Shishkina, I.G., and Belousov, Y.S. (2002). Minor groove binder-conjugated DNA probes for quantitative DNA detection by hybridisation-triggered fluorescence. BioTechniques *32*, 940–949.

Bassam, B.J., Allen, T., Flood, S., Stevens, J., Wyatt, P., and Livak, K.J. (1996). Nucleic acid sequence detection systems: Revolutionary automation for monitoring and reporting PCR products. Australasian Biotechnology *6*, 285–294.

Becker A., Reith, A., Napiwotzki, J., and Kadenbach, B. (1996). A quantitative method of determining initial amounts of DNA by polymerase chain reaction cycle titration using digital imaging and a novel DNA stain. Anal. Biochem. *237*, 204–207.

Bernard, P.S., Lay, M.J., and Wittwer, C.T. (1998). Integrated amplification and detection of the C677T point mutation in the methylenetetrahydrofolate reductase gene by Fluoresence Resonance Energy Transfer and probe melting curves. Anal. Biochem. *255*, 101–107.

Bustin, S.A. (2000). Absolute quantification of mRNA using real-time reverse transcriptase polymerase chain reaction assays. J. Mol. Endo. *25*, 169–193.

Broude, N.E. (2002). Stem-loop oligonucleotides: a robust tool for molecular biology and biotechnology. Trends Biotechnol. *20*, 249–256.

Cardullo, R.A., Agrawal, S., Flores, C., Zamecnik, P.C., and Wolf, D.E. (1988). Detection of nucleic acid hybridisation by non-radiative fluorescence resonance energy transfer. Proc. Natl. Acad. Sci. U.S.A. *85*, 8790–8794.

Clegg, R.M. (1992). Fluorescence resonance energy transfer and nucleic acids. In Methods in Enzmology, Lilley, D.M.J., and Dahlber, J.E., eds. (Academic Press, New York) *211*, 353–388.

Dubertret, B., Calame, M., and Libchaber, A. (2001). Single-mismatch detection using gold-quenched fluorescent oligonucleotides. Nat. Bio. *19*, 365–370.

French, D.J., Archard, C.L., Brown, T., and McDowell, D.G. (2001). Hybeacon™ probes: a new tool for DNA sequence detection and allele discrimination. Mol. Cell. Probes *15*, 363–374.

Gelfand, D.H. (1989). Taq D.N.A polymerase. In P.C.R Theory, Principles, and Application to D.N.A Amplification. (1988). In Erlich, H.A. (ed.) PCR Technology (Stockton Press, NY). p. 17–22.

Gelmini, S., Orlando, C., Sestini, R., Vona, G., Pinzani, P., Ruocco, L., and Pazzagli, M. (1997). Quantitative polymerase chain reaction-based homogeneous assay with fluorogenic probes to measure c-erB-2 oncogene amplification. Clin. Chem. *43*, 752–758.

Häner, R., Biner, S.M., Langeneger, S.M., Meng, T., and Malinovskii, V.L. (2010). A highly sensitive, excimer-controlled molecular beacon. Angew. Chem. *49*, 1–5.

Heller, M.J., and Jablonski, E.J. (1987). Fluorescent stokes shift probes for polynucleotide hybridisation assays. European patent application 86116652.8.

Heyduk, T., and Heyduk, E. (2002). Molecular beacons for detecting DNA binding proteins. Nat Biotech. *20*, 171–176.

Higuchi, R., Dollinger, G., Walsh, P.S., and Griffith, R. (1992). Simultaneous amplification and detection of specific DNA sequences. Biotechnology *10*, 413–417.

Higuchi, R., Fockler, C., Dollinger, G., and Watson, R. (1993). Kinetic PCR analysis: Real-time monitoring of DNA amplification reactions. Biotechnonology *11*, 1026–1030.

Holland, P.M., Abramson, R.D., Watson, R., and Gelfand, D.H. (1991). Detection of specific polymerase chain reaction product by utilising the 5′–3′ exonuclease activity of *Thermus aquaticus* DNA polymerase. Proc. Natl. Acad. Sci. U.S.A. *88*, 7276–7280.

Hwang, G.T., Seo, Y.J., and Kim, B.H. (2004). A highly discriminating quencher-free molecular beacon for probing DNA. J. Am. Chem. Soc. *126*, 6528–6529.

Isacsson, J., Cao, H., Ohlsson, L., Nordgren, S., Svanik, N., Westman, G., Kubista, M., Sjöback, R., and Sehlstedt, U. (2000). Rapid and specific detection of PCR products using light-up probes. Mol. Cell. Probes *14*, 321–328.

Ju, J., Ruan, C., Fuller, C.W.,Glazer, A.A., and Mathies, R.A. (1995). Fluorescence energy transfer dye-labeled primers from DNA sequencing and analysis. Proc. Natl. Acad. Sci. U.S.A. *92*, 4347–4351.

Kandimalla, E.R., and Agrawal, S. (2000). 'Cyclicons' as hybridisation-based primer-probes: synthesis, properties and application in real-time PCR. Bioorg. Med. Chem. *8*, 1911–1916.

Kong, D.M., Gu, L., Shen, H.X., and Mi, H.F. (2002). A modified molecular beacon combining the properties of Taqman probe. Chem. Commun. *8*, 854–855.

Koshkin, A.A., Rajwanshi, V.K., and Wengel, J. (1998). Novel convenient syntheses of LNA [2.2.1] bicyclo nucleosides. Tetrahedron Lett. *39*, 4381–4384.

Kutyavin, I.V., Afonina, I.A., Mills, A., Gorn, V.V., Lukhtanov, E.A., Belousov, E.S., Singer, M.J., Walburger, D.K., Lokhov, S.G., Gall, A.A., *et al.* (2000). 3′ minor groove binder-DNA probes increase sequence specificity at PCR extension temperatures. Nucleic Acids Res. *28*, 655–661.

Kyger, E.M., Krevolin, M.D., and Powell, M.J. (1998). Detection of the hereditary hemochromatosis gene mutation by real-time fluorescence polymerase chain reaction and peptide nucleic acid clamping. Anal. Biochem. *260*, 142–148.

Lee, L.G., Chen, C.H., and Chiu, L.A. (1986). Thiazole orange: a new dye for reticulocyte analysis. Cytometry. *7*, 508–517.

Lee, L.G., Connell, C.R., and Bloch, W. (1993). Allelic discrimination by nick-translation PCR with fluorogenic probes. Nucleic Acids Res. *21*, 3761–3766.

Lee, L.G., Livak, K.J., Mullah, B., Graham, R.J., Vinayak, R.S., and Woudenberg, T.M. (1999). Seven-colour, homogeneous detection of six PCR products. Biotechniques *27*, 342–349.

Lee, M.A., Brightwell, G., Leslie, D., Bird, H., and Hamilton, A. (1999). Fluorescent detection techniques for real-time multiplex strand specific detection of *Bacillus anthracis* using rapid PCR. J. Applied. Micro. *87*, 218–223.

Lee, M.A., Siddle, S.L., and Hunter, R.P. (2002). ResonSense®: Simple linear probes for quantitative homogeneous rapid polymerase chain reaction. Anal. Chimic. Acta. *457*, 61–70.

Lee, M.A., Leslie, D., and Squirrell, D. J. (2009). Internal and external controls for reagent validation. In Real-time PCR: Current Technology and Applications, Logan, J., Edwards, K., and Saunders, N., eds. (Caister Academic Press, Norfolk, UK), pp. 95–104.

Leone, G., Schijndel, H., Gemen, B., Kramer, F.R., and Schoen, C D. (1998). Molecular beacon probes combined with amplification by NASBA enable homogenous, real-time detection of RNA. Nucleic Acids Res. *26*, 2150–2155.

Letertre, C., Perelle, S., Dilasser, F., Arar, K., and Fach, P. (2003). Evaluation of the performance of LNA and MGB probes in 5′-nuclease PCR assays. Mol. Cell. Probes *17*, 307–3111.

Li, Q., Luan, G., Guo, Q., and Liang, J. (2002). A new class of homogeneous nucleic acid probes based on specific displacement hybridisation. Nucleic Acids Res. *30*, e5.

Livak, K.J., Flood, S.A.J., Marmaro, J., Giusti, W., and Deetz, K. (1995). Oligonucleotides with fluorescent dyes at opposite ends provide a quenched probe system useful for detecting PCR product and nucleic acid hybridisation. PCR. Meth. Appl. *4*, 357–362.

Logan, M.J., and Edwards, K. (2009a). An overview of PCR platforms. In Real-time PCR: Current Technology and Applications, Logan, J., Edwards, K., and Saunders, N., eds. (Caister Academic Press, Norfolk, UK), pp. 7–23.

Logan, M.J., and Edwards, K. (2009b) Performing real-time PCR. In Real-time PCR: Current Technology and Applications, Logan, J., Edwards, K., and Saunders, N., eds. (Caister Academic Press, Norfolk, UK), pp. 85–94.

Matayoshi, E.D., Wang, G.T., Krafft, G.A., and Erickson, J. (1990). Novel fluorogenic substrates for assaying retroviral proteases by resonance energy transfer. Science *247*, 954–958.

Morrison, L.E., Halder, T.C., and Stols, L.M. (1989). Solution-phase detection of polynucleotides using interacting fluorescent labels and competitive hybridisation. Anal Biochem. *183*, 231–244.

Morrison, T.B., Weis, J.J., and Wittwer, C.T. (1998). Quantification of low-copy transcripts by continuous SYBR®Green 1 monitoring during amplification. Biotechniques *24*, 954–962.

Mouritsen, P., Noerholm, M., Nielson, P.S., Jacobsen, N., Lornholt, C., Pfundheller, H.M., and Tolstrup, N. (2005). ProbeLibrary: A new method for faster design and execution of quantitative real-time PCR. Nat. Methods *2*, 313–316.

Mullis, K.B., and Faloona, F.A. (1987). Specific synthesis of DNA *in vitro* via a polymerase catalyzed chain reaction. Meth. Enzymol. *155*, 335–350.

Nazarenko, I.A., Bhatnagar, S.K., and Hohman, R.J. (1997). A closed tube format for amplification and detection of DNA based on energy transfer. Nucleic Acids Res. *25*, 2516–2512.

Newton, C.R., Graham, A., Heptinstall, L.E., Powell, S.J., Summers, C., Kalsheker, N., Smith, J.C., and Markham, A.F. (1989). Analysis of any point mutation in DNA. The amplification refractory mutation system (ARMS). Nucleic Acids Res. *17*, 2503–2516.

Nazarenko, I., Pires, R., Lowe, B., Obaidy, M., and Rashtchian. (2002). Effect of primary and secondary structure of oligodeoxyribonucleotides on the fluorescent properties of conjugated dyes. Nucleic Acids Res. *30*, 2089–2195.

Newton, C.R., Holland, D., Heptinstall, L.E., Hodgson, I., Edge, M.D., Markham, A.F., and Mclean, M.J. (1993). The production of PCR products with 5′ single-stranded tails using primers that incorporate novel phosphoramidite intermediates. Nucleic Acids Res. *21*, 1155–1162.

Nurmi, J., Wikman, M., Karp, M., and Lövgren, T. (2002). High-performance real-time quantitative RT-PCR using lanthanide probes and a dual temperature hybridization assay. Anal Chem. *74*, 3525–32.

Ogura, M., and Mitsuhahi, M. (1994). Screening method for a large quantity of polymerase chain reaction products by measuring YOYO-1 fluorescence on 96-well polypropylene plates. Anal Biochem. *218*, 458–459.

Ortiz, E., Estrada, G., and Lizardi, P.M. (1998). PNA molecular beacons for rapid detection of PCR amplicons. Mol. Cell. Probes *12*, 219–226.

Richard, J.A., Gerowaska, M., Shelbourne, M., Franch, D., and Brown, T. (2010). Six-colour HyBeacon probes for multiplex genetic analysis. Chembiochem *11*, 2530–2533.

Ririe, K.M., Rasmussen, P.R., and Wittwer, C.T. (1997). Product differentiation by analysis of DNA melting curves during the polymerase chain reaction. Anal. Biochem. *245*, 154–160.

Robinson, J.K., Mueller, R., and Filippone, L. (2000). New molecular beacon technology. Am. Lab. *32*, 30–24.

Saha, B.K., Tian, B., and Bucy, R.P. (2001). Quantitation of HIV-1 by real-time PCR with a unique fluorogenic probe. J. Virol. Meth. *93*, 33–42.

Saiki, R.K., Scharf, S., Faloona, F.A., Mullis, K.B., Horn, G.T., Erlich, H.A., and Arnheim, N. (1985). Enzymatic amplification of beta-globin genomic sequences and restriction site analysis for diagnosis of sickle cell anemia. Science *230*, 1350–1354.

Salmon, M.A., Vendrame, M., Kummert, J., and Lepoivre, P. (2002). Detection of apple chlorotic leaf spot virus using a 5′ nuclease assay with a fluorescent 3′ minor groove-DNA binding probe. J. Virol. Meth. *104*, 99–106.

Salvatore, A.E.M., Tyagi, S., and Kramer, F.R. (2006). Real-time assays with molecular beacons and other fluorescent nucleic acid probes. Clinica Chimica Acta *363*, 48–60.

Sanchez, J.A., Abramowitz, J.D., Salk, J.J., Reis, A.H., Rice, J.E., Pierce, K.E., and Wangh, L.J. (2006). Two-temperature LATE-PCR endpoint genotyping. BMC Biotech. *6*, 44.

Schofield, P., Pell, A.N., and Krause, D.O. (1997). Molecular Beacons: Trial of a fluorescence-based solution hybrization technique for ecological studies with ruminal bacteria. App. Env. Micro. *63*, 1143–1147.

Selvin, P.R. (1995). Fluorescence resonance energy transfer. Meth. Enzymol. *246*, 300–334.

Solinas, A., Brown, L.J., McKeen, C., Mellor, J.M., Nicol, J.T.G., Thelwell, N., and Brown, T. (2001). Duplex scorpion primers in SNP analysis and FRET applications. Nucleic Acids Res. *29*, E96.

Solinas, A., Thelwell, N., and Brown, T. (2002). Intramolecular TaqMan probes for genetic analysis. Chem. Commun. 2272–2273.

Svanik, N., Westman, G., Wang, D., and Kubista, M. (2000a). Light-Up probes: Thiazole orange-conjugated peptide nucleic acid for detection of target nucleic acid in homogeneous solution. Anal. Biochem. *281*, 26–35.

Svanik, N., Nygren, J., Westman, G., and Kubista, M. (2000b). Free-probe fluorescence of light-up probes. J. Am. Chem. Soc. *123*, 803–809.

Taylor, M.J., Hughes, M.S., Skuce, R.A., and Neill, S.D. (2001). Detection of Mycobacterium bovis in bovine clinical specimens using real-time fluorescence and fluorescence energy transfer probe rapid-cycle PCR. J. Clinic. Micro. *39*, 1272–1278.

Täpp, I., Malmberg, L., Rennel, E., Wik, M., and Syvänen, A.-C. (2000). Homogeneous scoring of single nucleotide polymorphisms: comparison of the 5′-nuclease TaqMan® assay and molecular beacon probes. Biotechniques *28*, 732–738.

Thelwell, N., Millington, S., Solinas, A., Booth, J and Brown, T. (2000). Mode of action and application of scorpion primers to mutation detection. Nucleic Acids Res. *28*, 3752–3761.

Tsourkas, A., Behlke, M.A., Rose, D., and Gang, B. (2003). Hybrisidisation kinetics and thermodynics of molecular beacons. Nucleic Acids Res. *31*, 1319–1330.

Tyagi, S., and Kramer, F.R. (1996). Molecular beacons: probes that fluoresce upon hybridisation. Nat. Biotech. *14*, 303–308.

Tyagi, S., Bratu, D.P., and Kramer, F.R. (1997). Multicolour molecular beacons for allele discrimination. Nat. Biotech. *16*, 49–53.

Vet, J.A.M., Majithia, A.R., Marras, S.A.E., Tyagi, S.., Dube, S., Poiesz, B.J., and Kramer, F.R. (1999). Multiplex detection of four pathogenic retroviruses using molecular beacons. Proc. Natl. Acad. Sci. U.S.A. *96*, 6394–6399.

Whitcombe, D., Brownie, J., Gillard, H.L., Mckechnie, D., Theaker, J., Newton, C.R., and Little, S. (1998). A homogeneous fluorescence assay for real-time PCR amplicons: its application to real-time, single tube genotyping. Clin. Chem. *44*, 918–923.

Whitcombe, D., Theaker, J., Guy, S.P., Brown, T., and Little, S. (1999a). Detection of PCR products using self-probing amplicons and fluorescence. Nat. Biotech. *17*, 804–807.

Whitcombe, D., Kelly, S., Mann, J., Theaker, J., Jones, C., and Little, S. (1999b). Scorpions™ primers – a novel method for use in single tube genotyping. Am. J. Hum. Genet. *65*, 2333.

Wittwer, C.T., Ririe, K.M., Andrew, R.V., David, D.A., Grundy, R.A., and Balis, U.J. (1997). The LightCycler™: a microvolume multisample fluorimeter with rapid temperature control. Biotechniques *22*, 176–181.

Wolffs, P., Knutsson, R., Sjöback, R., and Rådström, P. (2001). PNA-based light-up probes for real-time detection of sequence-specific PCR products. Biotechniques *31*, 766–771.

Xiao, Y., Plakos, K.J.I., Lou, X., White, R.J., Jiangrong, Q., Lacxco, P., K.W., and Soh, H.T. (2009). Fluorescence detection of single-nuleotide polymorphisms with a single self-complementary, triple-stem DNA probe. Angnew. Chem. Int. Ed. Engl. *48*, 4354–4358.

Yi, J., Zhang, W., and Zahng, D.Y. (2006). Molecular zipper: a fluorescent probe for real-time isothermal DNA amplification. Nucleic Acids Res. *34*, e81.

Young, D. (1996). Viral load quantitation: An integral part of future health management. Australasian Biotechnology *6*, 295.

Zahng, P., Beck, T., and Tan, W. (2001). Design of a molecular beacon DNA probe with two fluorophores. Angew. Chem. Int. Ed. Engl. *40*, 402–405.

Zahng, Y., Zahng, D., Wenquan, L., Chen, J., Peng, Y., and Cao, W. (2003). A novel real-time quantitative PCR method using attached universal template probe. Nucleic Acid Res. *31*, 20.

DOI: https://doi.org/10.21775/9781912530243.04

Instrumentation and Fluorescent Chemistries Used in Quantitative Polymerase Chain Reaction

4

Mathilde H. Josefsen, Charlotta Löfström, Trine Hansen, Eyjólfur Reynisson and Jeffrey Hoorfar

Abstract

The polymerase chain reaction has revolutionized the world of scientific research and its broad application has caused a tremendous development of versatile PCR instruments and chemistries to fit its purpose. This chapter provides the reader with a general introduction to the basics of real-time PCR instrumentation, including the thermal and optical systems and the software. Performance parameters such as temperature uniformity, accuracy and ramp speed as well as reaction format, optical systems, calibration of dyes, software and comparison between different real-time PCR platforms will be discussed from a user perspective leading to an instrument selection guide. Differences between fluorescent DNA binding dyes and target-specific fluorescently labelled primers or probes for detection of amplicon accumulation will be discussed, along with the properties and applications of the most frequently applied chemistries. The fluorophores and quenchers used for primer and probe labelling and their compatibility will be presented, and finally the future challenges and trends within the field of qPCR instrumentation will be discussed.

Introduction

In 1983, Dr Kary Mullis conceptualized PCR, a technique that would come to revolutionize the scientific research arena of the twentieth century (Lo and Chan, 2006). Also, in the field of applied microbiology, the sensitivity, specificity and rapidness of PCR has prompted its worldwide implementation and routine application. Before the discovery of the thermo-stable DNA polymerase and the development of thermal cycling instruments, running a PCR entailed moving the sample between water baths of different temperatures and supplying fresh DNA polymerase after each denaturation step. Since then, much effort has been devoted to developing the field of PCR and its application as well as developing and improving thermal cycling instruments. The first generation of thermal cyclers was only capable of DNA amplification, requiring subsequent detection of amplicons by gel electrophoresis. The major limitations of this end-point detection approach included lack of quantification, poor sensitivity and post PCR processing, which brought about the risk of carryover contamination. The development of the real-time PCR (Higuchi *et al.*, 1992) and the accompanying instrumentation (1996) presented the ability to monitor the accumulation of amplicons in real-time. Detection in real-time has the advantage of being more sensitive and specific, offering the possibility of quantification, and since the amplicons are not processed post PCR, the risk of carryover contamination is reduced. This chapter will provide the reader with a thorough description of the real-time PCR instrumentation, focusing on suitable applications and including state-of-the-art microchip-based PCR systems.

Independent of the level of prior experience with real-time PCR, selecting a thermal cycling instrument can be a difficult challenge. In the abundance of manufacturers and instrument models with different technical specifications, it can be difficult to identify a thermal cycler that meets the specific requirements of the laboratory. In the instrumentation section of this chapter the capacity, technical specifications and suitable application of the different thermal cyclers will be discussed, providing the reader with a guide to instrument selection.

The accumulation of amplicons can be monitored in real time either by incorporating fluorescent DNA binding dyes or target-specific fluorescently labelled primers or probes. The different approaches have their justification depending on the PCR assay, and advantages and drawbacks will be discussed in the following sections. A variety of different DNA binding dyes, primer chemistries and probe chemistries have been developed for real-time PCR, sharing the common property of producing a change in emission of fluorescence upon amplicon accumulation. The properties and applications of the alternative primer and probe chemistries will be discussed in this chapter.

Real-time PCR instrumentation

The number of commercially available real-time PCR instruments has increased remarkably since the introduction of the first instruments (ABI Prism 7700 from Applied Biosystems and the LightCycler from Roche) (Wittwer *et al.*, 1997a). The machines now cover the whole range from very simple one-channel instruments to sophisticated high-throughput systems capable of multiplexing up to six channels, as well as portable devices enabling point-of-care diagnostics (for details on available instruments see Tables 2.1–2.3). In the following sections, the most commonly used platforms for real-time PCR available on the market today will be introduced followed by a guide on how to select an instrument suitable for the needs of a particular laboratory. Finally, the future trends of the instrument development will be discussed, including some novel lab-on-a-chip technologies and devices for digital PCR.

Basics about real-time PCR instruments

Real-time PCR instrument systems consist of three basic parts (Javorschi-Miller and Delgado Orlic, 2011):

1. a thermal system to perform temperature cycling;
2. an optical system to emit light necessary for activation of the fluorophore(s) combined with a system to capture the generated fluorescence;
3. software to control the instrument operation, and collect and analyse the data generated.

For each of these points, there are several technical solutions available.

Thermal system

An important part of the thermocycler is the thermal system that automates the temperature cycling required for PCR. The majority of real-time PCR platforms employ blocks that rely on the Peltier effect for heat transfer. Alternative ways to transfer heat include the use of heated air (Roche's LightCycler 2.0), a resistive heater with air cooling and centrifugation (Qiagen's Rotorgene Q), a block filled with thermal conductive fluid agitated by the use of electromagnetic paddles (Illumina's Eco) or the use of direct heating in combination with an electrically conducting polymer (InSyte and SynChron from Biogene). See Table 2.1 for more platforms. Each of these techniques has its own properties and is suitable for different purposes. When evaluating thermal systems, factors such as uniformity, speed and reaction format can be important to take into consideration.

Temperature uniformity and accuracy

The uniformity of heat transfer and temperature accuracy has been identified as a crucial factor to obtain reliable PCR results, and differences in temperature performance between brands and types of cyclers have been demonstrated to influence PCR performance (Saunders *et al.*, 2001; Yang *et al.*, 2005; Kim *et al.*, 2008). For the highest accuracy, the temperature must remain uniform across the entire heat block, ensuring that all samples are processed equally. For demanding applications such as high-resolution melting (HRM) analysis, the instrument must be able to distinguish a temperature difference of 0.2°C (Wittwer *et al.*, 2003). Most instruments have thermal uniformity between ±0.2°C and ±0.5°C (Table 2.1). However, there are some exceptions;

the Qiagen Rotorgene Q demonstrates a uniformity of ±0.01°C (using air cooling) and the Illumina Eco which applies a block-based system with uniformity of ±0.1°C.

Ramp speed

Another aspect of the thermal system performance is speed, which is dependent on the ramping rate of the system. The ramping rates vary significantly depending on the type of heat block or system used. Conventional Peltier-based block cyclers are relatively slow, due to the thermal inertia in the solid block itself (Table 2.1). Furthermore, the plastic coating of the reaction vessels forms a layer of insulation that further slows down the heat transfer. Typically, these kinds of blocks have ramping rates of 1.5–2.5°C/s for heating, generally with cooling rates being slightly slower. Improvements leading to sample ramp rates of 2.5–8°C/s have been made using several different approaches: (i) silver and gold plated blocks (e.g. Illumina Eco, with a thermal conductive fluid in its silver hollow block), (ii) use of fast DNA polymerases and other reagents (e.g. ABI 7500 FAST using FAST reagents) and (iii) use of an heat-equalizing layer between the heating and cooling units (e.g. ThermaBase in Roche's LightCycler 480 and 1536). Alternative technologies to blocks are used by a number of platforms, generating ramp speeds as high as 40°C/s. These techniques include, for example: (i) a stationary turbulent air-heated glass capillary format (LightCycler 2.0), (ii) a centrifugal air-heated plastic tube format (Rotorgene), (iii) a high-thermal-conductivity ceramic heating plate plastic tube format (SmartCycler), and (iv) silicon heating elements (In-Check from STMicroelectronocs) (Table 2.1). To increase the uniformity between different parts of the block, a few platforms use multiple independent temperature blocks (e.g. StepOne Plus from Applied Biosystems and BioRad's CFX384 and CFX96) or even independent temperature control of each well (InSyte and SynChron from biogene).

Reaction format

There are a number of different reaction formats used for the different platforms. Several of the available instruments use a standard 96-well block format with 0.2-ml wells. Some of them also offer the possibility to use 384- and/or 1536-well blocks within the same system. In the 96-well blocks, it is also possible to use individual 0.2-ml tubes or racks of tubes if a lower throughput is needed. These blocks have the possibility to use tubes and racks available from many suppliers, which makes it cheaper compared with cyclers where instrument-specific vessels are used (e.g. LightCycler glass capillaries, SmartCycler and Rotorgene specific plastics). Furthermore, platforms with up to 48-wells (StepOne, Eco and MiniOpticon) have been introduced, offering a convenient solution between handling single tubes and using 96-well plates. Some of these allows for rapid cycling, thus making it possible to perform multiple runs in a short period of time. For laboratories that need a high-throughput, there are a number of platforms based on 384- and 1536-well blocks, as well as array-based formats available. At the other end of the scale, robust portable systems exist with a limited sample capacity (1–32 samples), which can perform in-field analysis within the hour (Table 2.3).

Optical systems

The optical system of the real-time PCR machines consists of two main parts: a light source for excitation and a detector to monitor the emitted fluorescence. For both of these, there are a range of options available.

Light source for excitation

The excitation light source can be classified into broad or narrow spectrum. In broad spectrum, light source filters can be used to narrow the light down to a suitable wavelength range for a particular fluorophore. This gives a wider choice of available fluorophores, but the sensitivity of the system might be reduced because the filters limit the light that can pass through the system. Another disadvantage is that the bulbs decay over time and need to be replaced. Examples of platforms that use this approach are the 7500 from Applied Biosystems and ExiCycler from Bioneer. Some platforms employ narrow spectrum light sources including laser (e.g. 7900HT from Applied Biosystems and InSyte from Biogene) and light-emitting diodes (LEDs) (e.g. StepOne from Applied Biosystems and MiniOpticon from Bio-Rad) (Table 2.2).

Table 2.1 Thermal system specifications of selected commercially available real-time PCR systems

Model	Company	Thermal control	Average heating ramp temperature (°C/s)	Accuracy (°C)	Uniformity (°C)	Temperature range (°C)
Mx3000p	Agilent	Peltier	2.5a	±0.25	±0.25	25–99
Mx3005p	Agilent	Peltier	2.5a	NA	±0.25	25–99
qTOWER	Analytik Jena	NA	12a	NA	NA	NA
7300	Applied Biosystems	Peltier	1.1	±0.25	±0.5	4–100
7500	Applied Biosystems	Peltier	2.5a	±0.25	±0.5	4–100
7500 Fast	Applied Biosystems	Peltier	3.5 (fast mode), 1.6 (standard mode)	±0.25	±0.5	4–100
7900HT Fast	Applied Biosystems	Peltier	3.0 (fast mode), 1.6 (standard mode)	±0.25	±0.5	4–100
OpenArray	Applied Biosystems	Flat Block	4	NA	≤0.5	35- 99.9
StepOne	Applied Biosystems	Peltier	2.2 (fast mode), 1.6 (standard mode)	±0.25	±0.5	4–100
StepOnePlus	Applied Biosystems	Peltier	2.2 (fast mode), 1.6 (standard mode)	±0.25	±0.5	4–100
ViiA 7	Applied Biosystems	Peltier	NA	NA	±0.5	NA
InSyte	Biogene	Integrated thermal control	10–20a	NA	NA	Ambient-100
SynChron	Biogene	Integrated thermal control	10–20a	±0.3	NA	Ambient-100
TOPtical	Biometra	Silver block	5	±0.1	±0.15 – 0.50	3–99

Exicycler 96	Bioneer	Peltier	2.5a	±0.3	±0.3	4.0–99.9
CFX384	Bio-Rad	Peltier	2	±0.2	±0.4	0–100
CFX96	Bio-Rad	Peltier	3.3	±0.2	±0.4	0–100
MiniOpticon	Bio-Rad	Peltier	2.5a	±0.2	±0.4	35–99
GeneXpert	Cepheid	Solid state heater	10a	±1.0	NA	50–95
SmartCycler II	Cepheid	Solid state heater, forced air cooling	10a	±0.5	NA	NA
BioMark	Fluidigm	Peltier	>2	NA	NA	4–99
LightScanner32	Idaho Technology	Heated air	NA	NA	NA	40–99
R.A.P.I.D	Idaho Technology	Heated air	20*a*	NA	NA	ambient-95
RAZOR EX	Idaho Technology	Heat plate	NA	±1.0	0.5–1.0	56–94
Eco	Illumina	Hollow silver block	5.5	±0.1	±0.1	35–100
Liat Analyser	IQuum	Heated segments	NA	NA	NA	NA
Rotorgene Q	QIAGEN	Centrifugal rotor with heated air	>15	±0.25	±0.01	ambient-99
LightCycler 1536	Roche	Peltier with Therma-Base	4.8	NA	NA	20–95
LightCycler 2.0	Roche	Heated air	~2.5	±0.3	±0.4	40–98
LightCycler 480	Roche	Peltier with Therma-Base	96-well: 4.4; 384-well: 4.8	NA	NA	20–95
In-Check	STMicroelectronics	Silicon heating elements	13	±0.2	NA	NA
SmartChip	Wafergen	NA	1.5	NA	±0.5	NA

NA, information not available.
[a]Maximum heating rate.

Table 2.2 Specifications of optical systems of selected commercially available real-time PCR systems

Model	Light source	Excitation spectrum	Detector	Channels (emission filters, nm)	Dye calibration
Mx3000p	QTH lamp	350–750 nm	Single PMT	4 (350–700)	Any dye compatible with filters
Mx3005p	QTH lamp	350–750 nm	Single PMT	4 (350–700)	Any dye compatible with filters
qTOWER	NA	5	NA	5 (520, 545, 580, 605, 670, + 4 FRET)	Any dye compatible with filters
7300	TH lamp	1	CCD camera	4	FAM, SYBR Green I, VIC, JOE, NED, TAMRA, ROX
7500	TH lamp	5	CCD camera	5	FAM, SYBR Green I, VIC, JOE, NED, TAMRA, ROX, Texas Red, Cy3, Cy6
7500 Fast	TH lamp	5	CCD camera	5	FAM, SYBR Green I, VIC, JOE, NED, TAMRA, ROX, Texas Red, Cy3, Cy5
7900HT Fast	Argon-iron laser	488 nm	CCD camera	Continuous 500–660	FAM, SYBR Green I, VIC, JOE, NED, TAMRA, ROX
OpenArray	3 LEDs	3	CCD camera	4 (510, 535, 560, 620)	FAM, VIC, SYBR Green I, ROX
StepOne	LED	3	Photodiodes	3	FAM, SYBR Green I, VIC, JOE, ROX
StepOnePlus	LED	4	Photodiodes	4	FAM, SYBR Green I, VIC, JOE, NED, TAMRA, ROX
ViiA 7	NA	6	NA	6 (500–720)	Any dye compatible with filters
InSyte	Solid state laser	473 nm	PMT	32 (510–710, no filters)	Any dye compatible with wavelength range
SynChron	NA	NA	NA	NA	NA
TOPtical	3 LEDs	3	CPM	6 (520, 545, 580, 605, 670, 705, + 4 FRET)	Any dye compatible with filters
Exicycler 96	Short arc lamp	5	CCD camera	5 (520, 550, 580, 610, 670)	Any dye compatible with filters
CFX384	5 LEDs	4	Photodiodes	4 (515–530, 560–580, 610–650, 675–690)	Any dye compatible with filters

CFX96	6 LEDs	5	Photodiodes	5 (510–530, 560–580, 610–650, 675–690, 705–730)	Any dye compatible with filters
MiniOpticon	LED	470–505 nm	Photodiodes	2 (523–543, 540–700)	FAM, SYBR Green I, TET, VIC
GeneXpert	LED	6	Photodiodes	6 (420–480, 510–535, 565–590, 606–650, 665–685, >700)	FAM, Alexa 532, Texas Red, Alexa 647
SmartCycler	LEDs	4	NA	4 (510–527, 565–590, 606–650, 670–750)	FAM, Cy3, TET, Texas Red, Cy5, Alexa 532, Alexa 647
BioMark	Xenon arc lamp	5	CCD camera	5 (525, 570, 645, 2 empty)	Any dye compatible with filters
LightScanner32	LED	470 nm	Fluorimeter	3 (530, 640, 705)	SYBR Green, hydrolysis probes, HybProbes
R.A.P.I.D	LED	470 nm	NA	3 (530, 640, 705)	NA
RAZOR EX	LED	470 ± 20 nm	NA	2 (530 ± 10, 690 ± 60)	TaqMan, Hybprobes
Eco	Dual LED	2	CCD camera	4 (505–545, 562–596, 604–644, 665–705)	Any dye compatible with filters
Liat Analyser	NA	4	Photometer	4	TaqMan, Molecular beacons
Rotorgene Q	LEDs	2–6	PMT	2–6 (460, 510, 557, 610, 660, 712)	Any dye compatible with filters
LightCycler 1536	Xenon lamp	2	CCD camera	2 (510 or 580)	Any dye compatible with filters
LightCycler 2.0	LED	470 ± 10 nm	Photodiodes	6 (530, 560, 610, 640, 670, 710)	Hybridization probes, hydrolysis probes, Molecular beacons, SYBR Green I, Simple probes, UPL
LightCycler 480	Xenon lamp	5	CCD camera	6 (488, 510, 580, 610, 640, 660)	Any dye compatible with filters
In-Check	NA	628 ± 40 nm	NA	692 ± 40	Any dye compatible with filters
SmartChip	QTH lamp	3	CCD camera	3 (550, 605, 625, 2 empty)	Any dye compatible with filters

QTH, quartz tungsten halogen; TH, tungsten halogen; CCD, charge-coupled device; PMT, photomultiplier tube; CPM, channel photomultiplier; LED, light-emitting diode; UPL, universal probelibrary; NA, information not available.

Table 2.3 Additional information about selected commercially available real-time PCR systems

Model	Format	Reaction volumes (µl)	Dimensions (cm)	Weight (kg)
Mx3000p	96-well plate/0.2 ml tubes	25 (optimal)	33 × 46 × 43	20
Mx3005p	96-well plate/0.2 ml tubes	25 (optimal)	33 × 46 × 43	20
qTOWER	96-well plate	5–20	NA	NA
7300	96-well plate/0.2 ml tubes	20–100	34 × 49 × 45	29
7500	96-well plate/0.2 ml tubes	20–100	34 × 49 × 45	34
7500 Fast	96-well (fast)	5–30	34 × 49 × 45	34
7900HT Fast	96-well (fast or standard), 384-wells, low density array	96-well: standard 25–100, fast 10–30, low-density array: 5–20	72 × 64 × 84	114
OpenArray	3 OpenArray Plates	33 nl	63 × 77 × 73	80
StepOne	48-well	10–30	25 × 51 × 49	24
StepOnePlus	96-well	10–30	25 × 51 × 49	24
ViiA 7	96, 384-well plates, TaqMan array microfluidic cards	96-well: standard 10–100, fast 5–30, 384-well: 5–20, cards: ~1	53 × 64 × 65	61
InSyte	96-well, independent temp. control of wells/8- strips	20–50	44 × 46 × 60	45
SynChron	96-well, independent temp. control of wells/special plate	10–50	37 × 53 × 43	25
TOPtical	96-well plate	5–80	28 × 38 × 43	16
Exicycler 96	96-well plate/0.2 ml tubes	20–50	36 × 54 × 47	30
CFX384	384-well	1–30 (5–20 recommended)	33 × 46 × 36	21
CFX96	96-well	1–50 (10–25 recommended)	33 × 46 × 36	21
MiniOpticon	48-well, 48 tubes, 6 × 8-tube strip	10–100 (20 recommended)	18 × 32 × 33	7
GeneXpert	1 cartridge	Depending on the sample type/ assay	11 × 34 × 34	NA
SmartCycler	16, 32, 48, 64, 80 or 96 tubes (special design)	25	31 × 31 × 25	10
BioMark	Integrated fluidic chip; up to 48 samples	>4 µl input volume, 0.85 nl reaction volume	49 × 24 × 33	NA
LightScanner32	32 capillaries	10	30 × 41 × 41	11
RAPID	32 capillaries	5–20	27 × 36 × 49	23
RAZOR EX	1 pouch (12 samples)	100	25 × 11 × 19	5
Eco	48-well	5–20	35 × 31 × 32	14
Liat Analyser	1 tube	10–200	8 × 17 × 15	2
Rotorgene Q	72 or 96 well rotor discs, tubes, strip tubes	10–30	37 × 28 × 42	14
LightCycler 1536	LC 1536 multiwell plates	0.5–2	57 × 59 × 50	56
LightCycler 2.0	32 capillaries	20, 100	45 × 30 × 40	22
LightCycler 480	LC 480 multiwell plate (96 or 384 wells)	5–20 (384-well), 10–100 (96-well)	57 × 59 × 50	56
In-Check	1 chip	25	50 × 30 × 15	10
SmartChip	SmartChip Nanowell Chip (up to 384 samples)	0.1	70 × 50 × 60	57

NA, information not available.

This approach might limit the number of available fluorophores that can be excited because of the narrow wavelength range of the emitted light. For example, a standard blue-green LED has a peak emission of 470 nm, corresponding to the excitation wavelength of FAM and SYBR Green dyes. Many of the other commonly used dyes cannot be excited at this wavelength (Table 2.4). However, one strategy to increase the number of excitable fluorophores is to use multiple LEDs with different wavelengths (e.g. TOPical from Biometra and CFX384 from Bio-Rad). Today, instruments with up to six different LEDs are available (e.g. CFX96 from Bio-Rad). The main advantage of the LEDs is that replacement is unnecessary and they keep the same intensity throughout the life-time of the diodes.

Detector

The second part of the optical system is the detector where the light emitted from the reporter fluorophores is recorded. Most systems use photo-multiplier tubes (PMTs) (e.g. Rotorgene Q from QIAGEN), photodiodes (e.g. StepOne from Applied Biosystems) or charge-coupled device (CCD) cameras (e.g. most models from Applied Biosystems; see Table 2.2). CCD cameras can record all wells simultaneously, leading to a more robust mechanical design, whereas both photodiodes and PMTs record one point at a time. Some platforms utilize multiple photodiodes or PMTs which reduces the number of moving parts (e.g. CFX384 and CFX96). The number of channels detected is between one and six. It is important to assure that crosstalk, i.e. the fluorescence from one channel being detected in another channel, is minimized. There are several recent studies describing the use of pentaplex (Molenkamp *et al.*, 2007; Kirchner *et al.*, 2010; Basuni *et al.*, 2011) and even hexaplex real-time PCR assays (Bahrdt *et al.*, 2010; Hopkins *et al.*, 2011).

Calibration of specific dyes

Also important to consider is if the platform is optimized for specific dyes. For example, some platforms are delivered pre-calibrated for certain dyes, and thus show optimal performance with these particular dyes, although dyes with similar excitation and emission wavelengths can be used (see Table 2.2). For some platforms, it is necessary to buy and run calibration kits before using the instruments.

Software

Another important part of a real-time PCR platform is the software used to control the instrument and analyse the data. This software should be as user-friendly as possible, and should be able to analyse data generated by the probe chemistry applied (see section on primer and probe chemistries). A critical part of the software is to specify from which channels that data should be recorded from. If the wrong channels and/or wells are selected, data might be lost. Some platforms therefore have the option to record data from all channels and wells simultaneously, making it impossible to lose data. For some specific applications, e.g. HRM, additional software is required that sometimes needs to be purchased at an additional cost.

Comparison of performances between different platforms

Performance factors, such as ramp speed and uniform heat distribution, are important to achieve reproducible results. Therefore, it is recommended that thermocyclers are calibrated on a regular basis as part of the quality assurance programme of the laboratory (Khan, 2011). Calibration has previously been noted as an important factor when assessing the performance of thermocyclers (Saunders *et al.*, 2001).

Differences in temperature performance, between brands and types of cyclers, have been demonstrated to influence PCR performance (Saunders *et al.*, 2001; Yang *et al.*, 2005; Kim *et al.*, 2008). Validation is important to ensure proper and reproducible results from PCR analysis, and also when transferring an assay between different thermocycler models (Schoder *et al.*, 2005). It can also reveal problems caused by poor thermocycler performance (Schoder *et al.*, 2005; Kim *et al.*, 2008).

There are a number of recent studies in the literature comparing the performances of different real-time PCR platforms analysing identical

Table 2.4 Compatibility between common fluorophores and quenchers commercially available (indicated by ✓). Adapted from Osterman (2009)

Fluorophore/ Quencher[a]	Emax	Dabcyl	BHQ-1	BHQ-2	BHQ-3	Cy5Q	Cy7Q	Iowa BlackFQ	Iowa BlackRQ	IRDye QC-1	QSY35	QSY7	QSY21	QXL520	QXL570	QXL610	QXL680	DDQ II
Coumarin	472	✓									✓							
Alexa Fluor® 488-X	517	✓	✓					✓			✓			✓				
Oregon Green448-X	518	✓	✓					✓			✓			✓				
Fluorescein (FAM)	520	✓	✓					✓		✓	✓	✓		✓				
Rhodamine Green-X	531	✓	✓					✓		✓		✓						
TET	536	✓	✓					✓		✓		✓						
Yakima yellow	549	✓	✓			✓		✓	✓	✓		✓						✓
AlexaFluor-532	553		✓			✓		✓	✓	✓		✓						✓
JOE/VIC/HEX	555		✓			✓		✓	✓	✓		✓						✓
Cy3	564		✓	✓		✓		✓	✓	✓		✓			✓			✓
BODIPY 558/568	568		✓	✓		✓		✓	✓	✓		✓	✓		✓			✓
Alexa Fluor-546	571		✓	✓		✓		✓	✓	✓		✓	✓		✓			✓
TAMRA	583		✓	✓				✓	✓	✓		✓	✓		✓			✓
ROX	608			✓				✓	✓	✓		✓	✓		✓	✓		✓
Alexa Fluor-594	616			✓				✓	✓	✓			✓			✓		✓
Texas Red	617			✓	✓			✓	✓	✓			✓			✓		✓
BODIPY 630/650	653			✓	✓		✓		✓	✓			✓					✓
Cy5	668			✓	✓		✓		✓	✓			✓				✓	✓
Alexa Fluor-647	670			✓	✓		✓		✓	✓			✓				✓	✓
IRDye 680	702				✓					✓								✓
IRDye 700DX	700				✓					✓								✓
Cy5.5	706				✓					✓								✓
Alexa Fluor 750	775									✓								
IRDye 800CW	789									✓								
IRDye 800	809									✓								

[a]Dyes and quencher are presented with commercial names.

samples. Although some studies report differences in results between cyclers (Reynisson *et al.*, 2006), most studies show a good agreement in results (Donald *et al.*, 2005; Wolff *et al.*, 2009; Allnutt *et al.*, 2010; Schlesinger *et al.*, 2010; Stephens *et al.*, 2010; Matero *et al.*, 2011). However, an important aspect to bear in mind is that although performances are similar, it is not always a straightforward procedure to transfer a real-time PCR assay from one platform to another. Cycling times and steps need to be adjusted when changing from a system with different ramping speeds, and the performance of the real-time PCR assay might be different using different systems (Bentley *et al.*, 2005; Lu *et al.*, 2010). For example, when using glass capillaries (e.g. LightCycler), additives such as bovine serum albumin are recommended to lower the DNA-binding effect on the glass walls (Teo *et al.*, 2002). The BIOMARK platform has been found incompatible with the commonly used dye SYBRGreen I, which absorbs to the material in the channels in the microfluidics system (Kubista, 2010). The use of fast PCR protocols has been reported to cause problems with lower sensitivity and increased variability (Hilscher *et al.*, 2005) when transferring assays from a 'traditional' PCR protocol.

Recent trends in the real-time PCR platform development

The recent development of novel techniques, including HRM and digital PCR, is also reflected in the instrument development and introduction of commercially available platforms. For example, the platforms available on the market that are able to perform HRM (i.e. with a thermal resolution <0.02°C/s and suitable software), have been steadily increasing during the last couple of years. However, the accuracy of HRM genotyping has been reported to vary between different instruments depending on the data acquisition, data density and thermal control systems (Herrmann *et al.*, 2006, 2007). Also noteworthy is that the number of instruments able to multiplex five or even six channels is increasing, and so do the number of PCR assays taking advantage of this possibility (Bahrdt *et al.*, 2010; Hopkins *et al.*, 2011).

Digital PCR

Digital PCR is a technique to quantify gene target numbers by PCR using limiting dilution (Sykes *et al.*, 1992; Vogelstein and Kinzler, 1999). The sample is diluted so that it only contains a very small number of target molecules and is then distributed into reaction containers that contain either zero or a single template molecule each. When amplified by PCR, the number of target molecules will correspond to the number of positive reactions. The implementation of this technique has until recently been limited due to the need for manual dispensation into 96- or 384-well plates. However, with the introduction of high-throughput instruments such as the OpenArray from Life Technologies, the LC1536 from Roche, the SmartChip from Wafergen and the BIOMARK from Fluidigm, it is now possible to perform digital PCR in an automated manner which has increased the applicability of this method.

In-field and point-of-care analysis

There is a growing demand for robust, portable instruments suitable for in-field or point-of-care analysis, for example analysing bioterror agents or hospital 'point-of-care' diagnostic assays. A number of platforms have been available for some years, including the RAZOR from Idaho Technology and the Liat Analyser from IQuum. The pre-PCR sample preparation has been a potential source of error and risk of contamination with this type of instrument, and therefore this step has been integrated in a number of novel platforms, such as the Enigma ML and XL instruments from Enigma Diagnostics, BD Max from BD Diagnostics, and GeneXpert from Cepheid. The number of samples that it is possible to analyse simultaneously varies between 1 and 48 for the different instruments, many of them delivering results within only 30–60 min.

Lab-on-a-chip (LOC) devices to perform real-time PCR have gone from being promising research devices into a commercially available product (Zhang and Ozdemir, 2009; Lee *et al.*, 2010). These devices have great potential also for point-of-care diagnostics where the main advantage is the speed; cycling times for 40 cycles have been reported as low as 6 min (Neuzil *et al.*, 2006; Lee *et al.*, 2010). New devices are currently being developed and will most probably be introduced on the market within a few years (Lee *et al.*, 2010; Markey *et al.*, 2010; Shen *et al.*, 2010; Sundberg *et al.*, 2010).

Instrument selection

It is not trivial to choose a PCR platform that is suitable for the specific needs of a laboratory. As discussed in the last section, there are many factors that need to be taken into consideration, including (Logan and Edwards, 2009; Javorschi-Miller and Delgado Orlic, 2011):

- throughput (small laboratory use vs. core facility/shared instrument/large project);
- multiplexing features/supported chemistries;
- HRM capabilities;
- sample format (plate vs. tubes/capillaries, sample volume available for analysis);
- instrument size;
- ease of use and robustness of software; and
- cost (including instrument and software, servicing, reagents, consumables).

In connection with the novel development of LOC technologies and miniaturized systems (Zhang and Ozdemir, 2009), an additional factor might be added regarding the need for portability and use in the field and/or at point-of-care. Table 2.5 summarizes the features of the different instruments. From the information given here, the user can make a decision on a suitable platform.

Fluorescent DNA-binding dyes

Real-time PCR was established about 20 years ago when Higuchi and co-workers reported the addition of ethidium bromide directly to a PCR tube and monitored the DNA accumulation externally in real-time by the fluorescent signal generated in the reaction (Higuchi *et al.*, 1992). Since then, real-time PCR technology has developed and other double stranded (ds) DNA dyes with lower toxicity and improved fluorescent properties have been presented.

SYBR Green

Among the fluorescent DNA binding dyes, SYBR Green I is used most widespread for dsDNA quantification in real-time PCR applications (Wittwer *et al.*, 1997b). When the dye binds to the minor groove of the DNA molecule, it acquires morphological alterations with energy transfer that produces fluorescence. Upon accumulation of dsDNA in the PCR process, increased fluorescence signal is generated and detected by the real-time PCR instrument. Post PCR, a dissociation curve is usually generated by a stepwise increase in temperature and continuous fluorescent reading. When the melting temperature of the dsDNA product is reached, the two dissociated strands will produce a decrease in fluorescent signal due to lower affinity of the dye to single stranded DNA (Ririe *et al.*, 1997). The reported drawbacks of SYBR Green I are its inhibitory effect on the PCR reaction at elevated concentrations (Monis *et al.*, 2005), limited dye stability (Monis *et al.*, 2005) and suggested preferential binding with high G + C areas (Giglio *et al.*, 2003).

Alternative DNA binding dyes

New dyes have entered the market in recent years with reported improvements compared with SYBR Green I. These include YO-PRO (Ishiguro *et al.*, 1995), BEBO (Bengtsson *et al.*, 2003), BOXTO (Lind *et al.*, 2006), LCGreen (Wittwer *et al.*, 2003), SYTO9 (Monis *et al.*, 2005) and EvaGreen (Mao *et al.*, 2007), to name the most recognized. Some of these dyes belong to a class called 'new generation saturation dyes' which are frequently used in HRM analysis for genotyping and mutation scanning based on the real-time PCR principle and precise melting temperature analysis. For these applications, HRM instruments are needed along with dyes that fulfil the 'saturation model'. They have the potential of being added in such high concentrations to the reagent mixture that all binding positions of the DNA molecule are occupied. Several studies have been made on comparison of these alternative dyes in real-time PCR. BOXTO was shown to have less PCR inhibitory effect than SYBR Green I, but otherwise performed similarly (Ahmad, 2007), and EvaGreen was stable and

Table 2.5 Instrument selection guide

Model	Automation	Gradient	Digital PCR	In-field robust	Cycling timea	HRM	Multiplex
Mx3000p	No	No	No	No	NA	No	4
Mx3005p	No	No	No	No	NA	No	5
qTOWER	No	NA	NA	NA	<25 min	NA	4
7300	No	No	No	No	<2 h	No	4
7500	No	No	No	No	<2 h	No	5
7500 Fast	No	No	No	No	30–37 min (fast), <2 h (standard)	Optional	5
7900HT Fast	Yes	No	No	No	33 min (96-well fast), 52 min (384-well fast), <2 h (others)	No	2
OpenArray	No	NA	Yes	No	<3 h	NA	Yes
StepOne	No	No	No	No	<40 min (fast), <2 h (standard)	Possible	3
StepOnePlus	No	Yes	No	No	<40 min (fast), <2 h (standard)	Possible	4
ViiA 7	Optional	No	No	No	~ 30 min (96/384-well)	Yes	Yes
InSyte	No	Yes	No	NA	NA	NA	Yes
SynChron	No	Yes	No	NA	20 min/45 cycles	NA	NA
TOPtical	No	Optional	No	NA	NA	NA	6
Exicycler 96	No	Yes	No	Optional	2 h	NA	5
CFX384	No	Yes	No	No	NA	Yes	4
CFX96	No	Yes	No	No	<30 min	Yes	5
MiniOpticon	No	Yes	No	NA	NA	No	2
GeneXpert	Yes	Yes	No	Yes	<1 h	NA	5
SmartCycler	No	Yes	No	Yes	NA	Yes	4
BioMark	NA	No	Yes	No	NA	NA	5
LightScanner32	No	NA	No	NA	<30 min	Yes	Yes
RAPID	No	No	No	Yes	<30 min	No	Yes
RAZOR EX	No	No	No	Yes	<30 min	No	No
Eco	No	No	No	NA	<40 min/40 cycles	Yes	4
Liat Analyser	Yes	No	No	Yes	12 min/30 cycles	No	Yes
Rotorgene Q	No	No	No	No	40 min/45 cycles	Yes	2, 5 or 6
LightCycler 1536	No	No	Yes	No	<50 min	Yes	1
LightCycler 2.0	No	NA	No	NA	30 min/35 cycles (20 µl), 60 min/35 cycles (100µl)	Yes	4
LightCycler 480	Yes	No	No	No	<40 min (40 cycles, 384-well)	Yes	6
In-Check	NA	NA	No	NA	<1 h	No	Yes
SmartChip	NA	No	Yes	No	~2 h (40 cycles)	No	NA

NA, information not available.
[a]Cycling time for a standard PCR program, as stated by the manufacturers.

compatible with different PCR instruments (Mao *et al.*, 2007). SYT09 was shown to provide a broader range of dye concentrations without causing PCR inhibition compared with SYBR Green I. More robust and consistently shaped DNA melting curves that were less affected by dye concentration were produced by SYT09. This dye was also more readily used for the detection of multiple amplicons by DNA melting curve analysis without any apparent bias (Monis *et al.*, 2005). Despite general encouragement for using the new dyes for HRM applications (Reed *et al.*, 2007), other scientists have reported that SYBR Green I is just as good for HRM analysis without the need for any modifications to analysis protocols (Price *et al.*, 2007; Pornprasert *et al.*, 2008).

DNA binding dyes versus probe-based detection

In recent years, probe-based real-time PCR applications have received more attention, and numerous reports on new detection systems have been published (see section on primer and probe chemistries). Although the probe-based systems have been shown to be more suitable for DNA quantification than the dsDNA dye systems (Malinen *et al.*, 2003; Ciglenecki *et al.*, 2008), the latter have the advantages of being more cost-effective and the possibility of post-PCR melting curve analysis enables verification of product specificity. Its use is further advised in the first steps of assay development when performance of primers and efficiency is being tested (Cirino *et al.*, 2007). Theoretically, dsDNA dye could provide more sensitivity to an assay compared with a probe system, since the number of fluorescent molecules per amplicon is higher and thus should produce an earlier detectable amplification signal than a corresponding probe system. In our experience, this is also the case in a relatively broad target concentration range, where dsDNA dye assays usually produce lower Ct values than corresponding probe assays. However, at low target concentrations, the risk of unspecific signals by SYBR Green I can hinder a proper linearity close to the detection limit. Combining a DNA binding dye to a TaqMan reaction has been reported where the non-specific detection by the DNA dye was used as a quality measure of the reaction, for estimation on primer-dimer formation or unspecific amplification (Lind *et al.*, 2006). For a multiplex diagnostic assay, however, a mix of probe based and dsDNA dye systems is not recommended (Cirino *et al.*, 2007).

Primer and probe chemistries

Measurement of amplicon accumulation in real-time is, regardless of the choice of chemistry, based on monitoring a change in emitted fluorescence. Besides the fluorescent DNA binding dyes described in the previous section, real-time PCR is commonly performed by applying fluorescently labelled oligonucleotides. Several different chemistries are available, and although they are adaptable to most real-time PCR thermal cyclers, the choice of chemistry should be considered before purchasing a real-time thermal cycler. Two basic categories of fluorescently labelled oligonucleotides exist; primers and probes, and examples of both will be described in the following section.

Primer-based chemistries

Monitoring the accumulation of PCR amplicons by means of a single primer can be performed by labelling the primer with a fluorophore that changes fluorescence properties upon annealing to nucleic acids and subsequent extension, as shown in Fig. 2.1, exemplified by the Light Upon eXtension LUX™ primer (Nazarenko *et al.*, 2002). Other primer chemistries, such as the AmpliFluor primers (Fig. 2.2) are labelled with both a reporter and a quencher dye, relying on a collisional quenching mechanism brought about by the hairpin conformation of the un-annealed primer (Nazarenko *et al.*, 1997).

A major advantage of the primer-mediated detection chemistries is the reduced cost compared with corresponding probes. In contrast to the unspecific DNA binding dyes, the primer chemistries also offer the possibility of multiplexing. However, since fluorescence is changed simply upon annealing and extension, unspecific amplicons and primer-dimers will also give rise to a PCR signal, rendering this approach less specific than probe-based chemistries (Mackay and Landt, 2007).

Light Upon eXtension (LUX™)

The principle of Light Upon eXtension (LUX™) primers is based on internal quenching of the fluorophore in the un-annealed primer by a nearby guanine base. The labelled primer has a hairpin conformation (Fig. 2.1), and only upon annealing

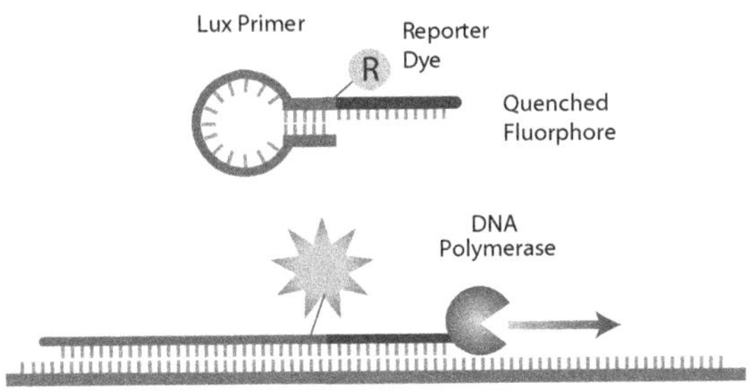

Figure 2.1 Fluorescence mechanism of the Light Upon eXtension (LUX™) primer (modified from Cepheid).

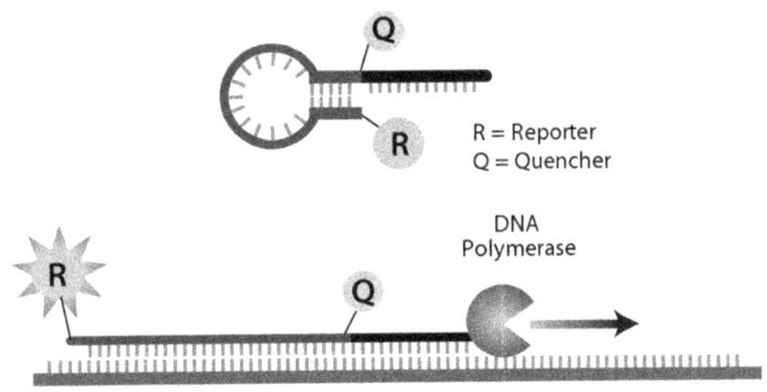

Figure 2.2 Fluorescence mechanism of the AmpliFluor® universal hairpin primer (modified from Cepheid).

and extension of this primer, can an up to 8-fold increase in fluorescence be detected (Nazarenko *et al.*, 2002). Comparison studies that include the LUX™ chemistry have been performed, and this primer-based system is reported to be a cost-effective detection alternative (McCrea *et al.*, 2007; Nordgren *et al.*, 2008; Buh Gasparic *et al.*, 2010; Chui *et al.*, 2010).

Plexor™

The Plexor™ technology distinguishes itself from other real-time PCR detection chemistries by reduced fluorescent signal upon product accumulation. One of the primers is synthesized with a modified nucleic acid coupled to a fluorophore in the 5′ end, and the reaction mixture contains complementary nucleotides that are modified to include a quencher. In subsequent PCR cycles, base pairing between the quencher-modified nucleotide and the fluorescently labelled one, on the newly synthesized DNA strand, will extinguish the fluorescent signal (Frackman *et al.*, 2005). Literature describing the application of the Plexor™ technology is limited (Atallah *et al.*, 2007; Buh Gasparic *et al.*, 2008; Lasalle *et al.*, 2011). For both the LUX™ and Plexor™ primer systems, analysis of dissociation curves can be performed to confirm the specificity.

AmpliFluor®

In the AmpliFluor® chemistry (formerly marketed as Sunrise™), three primers are applied, of which two are target specific and the third is a universal dual-labelled hairpin primer (Fig. 2.2). One of the target-specific primers has a recognition sequence for the hairpin primer incorporated, and becomes a template for this primer during the second cycle of PCR. Upon annealing of the hairpin primer and subsequent elongation, the hairpin stem will be pulled open eliminating the quenching and the fluorescence detection markedly enhanced (Nazarenko *et al.*, 1997). The application of AmpliFluor® primers for genotyping in single nucleotide polymorphism (SNP) and mutation detection studies has been described extensively, as differently labelled forward primers combined with one common reverse primer can identify sequences differing by a single nucleotide (Giancola *et al.*, 2006; Khripin, 2006; Fuhrman *et al.*, 2008).

Self-reporting DNA/PNA

Finally, a primer-based chemistry referred to as self-reporting DNA/peptide nucleic acid (PNA) primers (Fiandaca *et al.*, 2001) employ a tagged fluorophore-labelled target-specific primer, whose signal is silenced by a quencher-bearing PNA oligonucleotide. The PNA oligo is displaced upon primer annealing, and the subsequent increase in fluorescence detected. This system could be cost-effective and suitable for multiplexing, because the PNA oligo could quench the signal of any primer adapted for self-reporting by including the tag sequence and fluorophore labelling.

Probe-based chemistries

Compared with the primer-based detection chemistries, probe-based detection adds an additional level of specificity to the assay. Obtaining a fluorescent signal from the specifically designed probe ensures that the PCR amplicons accumulating are the correct ones, and that the increase in fluorescence is not caused by spurious binding of primers or primer-dimers.

Hydrolysis probes

The most widely used probe-based detection chemistry for real-time PCR remains to be the original hydrolysis probes, also known as 5′ nuclease or simply TaqMan® probes. In this chemistry, a linear, target-specific probe is designed to complement a sequence in between the forward and reverse primer, preferentially in close proximity to one of the primer binding sites. These probes are generally labelled at the 5′ end with a fluorescent reporter dye and at the 3′ end with a quencher. Hydrolysis probes are designed to have melting temperatures approximately 10°C higher than their corresponding primers, causing the probe to anneal to the DNA template prior to the primers. When the primers are subsequently extended and the probe is encountered, part of the probe will be hydrolysed by the 5′–3′ exonuclease activity of the DNA polymerase, separating the reporting dye from the quencher and allowing emission of fluorescence (Fig. 2.3).

Substituting part of the regular nucleotides of the TaqMan® probes with synthetic analogues such as locked nucleic acids (LNA) or peptide nucleic acids (PNA) allows for shorter probe designs. These analogues have a higher affinity for DNA and hence the thermal stability of the formed duplex is increased significantly (Egholm *et al.*, 1992; Kumar *et al.*, 1998). This is also the case for the next generation of TaqMan® probes, the minor groove binding (MGB) probes. In this probe type, a hyperstabilizing molecule attached to one end of the probe folds into the minor groove of the formed duplex, giving rise to increased thermal stability. Advantages of these short probe types are improved sensitivity towards single-base mismatches (specificity), ease of design and an improved signal–to-noise ratio, partly due to lower fluorescence from spurious binding and partly due to the reduced distance between the reporter and the quencher (Kutyavin *et al.*, 2000; Braasch and Corey, 2001). These advantages must be weighed against the additional cost of these more complex probe types; however, they have been reported to perform superior to other probe types and chemistries depending on the specific PCR assay (Costa *et al.*, 2004; Zhao *et al.*, 2005; Reynisson *et al.*, 2006; Josefsen *et al.*, 2009). These short probe types have been successfully applied for genotyping and detection of SNPs (Saito *et al.*, 2003; Petersen *et al.*, 2004; Li *et al.*, 2010). The advantages of the shorter probe types are not limited to the hydrolysis probe chemistries.

AllGlo™ probes are labelled with two identical reporter dyes capable of quenching each other and becoming de-quenched upon probe cleavage. This type of probe contains two signal-generating reporter dye molecules per oligo, therefore a stronger fluorescent signal can be expected. Furthermore, the cost of AllGlo™ probes is less due to the simplified manufacturing process (AlleLogic Biosciences). Use of this more recently developed probe type has not yet been reported widely.

The cycling probe technology (CPT) also belongs to the hydrolysis category, but here the probe nucleotides are substituted with modified RNA molecules. Following annealing of this DNA-RNA-DNA probe, the formed duplex is recognized and cut by RNaseH, thus separating the reporter and the quencher and releasing the fluorescence emission. This probe system does not, in contrast to the other hydrolysis probes, rely on the 5′–3′ exonuclease activity of the DNA polymerase (Duck *et al.*, 1990). The CPT was compared with eight other probe types for qualitative and quantitative applications in GMO detection and shown to have a moderate performance and a high cost, making them inappropriate for routine use (Buh Gasparic *et al.*, 2010).

The LionProbe™, in contrast, exploits the 3′–5′ exonuclease activity of a proofreading polymerase. The probe is labelled at both ends with a reporter and a quencher, respectively, and the 3′-end is designed to have the last few nucleotides mismatching the target sequence. The activity of the proofreading polymerase will cleave the reporter from the probe to correct the mismatch, and the probe now functions as a primer for the subsequent extension (Biotools).

Linear hybridization probes

In contrast to the fluorescence accumulation observed with the hydrolysis probe chemistries,

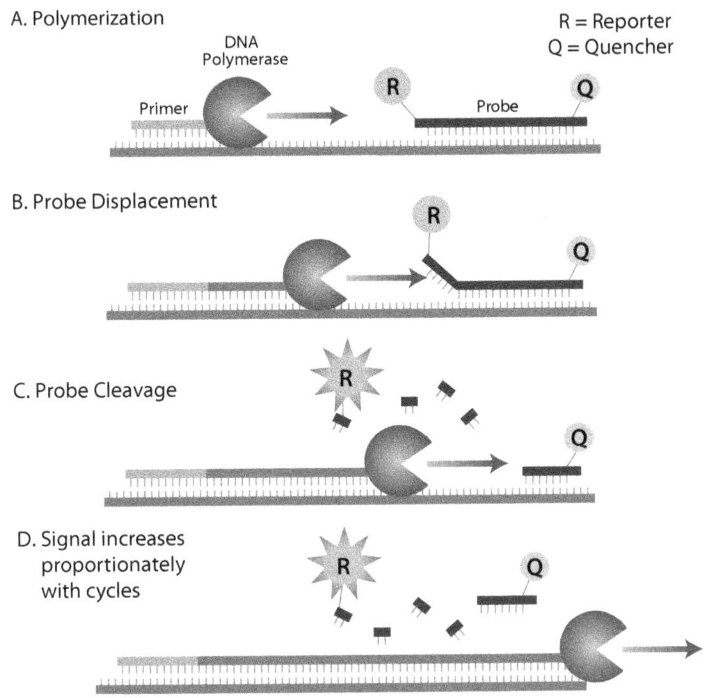

Figure 2.3 Hydrolysis of a TaqMan® probe during amplification (Cepheid).

the signal from hybridization probes is generated in each PCR cycle as the hybridization event occurs. While several probe types fall into the category of linear hybridization probes, the HybProbes® or dual hybridization probes, originally designed for application in the LightCycler instrument, are the most widely used today. In this approach, two single-dye labelled linear probes are designed to anneal adjacently on the DNA template in between the two primer binding sites. In most applications, one probe is labelled at the 3′-end with a donor dye and the other at the 5′-end with an acceptor dye. Upon hybridization of the two probes to the DNA template, the dyes are brought into close proximity and the acceptor dye will be excited by the donor dye, emitting fluorescence of a characteristic wavelength. The 5′-end labelled probe is modified and contains a PCR blocker in the 3′-end to prevent extension by the DNA polymerase. It can be challenging designing not only one, but two probes, if there is a lack of known conserved region. Likewise, the melting temperature of the probes should be identical, and in the range of 65–75°C in order to generate sufficient signal without clamping and thereby inhibiting the amplification process. An advantage of this chemistry is that it offers the possibility of performing subsequent melting curve analysis, allowing for multiplexing with only one dye, and revealing potential nucleotide polymorphisms in the DNA strand sequence under the hybridization probes (Lee *et al.*, 2009).

The displacement probes, or Yin-Yang probes as they are also referred to as, also belong to the class of linear hybridization probes (Li *et al.*, 2002). This is the probe-based counterpart of the self-reporting DNA/PNA primer described by Fiandaca *et al.* (2001). A reporter-labelled probe is silenced by a complementary, but shorter, quencher-bearing DNA strand. Upon PCR product amplification, a preferential sequence for the reporter-labelled probe will be made available for hybridization, eliminating the quenching of fluorescence (Li *et al.*, 2002).

LightUp™ Probes (Svanvik *et al.*, 2000) consists of PNAs linked to the fluorophore thiazole orange. Upon probe hybridization to a complementary sequence, the fluorescence emission will increase substantially. The SimpleProbe™ or HyBeacon™ probes for detection of mutations and genotyping, work in a similar manner to the LightUp™ probes, exhibiting enhanced fluorescence upon hybridization to a specific target (French *et al.*, 2001), allowing for identification of closely related sequences by melting point technology.

Conformational probes

Probably the most well-known types of conformational probes are the Molecular beacons (MB). The MBs are principally designed to form a secondary hairpin structure in which the fluorescence emission of the reporter is quenched due to the proximity of the quencher as shown in Fig. 2.4. The reporter and quencher-bearing stems of this conformational probe consist of G/C-rich complementary DNA strands of four to six nucleotides in length, while the probe sequence is contained in the loop region. Presence of the probe-specific target favours hybridization to this over returning to the hairpin conformation, causing the probe to unfold, and separating the reporter and the quencher, emitting detectable fluorescence (Tyagi and Kramer, 1996).

The MBs offer an exceptional specificity because the stability of the stem structure ensures that unfolding and hybridization only occurs if a perfect matching complementary annealing site is present, making them ideal for SNP analysis. In recent years, many structural variants of the original MBs have arisen such as the wavelength-shifting MB producing brighter fluorescence emission and simplifying and improving multiplex analysis, and the tripartite MB which has additional application possibilities and reduced cost of synthesis. Application of MBs is widely described and the possibilities are numerous (for a review, see Broude, 2002). A limitation of this chemistry is the high cost of synthesis, and it should be noted that MBs are unstable in the presence of DNA polymerases exhibiting exonuclease activity.

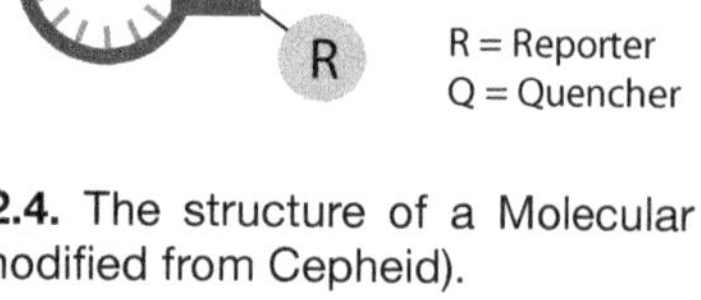

Figure 2.4. The structure of a Molecular beacon probe (modified from Cepheid).

The Scorpion™ probe system comes in two formats; uni- and bimolecular. The unimolecular Scorpion™ has a hairpin loop tail linked to the 5′-end of a primer via a PCR-blocker as shown in Fig. 2.5. This primer anneals to the template DNA, and the DNA polymerase extends it, creating a complementary probe binding site. Following the next denaturation, the probe element will fold up and anneal to this site, rather than entering the hairpin loop conformation again, because it will be more kinetically favourable. This event opens the hairpin loop, eliminating the quenching of the fluorophore and increasing the real-time PCR signal. The bi-molecular Scorpion consists of a fluorescently labelled probe coupled to a primer by a PCR-blocker. It is annealed to a complementary oligo bearing a quencher at the 3′-end. The mechanism of action is similar to that of the unimolecular in that the probe will anneal to the newly synthesized DNA template and will emit fluorescence since it is separated from the quencher. The intra molecular probing mechanism of the Scorpion™ system results in a rapid signal generation contributing to a reduced time of analysis (Whitcombe *et al.*, 1999; DxS Diagnostics). This probe system has also been used successfully for mutation detection and SNP analysis (Burucoa *et al.*, 2008; Sahoo *et al.*, 2011).

The IntraTaq probe presents a hydrolysis probe with the conformation of the Scorpion system relying on the exonuclease activity of the DNA polymerase to generate a fluorescent signal. This intra molecular probing mechanism has been shown to perform superior to equivalent linear TaqMan® probes (Solinas *et al.*, 2002).

As seen in Fig. 2.6, the MGB Eclipse™ probe also belongs to the grouping of conformational probes. An MGB moiety is attached to the 5′-end of the three-dimensional probe together with the quencher molecule and the reporter dye is linked to the 3′-end. As for the linear probes, the MGB Eclipse™ probes can be designed shorter due to the stabilizing action of the MGB molecule with similar advantages as mentioned earlier. In its non-annealed state, the short distance between the reporter and

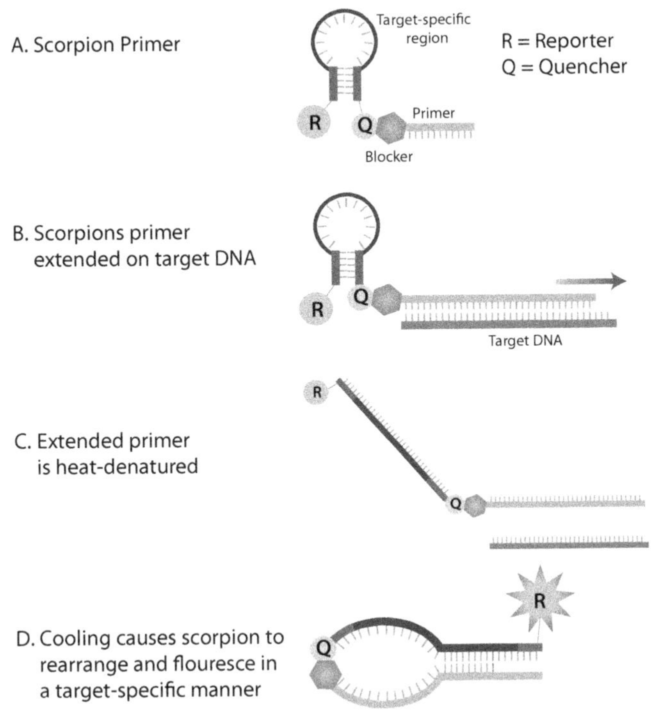

Figure 2.5 The Scorpion™ probe-based amplification process (Cepheid).

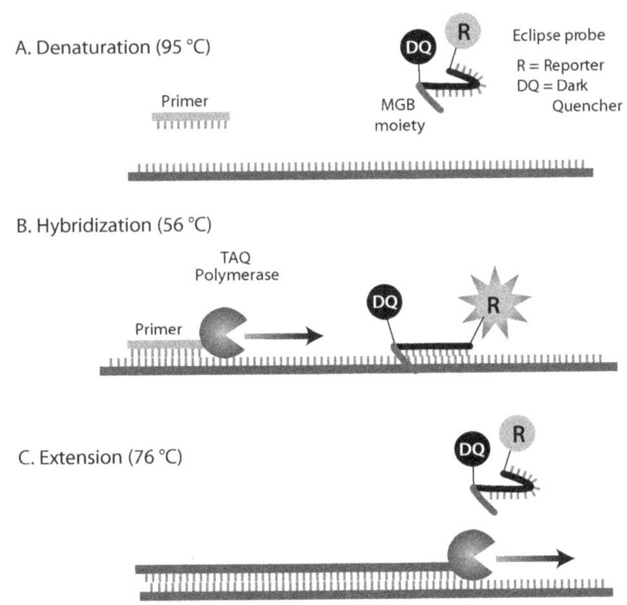

Figure 2.6 The Eclipse™ probe during the amplification process (Cepheid).

quencher will prevent emission of fluorescence, and only upon hybridization to target DNA and unfolding of the probe element will the quenching of the reporter be eliminated (Afonina *et al.*, 2002).

Fluorescent dyes and quenchers

Every probe-based real-time PCR application is dependent on molecular illumination through various types of fluorophores and quenchers

(Heid *et al.*, 1996; Cirino *et al.*, 2007). The principle of fluorescence is based on the ability of a substance to absorb excitation energy, which is usually a light source of a certain wavelength, and re-emit the energy as light at a another wavelength (Morrison, 2010). In real-time PCR, the fluorophore is in proximity to a quencher that absorbs the energy and eliminates illumination by the fluorophore upon excitation. The quenching mode can be accomplished in different manners, but in real-time PCR, it is usually either contact quenching (non-fluorescent ground state complex), which is the case with the Scorpion probe system and MBs, or energy transfer to another molecule (fluorescence resonance energy transfer, or FRET), which is the case in TaqMan probes (Morrison, 2010).

Fluorescent quenchers re-emit the energy as light (at another wave-length than the fluorophore) while other quenchers, so called dark quenchers, re-emit the energy as heat (Fig. 2.7) without affecting the temperature of the solution (May *et al.*, 2005).

Early on in real-time PCR development, the most frequent dyes were FAM quenched with TAMRA but today the dark quenchers are more commonly used (Heid *et al.*, 1996; Yang *et al.*, 2009). The major advantage of non-fluorescent quenchers is a lower fluorescent background, which improves the spectral discrimination and the ease of multiplexing. The possibility for multiplexing is further increased as all detectors available in the instrument can be used for targets of interest, rather than being occupied by light-emitting quenchers (Cirino *et al.*, 2007). Wide varieties of trademark fluorophores are available for real-time PCR but some of the most frequently applied are FAM, HEX, ROX, Cy5, Cy3, Yakima Yellow and Texas Red. Similarly, examples of dark quenchers are BHQ1–3 and Qxl, which are able to quench the entire visible spectrum, Dabcyl (green spectrum), Iowa quenchers (green-yellow-orange-red), IRDye QC-1 (visible to near infrared spectrum) and Deep Dark Quencher. Fluorophore–quencher combinations have to be chosen with respect to their absorbance and emission spectra, meaning that the absorbance spectra of the quencher must overlap with the emission wavelength of the fluorophore to ensure

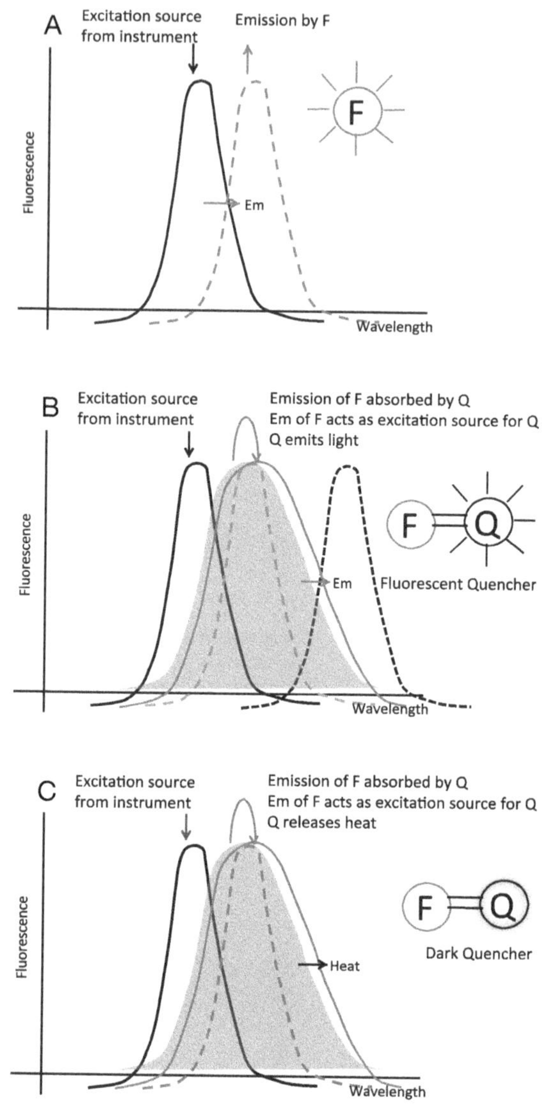

Figure 2.7 Principles of the fluorescent dyes used in a real-time PCR probe system. (A) When a fluorophore (F) is unquenched, the excitation source from the instrument results in a direct emission from the fluorophore. (B) When F is linked to a fluorescent quencher (Q) such as TAMRA, Q absorbs the emitted energy from F and results in a direct emission of Q at another wavelength than F. (C) When F is attached to a dark quencher, Q absorbs the emitted energy from F and releases it as heat.

efficient quenching (Peng *et al.*, 2009) (Table 2.4 and Fig. 2.7). The compatibility of fluorophores and instruments intended to be used must also be taken into consideration since the real-time PCR thermal cyclers are equipped with different detector channels (Table 2.4). The FAM reporter is still usually regarded as the first choice (Marras, 2006).

Multiplex real-time PCR

Multiplex real-time PCR systems offer an attractive opportunity to increase throughput and reduce reagent cost. In that manner, it is essential to carefully design the dye combinations based on the available detectors in the real-time PCR instrument to be used. The need for multiplex systems is also evident in routine diagnostics where it has been proposed to make internal amplification controls mandatory, to prevent false-negative results caused by PCR inhibitory substances (Hoorfar *et al.*, 2003). In other cases, the detection of multiple microorganisms such as different foodborne pathogens or food spoilage bacteria is advantageous (Haakensen *et al.*, 2008). Recently, an experimental strategy named 'multicolour combinatorial probe coding' (MCPC) was introduced which includes systematic labelling of large number of different probes with a limited number of differently coloured fluorophores. The identification of the 15 existing human papillomavirus types in one reaction was demonstrated, which is the maximum target number theoretically detectable in one reaction by MCPC in a four-colour channel instrument (Huang *et al.*, 2011).

Future challenges and trends

Despite the increasing number of primer and probe chemistries and real-time PCR instruments, the current applications are mostly limited to duplex detection and, in rare cases, triplex. This is far from the promised potential of real-time PCR for parallel testing of many targets in the same tube. Most scientists have alleviated this limitation by distributing the same sample in several parallel tubes, each containing different dyes. However, this is not what real-time was supposed to deliver. Instrument producers must develop true multiplex applications to support customers to take the full advantage of real-time PCR.

Scientists and end-users should develop new applications to exploit the enumeration capacity of real-time PCR. Most publications report new tests that merely detect presence or absence of certain genes. More intelligent sample preparation designs are needed to obtain quantitative data that provide more meaningful indication of the extent of the initial contamination or the extent of interventions that are necessary to deal with a situation. For instance, it has been shown to be possible to combine pre-enrichment with real-time PCR and still obtain quantitative data for foodborne pathogens (Josefsen *et al.*, 2004; Krämer *et al.*, 2011).

Interpretation of PCR results, in particular fluorescence readings close to baseline values, is very dependent on the software algorithms that come with the different instruments. The algorithms used to normalize output readings vary, in turn, among software producers. Some producers use sophisticated calculations to normalize PCR results for within sample controls (spontaneous fluorescence release, internal amplification control, etc.), sample-to-sample variations, coefficient of variations, baseline readings, and so on, while other instrument producers simply subtract a fixed value from all of the fluorescence readings to produce threshold cycle values for test samples. Thus, there is an urgent need to agree upon a standard algorithm in order to facilitate data comparison among current and future PCR platforms.

To add to the lack of standardization, the DNA or RNA standard dilutions used among different research groups to produce quantification curves could vary considerably in genome and gene copy numbers, depending on the purity, nucleic acid integrity and method of measurement. International standard organizations should agree upon a uniform guideline for the preparation and measurement of reference DNA dilutions.

There are certainly other issues that can be added to the abovementioned challenges. Nevertheless, new real-time PCR platforms and chemistries will continue to merge, in particular with connection of low-density microarray platforms that can result in true multiplex detection of hundreds of targets. To achieve this goal, scientists involved in real-life diagnostics must collaborate more closely with instrument producers, and instrument producers must welcome new application needs that are put forward by scientists.

Conclusion

The development of applications and instrumentation for real-time PCR presents the routine or research laboratory with a wealth of possibilities. Thorough considerations regarding fluorescent detection method and potential probe chemistry has to be made prior to implementing a real-time PCR assay. Selecting a real-time thermal cycler is not a straight forward task, and factors to consider include; cost (not only related to purchasing the instrument itself, but also service contracts, consumables and reagents), supported chemistries, multiplexing capacities, HRM capabilities, cycling speed, technical support, customer support, ease of use, throughput, format and footprint. In this respect it can be wise to make use of the possibility to 'try before you buy' that many companies provide today. Finally, advances in the area of fluorescent dyes and quenchers, adding to the sensitivity and spectral performance of real-time PCR assays, should be taken into consideration.

Web resources

http://www.gene-quantification.de

Manufacturers' homepages:

Idaho Technology: www.idahotech.com Illumina: www.illumina.com Roche: www.roche.com Bio-Rad: www.biorad.com BD: www.bd.com Cepheid: www.cepheid.com IQuum: www.iquum.com STMicroelectronics: www.st.com Applied Biosystems: www.appliedbiosystems.com Agilent: www.agilent.com Analytik Jena: www.analytik-jena.de Biogene: www.biogene.com Biometra: www.biometra.de Fluidigm: www.fluidigm.com Qiagen: www.qiagen.com TaKaRa: www.takara-bio.com Wafergen: www.wafergen.com

Acknowledgements

This work was supported in part by the EU-project AniBioThreat (Grant Agreement: Home/2009/ISEC/AG/191) with the financial support from the Prevention of and Fight against Crime Programme of the European Union, European Commission – Directorate General Home Affairs, in part by the European Union-funded Integrated Project BIOTRACER (Contract 036272) under the 6th RTD Framework, and in part by the Icelandic Technological fund with the era-SME programme through the FreshFish project (Contract RAN090914-1736).

References

Afonina, I.A., Reed, M.W., Lusby, E., Shishkina, I.G., and Belousov, Y.S. (2002). Minor groove binder-conjugated DNA probes for quantitative DNA detection by hybridization-triggered fluorescence. Biotechniques *32*, 940–949.

Ahmad, A.I. (2007). BOXTO as a real-time thermal cycling reporter dye. J. Biosci. *32*, 229–239.

AlleLogic Biosciences. AllGlo technology for real time PCR. Available at: http://www.allelogic.com/.

Allnutt, T.R., Ayadi, M., Berben, G., Brodmann, P., and Lee, D. (2010). Evaluation of different machines used to quantify genetic modification by real-time PCR. J. AOAC Int. 93, 1243–1248.

Atallah, Z.K., Bae, J., Jansky, S.H., Rouse, D.I., and Stevenson, W.R. (2007). Multiplex real-time quantitative PCR to detect and quantify Verticillium dahliae colonization in potato lines that differ in response to verticillium wilt. Phytopathology *97*, 865–872.

Bahrdt, C., Krech, A., Wurz, A., and Wulff, D. (2010). Validation of a newly developed hexaplex real-time PCR assay for screening for presence of GMOs in food, feed and seed. Anal. Bioanal. Chem. *396*, 2103–2112.

Basuni, M., Muhi, J., Othman, N., Verweij, J.J., Ahmad, M., Miswan, N., Rahumatullah, A., Aziz, F.A., Zainudin, N.S., and Noordin, R. (2011). A pentaplex real-time polymerase chain reaction assay for detection of four species of soil-transmitted helminths. Am. J. Trop. Med. Hyg. *84*, 338–343.

Bengtsson, M., Karlsson, H.J., Westman, G., and Kubista, M. (2003). A new minor groove binding asymmetric cyanine reporter dye for real-time PCR. Nucleic Acids Res. *31*, e45.

Bentley, H.A., Belloni, D.R., and Tsongalis, G.J. (2005). Parameters involved in the conversion of real-time PCR assays from the ABI prism 7700 to the Cepheid SmartCycler II. Clin. Biochem. *38*, 183–186.

Biotools. Retrieved 3/11/2011, 2011. Available at: http://www.biotools.eu/our_genomic.html.

Braasch, D.A., and Corey, D.R. (2001). Locked nucleic acid (LNA): fine-tuning the recognition of DNA and RNA. Chem. Biol. *8*, 1–7.

Broude, N.E. (2002). Stem-loop oligonucleotides: a robust tool for molecular biology and biotechnology. Trends Biotechnol. *20*, 249–256.

Buh Gasparic, M., Cankar, K., Zel, J., and Gruden, K. (2008). Comparison of different real-time PCR chemistries and their suitability for detection and quantification of genetically modified organisms. BMC Biotechnol. *8*, 26.

Buh Gasparic, M., Tengs, T., La Paz, J.L., Holst-Jensen, A., Pla, M., Esteve, T., Zel, J., and Gruden, K. (2010). Comparison of nine different real-time PCR chemistries for qualitative and quantitative applications in GMO detection. Anal. Bioanal. Chem. *396*, 2023–2029.

Burucoa, C., Garnier, M., Silvain, C., and Fauchere, J.L. (2008). Quadruplex real-time PCR assay using allele-specific scorpion primers for detection of mutations conferring clarithromycin resistance to Helicobacter pylori. J. Clin. Microbiol. *46*, 2320–2326.

Cepheid. SmartNotes Retrieved 3/11/2011, 2011, from http://www.cepheid.com/resources-and-support/support/smartnotes/

Chui, L., Couturier, M.R., Chiu, T., Wang, G., Olson, A.B., McDonald, R.R., Antonishyn, N.A., Horsman, G., and Gilmour, M.W. (2010). Comparison of Shiga toxin-producing *Escherichia coli* detection methods using clinical stool samples. J. Mol. Diag. (JMD) *12*, 469–475.

Ciglenecki, U.J., Grom, J., Toplak, I., Jemersic, L., and Barlic-Maganja, D. (2008). Real-time RT-PCR assay for rapid and specific detection of classical swine fever virus: Comparison of SYBR Green and TaqMan MGB detection methods using novel MGB probes. J. Virol. Methods. *147*, 257–264.

Cirino, N.M., Tavakoli, N.P., Madison-Anenucci, S., and Egan, C. (2007). Multiplex rtPCR in microbiology. In Real-Time PCR in Microbiology, Mackay, I.M., ed (Caister Academic Press, Norfolk, UK), pp. 183–192.

Costa, J.M., Ernault, P., Olivi, M., Gaillon, T., and Arar, K. (2004). Chimeric LNA/DNA probes as a detection system for real-time PCR. Clin. Biochem. 37, 930–932.

Donald, C., Qureshi, F., Burns, M., Holden, M., Blasic, J., and Woolford, A. (2005). An inter-platform repeatability study investigating real-time amplification of plasmid DNA. BMC Biotechnol. *5*, 15.

Duck, P., Alvarado-Urbina, G., Burdick, B., and Collier, B. (1990). Probe amplifier system based on chimeric cycling oligonucleotides. Biotechniques. *9*, 142–148.

DxS Diagnostics. Molecular diagnostic technologies. Scorpions reaction. Available at: http://www.dxsdiagnostics.com/(S(x1w3v045o5urvn45skrh2b45))/site/content/dxs/TheScorpionsReaction.aspx?Z=18

Egholm, M., Buchardt, O., Nielsen, P.E., and Berg, R.H. (1992). Peptide nucleic acids (PNA). Oligonucleotide analogs with an achiral peptide backbone. J. Am. Chem. Soc. *114*, 1895–1897.

Fiandaca, M.J., Hyldig-Nielsen, J.J., Gildea, B.D., and Coull, J.M. (2001). Self-reporting PNA/DNA primers for PCR analysis. Genome Res. *11*, 609–613.

Frackman, S., Ekenberg, S., Hoffmann, K., Krenke, B., Sprecher, C., and Storts, D. (2005). Plexor™ technology: a new chemistry for real-time PCR. Promega Notes *90*, 2.

French, D.J., Archard, C.L., Brown, T., and McDowell, D.G. (2001). HyBeacon probes: a new tool for DNA sequence detection and allele discrimination. Mol. Cell. Probes. *15*, 363–374.

Fuhrman, L.E., Shianna, K.V., and Aballay, A. (2008). High-throughput isolation and mapping of *C. elegans* mutants susceptible to pathogen infection. PLoS ONE *3*, e2882.

Giancola, S., McKhann, H.I., Berard, A., Camilleri, C., Durand, S., Libeau, P., Roux, F., Reboud, X., Gut, I.G., and Brunel, D. (2006). Utilization of the three high-throughput SNP genotyping methods, the GOOD assay, Amplifluor and TaqMan, in diploid and polyploid plants. Theor. Appl. Gen. *112*, 1115–1124.

Giglio, S., Monis, P.T., and Saint, C.P. (2003). Demonstration of preferential binding of SYBR Green I to specific DNA fragments in real-time multiplex PCR. Nucleic Acids Res. *31*, e136.

Haakensen, M., Dobson, C.M., Deneer, H., and Ziola, B. (2008). Real-time PCR detection of bacteria belonging to the Firmicutes Phylum. Int. J. Food Microbiol. *125*, 236–241.

Heid, C.A., Stevens, J., Livak, K.J., and Williams, P.M. (1996). Real-time quantitative PCR. Genome Res. *6*, 986–994.

Herrmann, M.G., Durtschi, J.D., Voelkerding, K.V., and Wittwer, C.T. (2006). Instrument comparison for DNA genotyping by amplicon melting. J. Assoc. Lab. Autom. *11*, 273–277.

Herrmann, M.G., Durtschi, J.D., Wittwer, C.T., and Voelkerding, K.V. (2007). Expanded instrument comparison of amplicon DNA melting analysis for mutation scanning and genotyping. Clin. Chem. *53*, 1544–1548.

Higuchi, R., Dollinger, G., Walsh, P.S., and Griffith, R. (1992). Simultaneous amplification and detection of specific DNA sequences. Biotechniques *10*, 413–417.

Hilscher, C., Vahrson, W., and Dittmer, D.P. (2005). Faster quantitative real-time PCR protocols may lose sensitivity and show increased variability. Nucleic Acids Res. *33*, e182.

Hoorfar, J., Cook, N., Malorny, B., Wagner, M., De Medici, D., Abdulmawjood, A., and Fach, P. (2003). Making internal amplification control mandatory for diagnostic PCR. J. Clin. Microbiol. *41*, 5835.

Hopkins, M.J., Moorcroft, J.F., Correia, J.B., and Hart, I.J. (2011). Using the full spectral capacity (six channels) of a real-time PCR instrument can simplify diagnostic laboratory screening and typing protocols for pandemic H1N1 influenza. Influenza Other Respi. Viruses 5, 110–114.

Huang, Q., Zheng, L., Zhu, Y., Zhang, J., Wen, H., Huang, J., Niu, J., Zhao, X., and Li, Q. (2011). Multicolor combinatorial probe coding for real-time PCR. PLoS One *6*, e16033.

Ishiguro, T., Saitoh, J., Yawata, H., Yamagishi, H., Iwasaki, S., and Mitoma, Y. (1995). Homogeneous quantitative assay of hepatitis C virus RNA by polymerase chain reaction in the presence of a fluorescent intercalater. Anal. Biochem. *229*, 207–213.

Javorschi-Miller, S., and Delgado Orlic, I. (2011). Real-time PCR instrumentation: an instrument selection guide. In PCR troubleshooting and optimization: The essential guide, Kennedy, S., and Oswald, N., eds (Caister Academic Press, Norfolk), pp. 119–138.

Josefsen, M.H., Jacobsen, N.R., and Hoorfar, J. (2004). Enrichment followed by quantitative PCR both for rapid detection and as a tool for quantitative risk assessment of food-borne thermotolerant campylobacters. Appl. Environ. Microbiol. *70*, 3588–3592.

Josefsen, M.H., Lofstrom, C., Sommer, H.M., and Hoorfar, J. (2009). Diagnostic PCR: comparative sensitivity of four probe chemistries. Mol. Cell. Probes. *23*, 201–203.

Khan, Z. (2011). Setup of a PCR Laboratory. In PCR Protocols, Park, D.J., ed. (Humana Press, New York), pp. 3–14.

Khripin, Y. (2006). High-throughput genotyping with energy transfer-labeled primers. Meth. Mol. Biol. *335*, 215–240.

Kim, Y.H., Yang, I., Bae, Y.S., and Park, S.R. (2008). Performance evaluation of thermal cyclers for PCR in a rapid cycling condition. Biotechniques *44*, 495–500.

Kirchner, S., Kramer, K.M., Schulze, M., Pauly, D., Jacob, D., Gessler, F., Nitsche, A., Dorner, B.G., and Dorner, M.B. (2010). Pentaplexed quantitative real-time PCR assay for the simultaneous detection and quantification of botulinum neurotoxin-producing clostridia in food and clinical samples. Appl. Environ. Microbiol. *76*, 4387–4395.

Krämer, N., Löfström, C., Vigre, H., Hoorfar, J., Bunge, C., and Malorny, B. (2011). A novel strategy to obtain quantitative data for modelling: Combined enrichment and real-time PCR for enumeration of salmonellae from pig carcasses. Int. J. Food Microbiol. *145*, S86–S96.

Kubista, M. (2010). Real-time PCR – where we are and where are we heading. Drug Discov. World Spring, 59–65.

Kumar, R., Singh, S.K., Koshkin, A.A., Rajwanshi, V.K., Meldgaard, M., and Wengel, J. (1998). The first analogues of LNA (locked nucleic acids): phosphorothioate-LNA and 2′-thio-LNA. Bioorg. Med. Chem. Lett. *8*, 2219–2222.

Kutyavin, I.V., Afonina, I.A., Mills, A., Gorn, V.V., Lukhtanov, E.A., Belousov, E.S., Singer, M.J., Walburger, D.K., Lokhov, S.G., Gall, A.A., *et al.* (2000). 3′-minor groove binder-DNA probes increase sequence specificity at PCR extension temperatures. Nucleic Acids Res. *28*, 655–661.

Lasalle, H.E., Duncan, G., and McCord, B. (2011). An analysis of single and multi-copy methods for DNA quantitation by real-time polymerase chain reaction. Forensic Sci. Int. Genet. *5*, 185–193.

Lee, D., Chen, P.J., and Lee, G.B. (2010). The evolution of real-time PCR machines to real-time PCR chips. Biosens. Bioelectron. *25*, 1820–1824.

Lee, M.A., Squirrell, D.J., Leslie, D.L., and Brown, T. (2009). Homogeneous fluorescent chemistries for real-time PCR. In Real-time PCR. Current technology and applications, Logan, J., Edwards, K., and Saunders, N., eds (Caister Academic Press, Norfolk), pp. 23–47.

Li, P., Ruel, T., Fujimoto, K., Hatano, H., Yukl, S., Eller, L.A., Liegler, T., Kamya, M., Gassasira, A., Dorsey, G., *et al.* (2010). Novel application of Locked Nucleic Acid chemistry for a TaqMan assay for measuring diverse human immunodeficiency virus type 1 subtypes. J. Virol. Methods *170*, 115–120.

Li, Q., Luan, G., Guo, Q., and Liang, J. (2002). A new class of homogeneous nucleic acid probes based on specific displacement hybridization. Nucleic Acids Res. *30*, E5.

Lind, K., Stalberg, A., Zoric, N., and Kubista, M. (2006). Combining sequence-specific probes and DNA binding dyes in real-time PCR for specific nucleic acid quantification and melting curve analysis. Biotechniques *40*, 315–319.

Lo, Y.M., and Chan, K.C. (2006). Introduction to the polymerase chain reaction. Meth. Mol. Biol. *336*, 1–10.

Logan, J., and Edwards, K., 2009. An overview of PCR platforms. In Real-time PCR: Current Technology and Applications, Logan, J., Edwards K., and Saunders, N., eds (Caister Academic Press, Norfolk, UK), pp. 7–22.

Lu, S., Smith, A.P., Moore, D., and Lee, N.M. (2010). Different real-time PCR systems yield different gene expression values. Mol. Cell. Probes *24*, 315–320.

McCrea, J.K., Liu, C., Ng, L.K., and Wang, G. (2007). Detection of the *Escherichia coli* pathogenic gene eae with three real-time polymerase chain reaction methods. Can. J. Microbiol. *53*, 398–403.

Mackay, J., and Landt, O. (2007). Real-time PCR fluorescent chemistries. Meth. Mol. Biol. *353*, 237–261.

Malinen, E., Kassinen, A., Rinttila, T., and Palva, A. (2003). Comparison of real-time PCR with SYBR Green I or 5′-nuclease assays and dot-blot hybridization with rDNA-targeted oligonucleotide probes in quantification of selected faecal bacteria. Microbiology *149*, 269–277.

Mao, F., Leung, W.Y., and Xin, X. (2007). Characterization of EvaGreen and the implication of its physicochemical properties for qPCR applications. BMC Biotechnol. *7*, 76.

Markey, A.L., Mohr, S., and Day, P.J. (2010). High-throughput droplet PCR. Methods *50*, 277–281.

Marras, S.A. (2006). Selection of fluorophore and quencher pairs for fluorescent nucleic acid hybridization probes. Meth. Mol. Biol. *335*, 3–16.

Matero, P., Hemmilä H., Tomaso, H., Piiparinen, H., Rantakokko-Jalava, K., Nuotio, L., and Nikkari, S. (2011). Rapid field detection assays for *Bacillus anthracis*, *Brucella* spp., *Francisella tularensis* and *Yersinia pestis*. Clin. Microbiol. Infect. *17*, 34–43.

May, J.P., Brown, L.J., van Delft, I., Thelwell, N., Harley, K., and Brown, T. (2005). Synthesis and evaluation of a new non-fluorescent quencher in fluorogenic oligonucleotide probes for real-time PCR. Org. Biomol. Chem. 3, 2534–2542.

Molenkamp, R., van der Ham, A., Schinkel, J., and Beld, M. (2007). Simultaneous detection of five different DNA targets by real-time TaqMan PCR using the Roche LightCycler480: Application in viral molecular diagnostics. J. Virol. Methods. *141*, 205–211.

Monis, P.T., Giglio, S., and Saint, C.P. (2005). Comparison of SYT09 and SYBR Green I for real-time polymerase chain reaction and investigation of the effect of dye concentration on amplification and DNA melting curve analysis. Anal. Biochem. *340*, 24–34.

Morrison, L.E. (2010). Basic principles of fluorescence and energy transfer applied to real-time PCR. Mol. Biotechnol. *44*, 168–176.

Nazarenko, I., Lowe, B., Darfler, M., Ikonomi, P., Schuster, D., and Rashtchian, A. (2002). Multiplex quantitative PCR using self-quenched primers labeled with a single fluorophore. Nucleic Acids Res. *30*, e37.

Nazarenko, I.A., Bhatnagar, S.K., and Hohman, R.J. (1997). A closed tube format for amplification and detection of DNA based on energy transfer. Nucleic Acids Res. *25*, 2516–2521.

Neuzil, P., Zhang, C., Pipper, J., Oh, S., and Zhuo, L. (2006). Ultra fast miniaturized real-time PCR: 40 cycles in less than six minutes. Nucleic Acids Res. *34*, e77.

Nordgren, J., Bucardo, F., Dienus, O., Svensson, L., and Lindgren, P.E. (2008). Novel light-upon-extension real-time PCR assays for detection and quantification of genogroup I and II noroviruses in clinical specimens. J. Clin. Microbiol. *46*, 164–170.

Osterman, H.L. (2009). The next step in near infrared fluorescence: IRDye® QC-1 dark quencher, LI-COR Biosciences, Lincoln, NE 68504.

Peng, X., Chen, H., Draney, D.R., Volcheck, W., Schutz-Geschwender, A., and Olive, D.M. (2009). A nonfluorescent, broad-range quencher dye for Förster resonance energy transfer assays. Anal. Biochem. *388*, 220–228.

Petersen, K., Vogel, U., Rockenbauer, E., Nielsen, K.V., Kolvraa, S., Bolund, L., and Nexo, B. (2004). Short PNA Molecular beacons for real-time PCR allelic discrimination of single nucleotide polymorphisms. Mol. Cell. Probes. *18*, 117–122.

Pornprasert, S., Phusua, A., Suanta, S., Saetung, R., and Sanguansermsri, T. (2008). Detection of alpha-thalassemia-1 Southeast Asian type using real-time gap-PCR with SYBR Green1 and high resolution melting analysis. Eur. J. Haematol. *80*, 510–514.

Price, E.P., Smith, H., Huygens, F., and Giffard, P.M. (2007). High-resolution DNA melt curve analysis of the clustered, regularly interspaced short-palindromic-repeat locus of *Campylobacter jejuni*. Appl. Environ. Microbiol. *73*, 3431–3436.

Reed, G.H., Kent, J.O., and Wittwer, C.T. (2007). High-resolution DNA melting analysis for simple and efficient molecular diagnostics. Pharmacogenomics *8*, 597–608.

Reynisson, E., Josefsen, M.H., Krause, M., and Hoorfar, J. (2006). Evaluation of probe chemistries and platforms to improve the detection limit of real-time PCR. J. Microbiol. Methods *66*, 206–216.

Ririe, K.M., Rasmussen, R.P., and Wittwer, C.T. (1997). Product differentiation by analysis of DNA melting curves during the polymerase chain reaction. Anal. Biochem. *245*, 154–160.

Sahoo, R., Harini, V.V., Babu, V.C., V Patil Okaly, G., Rao, S., Nargund, A., Venkataswamy, E., Rao, R., and Kumar, B.S. (2011). Screening for EGFR mutations in lung cancer, a report from India. Lung Cancer. *73*, 316–319.

Saito, K., Miyake, S., Moriya, H., Yamazaki, M., Itoh, F., Imai, K., Kurosawa, N., Owada, E., and Miyamoto, A. (2003). Detection of the four sequence variations of MDR1 gene using TaqMan MGB probe based real-time PCR and haplotype analysis in healthy Japanese subjects. Clin. Biochem. *36*, 511–518.

Saunders, G.C., Dukes, J., Parkes, H.C., and Cornett, J.H. (2001). Interlaboratory study on thermal cycler performance in controlled PCR and random amplified polymorphic DNA analyses. Clin. Chem. *47*, 47–55.

Schlesinger, J., Tönjes, M., Schueler, M., Zhang, Q., Dunkel, I., and Sperling, S.R. (2010). Evaluation of the LightCycler™ 1536 Instrument for high-throughput quantitative real-time PCR. Methods *50*, S19-S22.

Schoder, D., Schmalwieser, A., Schauberger, G., Hoorfar, J., Kuhn, M., and Wagner, M. (2005). Novel approach for assessing performance of PCR cyclers used for diagnostic testing. J. Clin. Microbiol. *43*, 2724–2728.

Shen, F., Du, W., Kreutz, J.E., Fok, A., and Ismagilov, R.F. (2010). Digital PCR on a SlipChip. Lab Chip. *10*, 2666–2672.

Solinas, A., Thelwell, N., and Brown, T. (2002). Intramolecular TaqMan probes for genetic analysis. Chem. Comm. *19*, 2272–2273.

Stephens, K.W., Hutchins, R.J., and Dauphin, L.A. (2010). Cross-platform evaluation of commercial real-time reverse transcription PCR master mix kits using a quantitative 5′nuclease assay for Ebola virus. Mol. Cell. Probes. *24*, 370–375.

Sundberg, S.O., Wittwer, C.T., Gao, C., and Gale, B.K. (2010). Spinning disk platform for microfluidic digital polymerase chain reaction. Anal. Chem. *82*, 1546–1550.

Svanvik, N., Westman, G., Wang, D., and Kubista, M. (2000). Light-up probes: thiazole orange-conjugated peptide nucleic acid for detection of target nucleic acid in homogeneous solution. Anal. Biochem. *281*, 26–35.

Sykes, P.J., Neoh, S.H., Brisco, M.J., Hughes, E., Condon, J., and Morley, A.A. (1992). Quantitation of targets for PCR by use of limiting dilution. Biotechniques *13*, 444–449.

Teo, I.A., Choi, J.W., Morlese, J., Taylor, G., and Shaunak, S. (2002). LightCycler qPCR optimisation for low copy number target DNA. J. Immunol. Methods *270*, 119–133.

Tyagi, S., and Kramer, F.R.,1996. Molecular beacons: probes that fluoresce upon hybridization. Nat. Biotechnol. *14*, 303–308.

Vogelstein, B., and Kinzler, K.W. (1999). Digital PCR. Proc. Natl. Acad. Sci. U.S.A. *96*, 9236–9241.

Whitcombe, D., Theaker, J., Guy, S.P., Brown, T., and Little, S. (1999). Detection of PCR products using self-probing amplicons and fluorescence. Nat. Biotechnol. *17*, 804–807.

Wittwer, C.T., Ririe, K.M., Andrew, R.V., David, D.A., Gundry, R.A., and Balis, U.J. (1997a). The LightCycler: A microvolume multisample fluorimeter with rapid temperature control. Biotechniques *22*, 176–181.

Wittwer, C.T., Herrmann, M.G., Moss, A.A., and Rasmussen, R.P. (1997b). Continuous fluorescence monitoring of rapid cycle DNA amplification. Biotechniques *22*, 130–138.

Wittwer, C.T., Reed, G.H., Gundry, C.N., Vandersteen, J.G., and Pryor, R.J. (2003). High-resolution genotyping by amplicon melting analysis using LCGreen. Clin. Chem. *49*, 853–860.

Wolff, D.J., Heaney, D.L., Neuwald, P.D., Stellrecht, K.A., and Press, R.D. (2009). Multi-site PCR-based CMV viral load assessment-assays demonstrate linearity and precision, but lack numeric standardization: a report of the association for molecular pathology. J. Mol. Diag. *11*, 87–92.

Yang, G.P., Erdman, D.D., Tondella, M.L., and Fields, B.S. (2009). Evaluation of tetramethylrhodamine and black hole quencher 1 labeled probes and five commercial amplification mixes in TaqMan real-time RT-PCR assays for respiratory pathogens. J. Virol. Methods. *162*, 288–290.

Yang, I., Kim, Y.H., Byun, J.Y., and Park, S.R. (2005). Use of multiplex polymerase chain reactions to indicate the accuracy of the annealing temperature of thermal cycling. Anal. Biochem. *338*, 192–200.

Zhang, Y., and Ozdemir, P. (2009). Microfluidic DNA amplification – A review. Anal. Chim. Acta *638*, 115–125.

Zhao, J.R., Bai, Y.J., Zhang, Q.H., Wan, Y., Li, D., and Yan, X.J. (2005). Detection of hepatitis B virus DNA by real-time PCR using TaqMan-MGB probe technology. World J. Gastroenterol. *11*, 508–510.

DOI: https://doi.org/10.21775/9781912530243.05

Quantification Strategies in Real-time Polymerase Chain Reaction

5

Michael W. Pfaffl

Abstract

The present chapter describes the quantification strategies used in real-time RT-PCR (RT-qPCR), focusing on the main elements that are essential to fulfil the MIQE guidelines. The necessity of initial proper data adjustment and background correction is discussed to allow reliable quantification. The advantages and disadvantages of the absolute and relative quantification approaches are also described. In conjunction with relative quantification, the importance of an amplification efficiency correction is shown, and software tools that are available to calculate relative expression changes are presented.

Introduction

In general, two quantification strategies can be applied to analyse real-time RT-PCR (RT-qPCR) expression data. The levels of expressed genes may be measured by absolute quantification, using a standard curve, or relative quantification, using validated reference genes in RT-qPCR.

Absolute quantification relates the PCR signal to input copy number using a calibration curve with known concentrations of the amplified gene of interest (Fig. 3.1). Relative quantification instead measures the expression changes between two or more analysed experimental groups. The reliability of an absolute RT-qPCR assay depends mainly on the validity of the 'calibration curve' used and the 'identical' amplification efficiencies for both the native target in the biological sample and the artificial standard material used in the calibration curve, in the RT and qPCR reactions (Souaze *et al.*, 1996; Bustin *et al.*, 2005).

From the handling perspective, the relative quantification approach seems easier to perform than the absolute one, because a calibration curve is not needed. But it bears other essential problems, such as the changing amplification efficiencies of the analysed target and reference genes, and lacks of amplification variability at low transcript concentrations. It is based on the expression levels of a gene of interest versus one (or more) reference gene(s), whereby the reference gene selection is not trivial (Vandesompele *et al.*, 2002, 2009). To guarantee optimal quantification results, the recently published MIQE guidelines – Minimum Information for Publication of Quantitative Real-Time PCR Experiments – are highly recommended and a prerequisite for any quantitative PCR study (Bustin *et al.*, 2009).

Data adjustment

Prior to any quantification, a proper data adjustment is necessary so that the biological samples and the standards are comparable. Between reaction vessel differences are obvious, which are introduced by the operator, by the handling errors or are derived from the complex enzymatic reaction steps. Furthermore, some instruments excite all reaction vials individually and detected the fluorescence independently, which also may introduce some variability. To overcome this potential variation between reaction set-ups and replicates, it is possible to incorporate a passive reference dye, such as ROX or Fluorescein (FL), in each reaction mix. The passive reference signal is monitored during the course of the amplification, and each

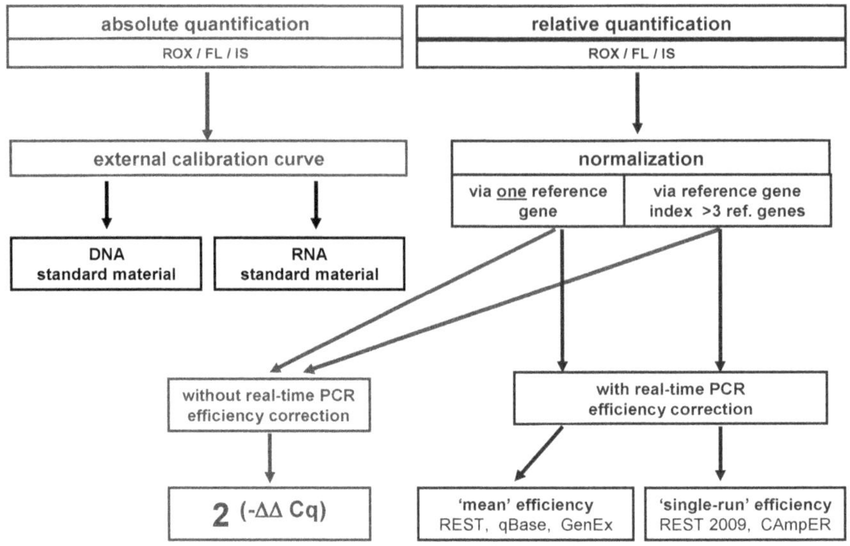

Figure 3.1 Quantification strategies in quantitative real-time RT-PCR.

raw data fluorescence reading is then corrected and equalized using the unchanging reference signal (ROX or FL). An advantage of the data normalization strategy is that the impact of any optical variability across the instrument block will be minimized, but as an additional fluorescence channel is introduced, this adds additional variability in the quantification workflow. The use of a passive reference dye may however help to remove background fluorescence and reduce the noise in the background phase from the measurements. This can be done by the software using the 'automatic background correction' option. An optimal fluorescence history shows low initial background signal, a significant and exponential fluorescence increase resulting from a stable PCR amplification, and a stable plateau (Keer, 2008; Pfaffl *et al.*, 2009).

After proper data adjustment, all samples under investigation are comparable and can be analysed dependent on the applied quantification strategy selected according to the scientific or analytical question. Either an absolute quantification approach, which results in defined units, e.g. copies per defined nucleic acid concentration, or a relative quantification, leading to gene expression ratios between different treatment groups, is used. In this later case, absolute units remain unknown.

Absolute quantification

The so-called 'absolute quantification' is misleading because the method is never 'absolute', as such, it is relative to the created calibration or standard curve. Hence everything depends on the reliability and validity of the standard curve and the used standard material. In particular, the external calibration curve model has to be thoroughly validated as the fidelity of 'absolute quantification' in real-time RT-qPCR depends entirely on the accuracy and quality of the standard used. It should be pure, not degraded, stable, and for this reason yield highly reproducible results over time and the entire quantification range. The standard curve model should allow the generation of highly specific, sensitive, and reproducible quantification data (Reischl and Kochanowsky, 1995; Bustin, 2000; Pfaffl and Hageleit, 2001; Reiter *et al.*, 2011). The standard composition, sequence design, purity, determination of the exact concentration, and stability over long storage time is not straightforward and can be problematic. In addition, the matrix of the biological samples is important. Some biological samples under investigation bear inhibitors for the enzymatic RT and/or qPCR reaction. Therefore, the standard curve itself should reflect as well the biological sample which is the focus for quantification. What is the real benefit of a suboptimal quantification model if the standard curve and the biological do not show the same amplification behaviour? Often the standard curve shows perfect properties with pure standard material and a perfect amplification history whereas the biological samples demonstrate high background, low amplification efficiency and seem to be inhibited by any unknown substance. Very importantly, both the standard curve dilutions and the unknown biological samples have to amplify in the same manner and with the same amplification characteristics. Elsewise the quantification is NOT reliable and will result in wrong quantification findings (Pfaffl, 2004).

What are the characteristics of a good standard curve? The dynamic range of the performed calibration curve should be as broad as possible. It can be up to nine orders of magnitude, starting at one theoretical molecule in the lowest, up to billions of molecules in the highest concentration. If plotted, all standard points and their replicates should be equally spread over the quantification range. All standard points will be summarized in a linear regression which should be linear with low variation, expressed in the confidence interval with a high correlation coefficient (r^2), and a slope, ideally –3.33, which corresponds to an amplification efficiency of 100% or a perfect doubling in each cycle.

The quantification range depends much on the applied standard material stability whereas the biggest difference appears between RNA and DNA (Pfaffl and Hageleit, 2001; Reiter *et al.*, 2011). The analytical question determines the type of the standard material, either DNA or RNA. If the focus is on the quantification of DNA, e.g. quantification of microorganisms in a sample, the calibration curve should be made out of DNA. If the interest is on the quantification of expressed transcripts or to measure RNA viruses concentrations, the standard substance must be RNA based.

For a DNA standard curve, the type of standard material is always based on known concentrations of double- or single-stranded DNA molecules. In most cases, serial dilutions with 10-fold dilution steps are sufficient (Fig. 3.2). Furthermore, the stability and reproducibility of the absolute quantification curve depends on the type of standard material applied, e.g. genomic DNA, RT-PCR product, recombinant plasmid DNA, commercially synthesized oligo-nucleotide or artificial genes (Morrison *et al.*, 1998, Pfaffl, 2004). Cloned

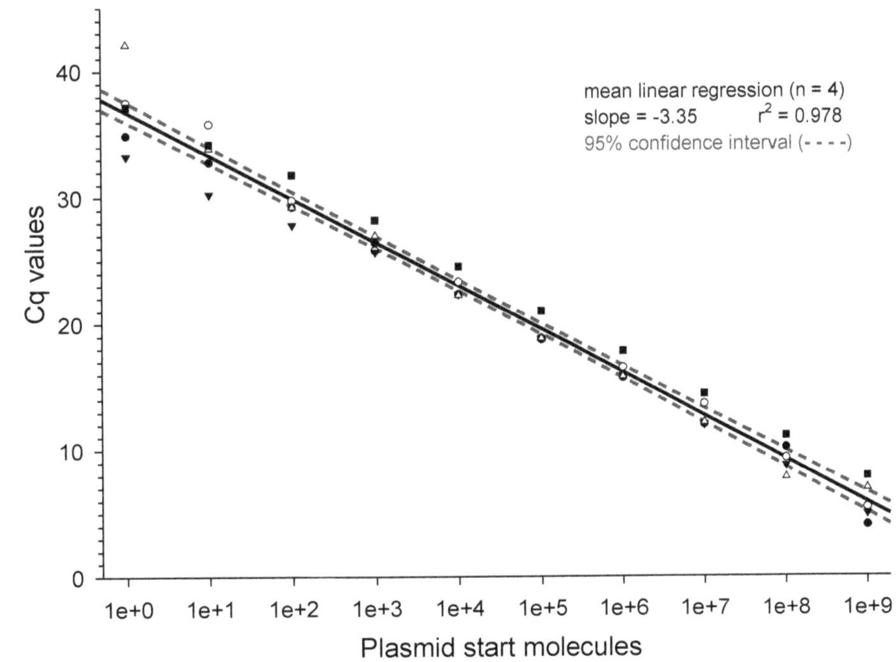

Figure 3.2 DNA standard curve.

recombinant DNA, artificial genes, and genomic DNA are very stable and generate highly reproducible standard curves even after a long storage time up to several years in a –20°C freezer (own findings).

RNA standards are much more sensitive to RNAse activity and degradation, especially at low concentration. Owing to the stability problem, it will introduce variability in the quantification of low abundant transcripts, and low concentration cannot be measured with high confidence. Furthermore, the longer templates derived from artificial genes mimic the average native length of messenger RNA of roughly 2 kb better than shorter templates or oligo-nucleotides. These longer standards have longer overhanging ends and are hence more resistant against unspecific enzymatic cleavage during RT or PCR reaction (Rasmussen, 2004).

A major problem with DNA-based calibration curves is that they can only be applied to quantify at the DNA level, and they are not suitable to quantify any RNA. DNA is subject to the PCR step only, unlike the unknown mRNA samples that must first be reverse transcribed to convert them to complementary DNA (cDNA). The RT step efficiency is usually around 30–40% and bears high variability between technical RT replicates (additional 30%) (Stahlberg *et al.*, 2004a,b). This decreases the amount of synthesized cDNA dramatically; pushes the variability in the quantitative RT-qPCR study and the derived quantification results may not reflect the reality of RNA initially present in the biological sample. If RNA is in the focus of the quantification, a recombinant RNA standard material *in vitro* transcribed from a plasmid, or artificially synthesized is mandatory (Pfaffl *et al.*, 2001; Fronhoffs *et al.*, 2002; Reiter *et al.*, 2011).

Independent of the standard material used, an absolute quantification model requires careful optimization of its precision. The reverse transcription efficiencies for both, standard curve material and native biological sample, must be tested thoroughly and should be confirmed by multiple technical RT replicates (Tichopad *et al.*, 2003b). We recommend performing replicates in the same RT reaction run, to get information about the RT assay variability, and replicates in separate runs, to get the inter-assay variation or reproducibility. This will help to find the detection and quantification limitations within the given absolute quantification procedure, and will show exactly the technical variance (Pfaffl and Hageleit, 2001; Tichopad *et al.*, 2009).

A special focus should be on the existence of inhibitors in the biological sample, which hampers either RT and/or PCR reactions. Inhibitors may be present in the biological sample but not in the standard dilutions used in the calibration curve. Known and unknown biological matrix compounds interact with the enzymatic reaction and therefore impact the validity and trueness. If present, a strong bias is introduced, leading to an overestimation of the standard curve and an underestimation of the unknown sample concentration. This will lead to completely wrong quantification results. The presence of inhibitors in the biological matrix should always be tested first. Many factors present in samples as well as exogenous contaminants have been reported to inhibit enzymatic reactions, either RT or PCR. For example, the presence of humic acids, haemoglobin, fatty acids, glycogen, cell components, calcium ions, and nucleic acid binding proteins are known inhibiting compounds (Rossen *et al.*, 1992; Wilson, 1997; Tichopad *et al.*, 2004). Additionally, exogenous contaminants such as glove powder, DNA or RNA degrading decontamination chemicals, and ethanol or phenolic compounds from the extraction process will also have an inhibitory effect (Rossen *et al.*, 1992; Wilson, 1997). Furthermore, low-quality and impure plasticware with high nucleic acid-binding capacity interact with the quantification procedure and should be avoided. To test for such inhibitors, a SPUD assay can be performed, which allows determining the degree of inhibition in the biological sample (Nolan *et al.*, 2006).

To get the most correct and valid results in an absolute quantification approach, either the user must take actions to get rid of any interfering compounds in the unknown biological samples during nucleic acid purification, or the calibration curve must be adapted. This means that the calibration curve amplification behaviour and the overall RT and qPCR efficiency must be brought to the same level as the biological sample material. In the end, both the standard curve and the biological samples should show most comparable amplification behaviour. The so called 'biological matrix background effect' in the standard curve model should be obtained from the sample under investigation. Eventually, identical background levels in the calibration curve and the biological samples should result in comparable levels of inhibition and therefore leads to a more realistic and valid quantification (Pfaffl, 2004).

Regardless on how accurate the concentration of the standard material is determined, the final result is always reported compared with a defined unit of interest, e.g. copies per defined RNA or DNA concentration, copies per genome, copies per solitaire-cell, copies per tissue unit, copies per millilitre blood, copies per gram of soil, etc. If absolute changes in copy number are important then the denominator must still be shown to be fully stable across the entire study. The applicability depends strongly on the adherence to the MIQE guidelines, the general 'good laboratory practice' and the personal skills of the researcher (Bustin *et al.*, 2009; Health Protection Agency, 2010). Failure to comply with the mentioned rules may introduce enormous additional variability to the 'absolute' quantification system and will result in artificial quantification results.

Relative quantification

Relative quantification determines the relative changes in steady-state messenger RNA (mRNA) levels of an expressed gene across multiple biological samples. Today the use of unregulated and stable expressed reference genes for normalization of transcript levels is certainly the gold standard (Vandesompele *et al.*, 2002). It is adequate for most purposes to investigate even low expression changes of minute amounts of mRNA or microRNA. The units used to express relative quantities are irrelevant, and the relative quantities can be compared across multiple real-time RT-qPCR experiments (Vandesompele *et al.*, 2009). These controls are internally grown, named reference genes (formerly known as housekeeping genes), and can be co-amplified in a multiplex assay or can be amplified in separate tubes (Wittwer *et al.*, 2001). Therefore, relative quantification does not require standards with known concentrations. The controls can be one or more reference gene(s), which can be any transcript as long as it is expressed in all biological samples and properly validated according to the MIQE guidelines (Bustin *et al.*, 2009).

To calculate the expression difference of a target gene in relation to an adequate reference gene, two general mathematical calculation models are established in the scientific qPCR community. The so called 'delta-delta Cq model', developed by Livak and Schmittgen (2001) or the efficiency corrected model, developed by Pfaffl (2001). Both calculations models are based on the comparison of the distinct cycle of quantification (Cq). The most applied and common calculation model is the 'delta-delta Cq method' which is assuming optimal amplification efficiency of 100% in any cycle, over the entire PCR cycles, with all target transcripts. It does not take any suboptimal PCR efficiency or its variation into account, derived from PCR inhibition or handling errors, either for the target or reference genes. As described in the previous 'absolute quantification' section, inhibitors or suboptimal reaction conditions may interfere with the reaction performance and hence the reaction efficiency is never exactly 100%. Therefore, a model was modified by Pfaffl and co-workers in 2001 by adding a real-time PCR efficiency correction. This correction is applied separately for each gene studied, e.g. the target gene (E_{target}) and the reference gene (E_{ref}), as shown in the equation:

$$\text{ratio} = \frac{(E_{target})^{\Delta CP_{target}(control - sample)}}{(E_{ref})^{\Delta CP_{ref}(control - sample)}} \qquad 3.1$$

This efficiency corrected relative quantification model was further developed, driven by innovations from other groups. It was expanded by adding multiple reference genes for normalization (summarized by Vandesompele *et al.*, 2009), by applying the model to group measurements (Pfaffl *et al.*, 2002), or by adding new statistical analysis tools to get more significant valid relative expression results (Qiagen, 2009).

We now know that reference genes have to be carefully selected and validated, which can be done using free available software applets, e.g. geNorm (Vandesompele, 2002), BestKeeper (Pfaffl *et al.*, 2004) or NormFinder (Andersen *et al.*, 2004). The main goal of the normalization procedure is to remove experimental induced variation, each with their own advantages and considerations (Huggett *et al.*, 2005). While most of these methods cannot completely reduce all sources of variation, they have been shown to be very important to control most sources of variation along the entire workflow of PCR-based gene expression analysis. If one does not meticulously try to standardize each step, variation can and will be introduced in the results which cannot be eliminated by applying a final normalization (Stahlberg *et al.*, 2004b). It is thus recommended to ensure that similar biological sample sizes and extraction volumes are used, and also to standardize the workflow. The amount of time invested in DNase treatment and reverse transcription optimization is important and can impact the cDNA conversion efficiency. Furthermore, artificial RNA molecules can be spiked into the sample prior to extraction or to the RNA extract prior to reverse transcription (Smith *et al.*, 2003; Gilsbach *et al.*, 2006). This will give an indication of the RT and qPCR efficiencies, and will reveal any inhibition. We however need to keep in mind that the spiked artificial RNA may not be extracted with the same yield as natural mRNAs and will not fully control for the final amount of input material in the reaction (Stahlberg *et al.*, 2004a).

Summarizing the qPCR literature from the last decade, it has been agreed that the reference gene concept is the currently preferred way of normalizing real-time PCR data for relative expression profiling (Huggett *et al.*, 2005; Vandesompele *et al.*, 2009). The reference gene concept is particularly attractive because the reference genes are grown internal controls that are affected by all sources of variation during the experimental workflow, from sampling, storage, extraction, nucleic acid stabilization, and enzymatic reaction (RT, DNAse digestion, and PCR) in the same way as the genes of interest. Furthermore, PCR based quantification findings for a gene of interest are best normalized using a factor that is measured using the same quantitative methodology set-up and chemicals.

After selecting the right references, the relative expression difference of the target genes can be calculated by multiple software tools. Either a software application is already implemented in the real-time cycler software package or the data can be exported and further analysed using proper quantification tools. The oldest available software is the REST (Relative Expression Software Tool), established by Pfaffl and colleagues (Pfaffl *et al.*, 2002; http://REST.gene-quantification.info), qBase and qBase plus, developed around the group of Vandesompele and co-workers (Hellemans *et al.*, 2007; http://www.biogazelle.com), or GeneEx set up by Kubista and colleagues (Bergkvist *et al.*, 2010; http://www.multid.se). In these software, the target gene expression is normalized by multiple validated non-regulated reference gene expressions, and even more advanced applications are implemented to evaluate if the target gene is expressed as part of a gene cluster, via cluster analysis, principal component analysis or neural networks (summarized in Pfaffl *et al.*, 2009; Bergkvist *et al.*, 2010). This way, relative quantification can generate useful and biologically relevant information when used appropriately according to the MIQE guidelines (Bustin *et al.*, 2009).

PCR amplification efficiency

Constant PCR amplification efficiency in all investigated biological samples is one important criterion for reliable comparison between samples and genes. Individual reaction set-ups can however generate different fluorescence histories in quantitative real-time PCR. The shapes of amplification curves differ in the background fluorescence level, the steepness of fluorescence increase, and in the absolute fluorescence levels at plateau. PCR efficiency has a major impact at each cycle and finally in the overall cycle fidelity or accuracy of the assay. Therefore, efficiency evaluation is an essential and requested criterion in gene quantification procedure (Tichopad *et al.*, 2003a; Bustin *et al.*, 2009). Some quantification models assume that the analysed genes amplify optimally with 100% efficiency (Livak and Schmittgen, 2001) or at least with similar efficiency. But this is often not the case. As discussed, sporadic RT and PCR inhibitors, sample impurity in the reaction set-up, or nucleic acid degradation may reduce the efficiency of the RT or the PCR reaction (Stahlberg *et al.*, 2004b; Becker *et al.*, 2010). To determine the level of inhibition, a dilution series can be performed and the inhibitory factors can often be diluted out, causing a non-linear standard curve (Morrison *et al.*, 1998; Rasmussen, 2001).

A correction for PCR efficiency and its variation, as performed in efficiency corrected mathematically models, is strongly recommended and is one of the key element of the MIQE guidelines (Bustin *et al.*, 2009). It results in a more reliable estimation of the 'real expression ratio' compared with no efficiency correction in the relative quantification model (Pfaffl, 2001). Small efficiency differences between target and reference genes, or between biological samples, generate false expression ratio and results in wrong scientific findings. Overall, the researcher over- or under-estimates the 'real and initial' mRNA amount present in the biological sample (LightCycler Relative Quantification Software; Roche Applied Science, 2001). This is why efficiency quantification corrections should be included in the calculation procedure, and is a major goal to implement it in any real-time qPCR cycler relative quantification software.

The most applied method to evaluate the PCR efficiency is from the slopes of the calibration curve or dilution row from a pool of biological samples. From the regressed expression data versus the concentration, the PCR efficiency (E) can be calculated according to the equation $E = 10^{[-1/\text{slope}]}$, as described earlier (Higuchi *et al.*, 1993; Rasmussen, 2001). Efficiency determination by the dilution method can be performed sample by sample, but this is very time consuming and needs lots of RNA. If not available, the efficiency can be evaluated using a pool of all starting total RNAs to accumulate all possible positive and negative 'impacts' on kinetic PCR efficiency estimation. Usually, this so called 'mean PCR efficiency' (Fig. 3.2) vary around the ideal regression line, behave linearly ($0.980 > r^2 > 0.999$), and range from $E = 1.75$ to maximal values up to $E = 2.10$ for cDNA input ranges over at least three orders of magnitude (Wittwer *et al.*, 1991; Pfaffl *et al.*, 2002). Typically, the relationship between Cq and the logarithm of the starting copy number of the target sequence should remain linear in the calibration curve as well as in the native sample RNA (Muller *et al.*, 2002).

Over the years further methods were established, for a more realistic estimation of the real-time PCR efficiency. These models are based on the analysis of all available data points in the fluorescence history. Raw fluorescence data reading points were fitted according to a given mathematical model. Often, sigmoidal, logistic or exponential curve model were used in different approaches and even multiple fitting models were established (Liu and Saint, 2002a,b; Tichopad *et al.*, 2003a; Zhao and Fernald, 2005). The advantage of such models is that all data points will be included in the calculation process and the 'single run' efficiency can be estimated from each biological sample individually. No work consuming and cost intensive dilution row has to be performed, and no background subtraction is necessary (Raeymaekers, 2000; Tichopad *et al.*, 2004). It can be measured at different points, either at the point of inflexion or at the absolute maximum fluorescence increase which equals the 'first derivate maximum' point. Alternatively, it can be measured earlier than or at the so called 'second derivate maximum' point, where the optimal reaction point is defined. However, the derived slope parameters generated by such more advanced models are in the eyes of a small group of researchers more reliable and 'directly' comparable with the real PCR efficiency (Liu and Saint, 2002a,b; Tichopad *et al.*, 2003a). These 'single run' efficiency methods can be fully automated, are easier to perform for the operator, but bear the problem of higher variability in a set of biological samples. Thus, PCR efficiency is different according to biological samples and has a variability which should be taken into account in the quantification model. To cope with the knowledge of individual 'single-run' PCR efficiencies in each biological sample, new software tools were established, e.g. REST 2009 (Pfaffl *et al.*, 2002; Qiagen, 2009) and Champer (Blom, 2006).

Conclusion

Any calculation model, any applied algorithm, and any quantification strategy are making assumptions which try to describe the nature of the qPCR reaction at its best. Which standard curve material is the most stable? Which biological background best mimics the biological reality? Which normalization approach is the most reliable? Which PCR efficiency calculation method is 'the most correct one'? In the end, the quantification model that shows the most valid expression results is still unknown. With the latest progress, we are however in a good position to find it in the near future.

References

Andersen, C.L., Jensen, J.L., Ørntoft, T.F. (2004) Normalization of real-time quantitative reverse transcription-PCR data: a model-based variance estimation approach to identify genes suited for normalization, applied to bladder and colon cancer data sets. Cancer Res. *64*, 5245–5250.

Becker, C., Hammerle-Fickinger, A., Riedmaier, I, and Pfaffl, M.W. (2010) mRNA and microRNA quality control for RT-qPCR analysis. Methods *50*, 237–243.

Bergkvist, A., Rusnakova, V., Sindelka, R., Garda, J.M., Sjögreen, B., Lindh, D., Forootan, A., and Kubista, M. (2010) Gene expression profiling – Clusters of possibilities. Methods *50*, 323–335.

Blom, J. (2006) Champer software http://camper.cebitec.uni-bielefeld.de.

Bustin, S.A. (2000) Absolute quantification of mRNA using real-time reverse transcription polymerase chain reaction assays. J. Mol. Endocrinol. *25*, 169–193.

Bustin, S.A., Benes, V., Nolan, T., and Pfaffl, M.W. (2005) Review: quantitative real-time reverse transcription polymerase chain reaction – A perspective. J. Mol. Endocrinol. *34*, 597–601.

Bustin, S.A., Benes, V., Garson, J.A., Hellemans, J., Huggett, J., Kubista M., Mueller R., Nolan, T., Pfaffl, M.W., Shipley, G.L., *et al.* (2009) The MIQE guidelines: minimum information for publication of quantitative real-time PCR experiments. Clin. Chem. *55*, 611–622.

Fronhoffs, S., Totzke, G., Stier, S., Wernert, N., Rothe, M., Bruning, T., Koch, B., Sachinidis, A., Vetter, H., and Ko, Y., (2002) A method for the rapid construction of cRNA standard curves in quantitative real-time reverse transcription polymerase chain reaction. Mol. Cell Probes *16*, 99–110.

Gilsbach, R., Kouta, M., Bönisch, H., and Brüss, M. (2006) Comparison of *in vitro* and *in vivo* reference genes for internal standardization of real-time PCR data. Biotechniques *40*, 173–177.

Health Protection Agency (2010). Good laboratory practice when performing molecular amplification assays. National Standard Method QSOP 38 Issue 4.

Hellemans, J., Mortier, G., De Paepe, A., Speleman, F., and Vandesompele, J. (2007) qBase relative quantification framework and software for management and automated analysis of real-time quantitative PCR data. Genome Biol. *8*, R19.

Higuchi, R., Fockler, C., Dollinger, G., and Watson, R. (1993) Kinetic PCR analysis: Real-time monitoring of DNA amplification reactions. Biotechnology *11*, 1026–1030.

Huggett, J., Dheda, K., Bustin, S., and Zumla A. (2005) Real-time RT-PCR normalization – strategies and considerations. Genes Immun. *6*, 279–284.

Keer, J.T. (2008) Essentials of Nucleic Acid Analysis: A Robust Approach (Royal Society of Chemistry, Cambridge, UK), pp. 132–166.

Liu, W., and Saint, D.A. (2002a), Validation of a quantitative method for real time PCR kinetics. Biochem. Biophys. Res. Commun. *294*, 347–353.

Liu, W., and Saint, D.A., (2002b) A new quantitative method of real time reverse transcription polymerase chain reaction assay based on simulation of polymerase chain reaction kinetics. Anal. Biochem. *302*, 52–59.

Livak, K.J., and Schmittgen, T.D. (2001) Analysis of relative gene expression data using real-time quantitative PCR and the 2^[-delta delta C(T)] method. Methods *25*, 402–408.

Morrison, T.B., Weis, J.J., and Wittwer, C.T. (1998) Quantification of low-copy transcripts by continuous SYBR Green I monitoring during amplification. Biotechniques *24*, 954–962.

Muller, P.Y., Janovjak, H., Miserez, A.R., and Dobbie, Z. (2002) Processing of gene expression data generated by quantitative real-time RT-PCR. Biotechniques *32*, 1372–1378.

Nolan T., Hands R.E., Ogunkolade W., and Bustin S.A. (2006) SPUD: a quantitative PCR assay for the detection of inhibitors in nucleic acid preparations. Anal. Biochem. *351*, 308–310.

Pfaffl, M.W. (2001) A new mathematical model for relative quantification in real-time RT-PCR. Nucleic Acids Res. *29*, e45.

Pfaffl, M.W. (2004) Quantification strategies in real-time RT-PCR. In The Real-time PCR Encyclopaedia A–Z of Quantitative PCR (1st edition), Bustin, S.A., ed (International University Line, La Jolla, CA), pp. 87–120

Pfaffl, M.W., and Hageleit, M. (2001) Validities of mRNA quantification using recombinant RNA and recombinant DNA external calibration curves in real-time RT-PCR. Biotechnol. Lett. *23*, 275–282.

Pfaffl, M.W., Lange, I.G., Daxenberger, A., and Meyer, H.H.D. (2001) Tissue-specific expression pattern of oestrogen receptors (ER): quantification of ER alpha and ER beta mRNA with real-time RT-PCR. APMIS 109(5), 345–355.

Pfaffl, M.W., Horgan, G.W., and Dempfle, L. (2002) Relative expression software tool (REST) for group-wise comparison and statistical analysis of relative expression results in real-time PCR. Nucleic Acids Res. 30(9), e36.

Pfaffl M.W., Tichopad A., Prgomet C., and Neuvians T.P. (2004) Determination of stable housekeeping genes, differentially regulated target genes and sample integrity: BestKeeper – Excel-based tool using pair-wise correlations. Biotechnol. Lett. *26*, 509–515.

Pfaffl, M.W., Vandesompele, J., and Kubista, M. (2009) Real-time PCR technology. In Data Analysis Software, Logan, J., Edwards, K., and Saunders, N., eds (Caister Academic Press, Norwich), pp. 65–83.

Qiagen (2009) REST 2009 Software – for gene expression analysis using real-time PCR data from the Rotor-Gene Q and other cyclers. REST 2009 Software. Available at: http://www.REST.de.com

Raeymaekers, L., (2000) Basic principles of quantitative PCR. Mol Biotechnol. *15*, 115–122.

Rasmussen, R. (2001) Quantification on the LightCycler. In Rapid Cycle Real-time PCR, Methods and Applications, Meuer, S., Wittwer, C., and Nakagawara, K., eds (Springer Press, Heidelberg), pp. 21–34.

Reischl, U., and Kochanowski, B. (1995) Quantitative PCR. A survey of the present technology. Mol. Biotechnol. *3*, 55–71.

Reiter, M., Kirchner, B., Müller, H., Holzhauer, C., Mann, W., and Pfaffl, M.W. (2011) Quantification noise in single cell experiments. Nucleic Acids Res. *39*, e124.

Rossen, L., Norskov, P., Holmstrom, K., and Rasmussen, F.O. (1992). Inhibition of PCR by components of food sample, microbial diagnostic assay and DNA-extraction solutions. Int. J. Food Microbiol. *17*, 37–45.

Smith, A.B., Mock, V., Melear, R., Colarusso, P., and Willis, D.E. (2003) Rapid detection of influenza A and B viruses in clinical specimens by Light Cycler real time RT-PCR. J. Clin. Virol. *28*, 51–58.

Souaze, F., Ntodou-Thome, A., Tran, C.Y., Rostene, W., and Forgez, P. (1996) Quantitative RT-PCR: limits and accuracy. Biotechniques *21*, 280–285.

Stahlberg A., Kubista M., and Pfaffl M.W. (2004a) Comparison of reverse transcriptases in absolute and relative gene expression analysis. Clin. Chem. *50*, 1678–1680.

Stahlberg, A., Håkansson, J., Xian, X., Semb, H., and Kubista, M. (2004b) Properties of the reverse transcription reaction in mRNA quantification. Clin. Chem. *50*, 509–515.

Tichopad, A., Dilger, M., Schwarz, G., and Pfaffl, M.W. (2003a) Standardized determination of real-time PCR efficiency from a single reaction setup. Nucl. Acids Res. *31*, e122.

Tichopad, A., Dzidic, A., and Pfaffl, M.W. (2003b) Improving quantitative real-time RT-PCR reproducibility by boosting primer-linked amplification efficiency. Biotechnol. Lett. *24*, 2053–2056.

Tichopad, A., Didier, A., and Pfaffl, M.W. (2004) Inhibition of real-time RT-PCR quantification due to tissue specific contaminants. Mol. Cell. Probes *18*, 45–50.

Tichopad, A., Kitchen, R., Riedmaier, I., Becker, C., Ståhlberg, A., and Kubista M. (2009) Design and optimization of reverse-transcription quantitative PCR experiments. Clin. Chem. *55*, 1816–1823.

Vandesompele, J., De Preter, K., Pattyn, F., Poppe, B., Van Roy, N., De Paepe, A., and Speleman, F. (2002) Accurate normalization of real-time quantitative RT-PCR data by geometric averaging of multiple internal control genes. Genome Bio. 3(7), RESEARCH0034.

Vandesompele, J., Kubista, M., and Pfaffl, M.W. (2009) Real-Time PCR Technology. In Reference Gene Validation Software for Improved Normalization, Logan, J., Edwards, K., and Saunders, N., eds (Caister Academic Press, London), pp. 47–64.

Wilson, I.G. (1997). Inhibition and facilitation of nucleic acid amplification. Appl. Environ. Microbiol. *63*, 3741–3751.

Wittwer, C.T., and Garling, D.J. (1991) Rapid cycle DNA amplification: Time and temperature optimization. Biotechniques *10*, 76–83.

Wittwer, C.T., Herrmann, M.G., Gundry, C.N., and Elenitoba-Johnson, K.S. (2001) Real-time multiplex PCR assays. Methods *25*, 430–442.

Zhao, S., and Fernald, R.D. (2005) Comprehensive algorithm for quantitative real-time polymerase chain reaction. J. Comput. Biol. *12*, 1047–1064.

DOI: https://doi.org/10.21775/9781912530243.06

The Extraction and Purification of Nucleic Acids for Analysis by PCR

6

Chaminda Salgado and Waqar Hussain

Abstract

A myriad of methods for the extraction and purification of nucleic acids prior to PCR are currently used throughout the community. While these methods have many unique and bespoke aspects, they broadly follow a sequence of lysis, isolation, washing and elution to get from a complex biological sample to purified nucleic acid that can be used in a PCR reaction. Various common methods available for each stage are described and potential sequences for particular sample types can be discerned. The potential for these methods to be automated are discussed and the process options summarized with respect to the speed of the methods, technical skill required and the resultant purity and yield that can be expected.

Introduction

In this chapter some of the more common methods for nucleic acid isolation are broadly summarized and compared. To describe all of the possible chemistries and interactions involved would warrant a dedicated book and even then many of the mechanisms and interactions would be postulated from theory as these are rarely understood well. This chapter is therefore intended to guide the reader through the common options available and to aid and inform the analyst of potential isolation routes depending on their sample type and desired target. Detailed protocols and method descriptions can be found by following the references supplied within this chapter.

Regardless of technique used all the methods undergo four basic stages. The first stage is some form of lysis, in order to gain access to the nucleic acids, within the sample. Lysis tends to be either mechanical, such as by sonication (Engler *et al.*, 1996) or bead-mill homogenization (Ho *et al.*, 2008); chemical, such as by detergents (Schüpbach *et al.*, 2006) or potassium ethyl xanthogenate (Leuko *et al.*, 2008) or biological, or a combination of the three. The choice is dependant on the sample type and the nucleic acid species under isolation. The second stage is the isolation, and forms the bulk of this review. Isolation tends to fall under two general categories of separation and capture. The isolation method tends to dictate the complexity of the subsequent steps with significant implications on the propensity of the process to automation or on the requirement for highly skilled technicians to perform the task. The choice isolation method can also entail compromises between robustness, yield, or purity and is therefore a critical process choice. The third stage are the wash steps, which depend not only on the choice of isolation used, but also the nucleic acid species desired. The wash steps can be amenable to the inclusion of a DNA or RNA digestion step to specifically select for either DNA or RNA offering potential for high levels of target purity. The forth and final stage is the elution. For compatibility with PCR there is little choice in the elution buffer as inhibition of DNA polymerases is a concern. The choice of elution buffer can play a key role in yield from the isolation method and so this step is often a compromise. Subtle changes, such as the heating of the elution buffer, can be made to improve yield and to help prevent enzymatic degradation of the nucleic acid.

This chapter will examine each of these stages and choices in turn, paying particular attention to the isolation step. A further issue to consider however is sample quality and storage. For example, for preservation of tissue samples from biopsies and to prevent degradation by proteases, DNases, and RNases, snap freezing in liquid nitrogen is the recommended technique, followed by storage at $\leq -70°C$. In the absence of such storage facilities, reagents such as RNA*later*©, available from Sigma, Qiagen or Life Technologies, should be used to preserve such samples at higher temperatures. Preservation by formalin fixation, as commonly used in histopathology techniques, however, is a strong inhibitor of enzymes, and should be avoided for enzyme dependent techniques such as PCR.

Sample lysis

The first issue to be tackled is gaining access to the nucleic acids within a given sample. This section explores the available options, starting from the toughest sample types, to the more delicate. Plant tissue samples, animal tissue samples and bacterial spores are among the toughest sample types from which to access the nucleic acid. As an initial step, mechanical disruption is commonly used to homogenize such sample types, followed by enzymatic lysis. The concept behind mechanical homogenization is to allow better access to the cells for subsequent lysis methods. Before entering this step of the lysis process, the analyst can take steps to help the homogenization of the sample. During sample selection, from plants or animal tissues, simply chopping the sample into smaller pieces can improve the success of the lysis process. Bead beating (Ogram *et al.*, 1987; Moré *et al.*, 1994; Yeates *et al.*, 1997; Cullen and Hirsch, 1998; Kuske *et al.*, 1998; Miller *et al.*,1999) is one of the more common mechanical disruption methods. Here beads such as, glass, ceramic, or steel beads are added to the sample, and then agitated at a high frequency in an attempt to shear the samples to a consistent homogenate. Issues to be aware of with this method are the frothing of the samples, causing difficulty in downstream processing of the samples, and heat generation, which may cause damage to the nucleic acids (particularly RNA), and result in a reduction in yield. Excessive periods of bead beating should therefore be avoided, and alternatives sought such as pulse beating, with pauses to avoid heat build up. Sonication (Porteous *et al.*, 1997) using ultrasound is another mechanical method used to break down the integrity of bacterial spores but heat generation is again a concern, as well as the generation of free radicals that are potentially damaging to nucleic acids,. Other crude methods are the use of a pestle and mortar (Volossiouk *et al.*, 1995) to physically grind snap frozen samples. A common issue to consider with all of these methods is the release of DNases and RNases from the cells which may now act upon the target molecule, reducing yield. This risk can be mitigated by homogenization at low temperatures (liquid nitrogen) or in the presence of a lysis buffer designed to inhibit or denature proteins, that will inhibit action of these enzymes.

Following homogenization which compromises the integrity of the cellular boundary the nucleic acid can be accessed by one of the following commonly used methods. Samples such as cultured cells, bacterial cells or crude virus samples, may not require homogenization and can proceed directly to enzymatic or chemical lysis. Enzymatic lysis may use proteases like Proteinase K (Gross-Bellard *et al.*, 1973) and/or lysozyme (Clewell and Helinski, 1970). Proteinase K is used to lyse the cell walls or viral capsid, but also plays a role in denaturing nucleases and proteins with a high affinity to nucleic acids. It is also a robust enzyme which remains active in the presence of several dilute detergents [i.e. sodium dodecyl sulfate (SDS)] used in cell lysis, enabling simultaneous chemical and enzymatic lysis. Lysozyme is commonly used for lysing bacterial cell walls, particularly gram negative bacterial walls, by digesting peptidoglycan in the cell walls. Chemical lysis methods in nucleic acid isolation use ionic detergents such as SDS to both denature proteins, and disrupt the lipid barrier. If chemical lysis is used alone it is commonly a strong detergent solution such as sodium hydroxide and SDS. Timing is crucial as this method not only denatures protein but also nucleic acids and, if left too long, will cause irreversible damage to the nucleic acids. As a method to isolate plasmid from genomic DNA, neutralization of the solution allows the plasmid to re-anneal but the large genomic DNA cannot, allowing for separation by centrifugation. Under centrifugal force the genomic DNA, still bound to protein, is spun down out of solution while the plasmid remains in solution. Heat denaturation by boiling or microwave (Marra *et al.*, 1999) is a method to denature the proteins encapsulating the nucleic acids, but again heat is a concern for RNA. Freeze-thaw (Erb and Wagner-Döbler, 1993) is another method for lysing cells. It is the freezing that disrupts the walls/capsid through the formation of ice crystals. Crystal formation however can also damage the nucleic acids, and reduce the yield.

Following one, or a combination of the above steps, the analyst should be left with a crude preparation of nucleic acid, proteins and lipids.

Nucleic acid binding and purification

Phenol–chloroform isolation

Phenol–chloroform isolation is a method to isolate RNA in the presence of a chaotropic denaturing solution, or DNA in the absence of this solution (Chomczynski and Sacchi, 1987). In summary, when the lysed sample is mixed with phenol and chloroform, protein is denatured and dissolved in the phenol/chloroform. The nucleic acids, being a more polar molecule, dissolves in the water and the solution will separate, with the nucleic acid in the upper aqueous phase and the protein and other components in the interphase and lower organic phase. The aqueous phase, containing the nucleic acid of interest can then be carefully siphoned off using a pipette into a separate tube. This step can be tricky, the more of the aqueous phase that can be extracted the higher the recovery; however, any co-extraction of the organic phase will lead to contaminants that could inhibit the PCR. Products such as Phase Lock gel, are designed to make this step easier by creating greater separation between the organic and aqueous phase. The tube containing the aqueous phase can then be centrifuged to pellet the nucleic acid before precipitation by isopropanol or ethanol. Commercial kits using the above principles are those available from Invitrogen.

Caesium chloride ultracentrifugation

Caesium chloride ultracentrifugation (Meselson *et al.*, 1957; Radloff *et al.*, 1967; Maniatis *et al.*, 1982) is a method for isolating specific species of double-stranded DNA. The lysed sample is mixed with a DNA stain, such as ethidium bromide, and CsCl. The solution is then ultracentrifuged for at least 10 hours where the DNA separates due to relative density and physical form (e.g. linear or circular plasmid molecules) in a manner analogous to migration through agarose (acting as a molecular sieve) during agarose gel electrophoresis. Under ultraviolet light, the DNA bands can be visualized. To extract the species of interest a needle and syringe is used to puncture the side of the tube and extract the band. The DNA itself is then purified using isopropanol or ethanol precipitation.

Isopropanol or ethanol purification

Following isolation the nucleic acid maybe contaminated with salts (Zeugin and Hartley, 1985). The addition of sodium acetate and two to three volumes of 95% ethanol, creating a low polarity environment, causes exposure of the negatively charged phosphate back bone of nucleic acids to the positively charged Na^+ ions of sodium acetate. The nucleic acid is therefore neutralized making it much less hydrophilic and therefore less soluble in water. Centrifugation is then used to pellet the precipitated nucleic acid. A typical protocol then involves decanting the supernatant and resuspending the pellet in 70% ethanol by vortexing. Then once again centrifuge into a pellet, decant the supernatant and allow the pellet to air dry. If isopropanol is used instead of ethanol only a 1:1 volume is required, but a high centrifuge speed and longer drying time is needed, as isopropanol is less volatile. Over drying may cause a decreased yield. Finally resuspend the pellet in water or 1 × TE (Tris-EDTA) buffer.

Owing to the requirement to centrifuge and visualize bands, phase separation or pellets, the above techniques do not lend themselves well to simple, walk away, automated procedures.

Silica adsorption

The exact mechanism as to how nucleic acids adsorb to silica appears to be poorly understood and is still a matter of debate (Boom *et al.*, 1990). Particles of diatomaceous earth and other amorphous forms of silica dioxide are highly porous, thus presenting a very high surface area (800 m^2 per gram). As the lysate migrates through the micro pores in silica, in the presence of chaotropic salts, it interacts with the silica, binding to it and allowing the sample impurities to be washed away. Both the nucleic acid phosphorous backbone and silica have a negative charge and should, therefore, repel each other. There are several theories of how the nucleic acid adsorb onto the silica dioxide in the presence of chaotropic salts. One theory is that water forms a hydration shell around nucleic acid, preventing it from interacting with the silica dioxide. In the presence of chaotropic salts, this shell is disrupted allowing the formation of hydrogen bonds between the nucleic acid and the silica dioxide surface. Alternatively, following disruption of the hydration shell, cation bridges form between the nucleic acid and the silica dioxide surface. The ionic strength of the buffer can selectively cause adsorption that is specific to the nucleic acid species involved, so plasmid or genomic DNA can be selected by carefully optimized control of the ionic strength of the binding buffer containing chaotropic salts. For this reason, it is possible to have protocols and kits containing optimized binding and wash buffers, that are specifically designed to purify certain species of nucleic acids, even though the silica chemistry involved may be identical for all these protocols and kits. Other interactions are also possible; for example smaller nucleic acid species are able to migrate through the pores of the silica, whereas the larger genomic DNA will have more restricted access to such spaces. Ultimately, once binding has been established the excess chaotropic salts are washed away using ethanolic buffers with increasingly diluted ethanol throughout the protocol. Finally the nucleic acid is eluted using a low salt (i.e. Tris-EDTA) buffer or water. This elution can be understood in the context of the postulated hydration shell mechanism where the lower salt buffers allow reformation of the hydration shell, breaking the interaction between the silica and nucleic acid.

Silica adsorption technology is a cost-effective solution to provide purified nucleic acid for many downstream applications such as PCR and sequencing. However the purity achieved is not sufficient for more sensitive applications, such as transfecting into endotoxin-sensitive cells. The chemistry has also been criticized for producing low yields of typically 25–50%. It is generally accepted that this is not an adsorption issue, but an elution issue, due to the strength of the interaction between the nucleic acid and silica. Using elution buffer at elevated temperature (> 50°C) has been shown to improve nucleic acid yields. There has been recent advances in this chemistry which appear to show marked improvement in the yields, close to the yields often reserved to precipitation techniques.

From a practical point of view, the first iteration (Boom *et al.*, 1990) of using silica for purifying nucleic acids had only a slight advantage over isopropanol isolation, in that a phase separation step was not required. But the subsequent washing and centrifugation steps still required a high degree of manual intervention. Over time the method evolved to simplify the manual intervention by packing the silica into a column, priming the column, and then adding homogenized sample with binding buffer, to initiate adsorption. Gravity was used to allow passage of sample and buffers through the column. However, the time to achieve purification using gravity feed was not efficient. Eventually efficiency was addressed by further evolution to the current use of kits available. In kit form there appear to be two common technologies that use silica adsorption, beads and centrifuge columns. Centrifuge column technology was the first to market where the homogenized sample is mixed with binding buffer (optimized chaotropic salt solution) before adding the mix to columns containing a membrane (possibly also used in size exclusion) sandwiching the silica resin chemistry. The column is then centrifuged, and it is the centrifugal force that pushes the sample through the resin, and representing the first adsorption step described above. An alternative to centrifugation, which was designed to ease the automation of column technology, is the use of a vacuum manifold. In this application a vacuum is created to force the sample and buffers through the column. However, due to the physics of a disc (membrane/resin), the force exerted on the area of the membrane by a vacuum is concentrated on the centre, even more so than centrifugal force. Therefore the edges of the membrane may trap more of the sample or buffers. Another issue with column technology is that samples containing non/low-soluble PCR inhibitors can remain on the membrane, and some may be co-eluted with the nucleic acid in the final stages causing partial or complete inhibition of PCR. Automation, using columns, must involve liquid handling, due to specific volumes that need to be added to the columns. In a regulated environment this brings the additional burden of validating the pipetting performance, depending on the liquid type used. The DNeasy, and RNeasy family of kits, as sold by Qiagen, are an example of column kits using silica adsorption.

Silica magnetite beads are a more recent technology that utilize an iron core coated in silica. This technology was developed for the specific need of automating the silica binding chemistry. Here, the homogenized sample is again mixed with a binding buffer which is then mixed with the beads to initiate the adsorption step in the protocol. Bead technology can use liquid handling, by immobilizing the beads on the bottom of the tube/plate with a magnet under the tube/plate, and using the liquid handler to add and remove the wash and elution buffers. However a much simpler and neater technique is to move the beads themselves by using automation with simple magnetic rods to pickup and release the beads into the different wash buffers, leading ultimately to the elution buffer. Such automation is much simpler to maintain as the need to validate any liquid handling performance is negated. Another advantage that beads have over columns is the fact that the non/low-soluble inhibitors are not forced to interact with the chemistry, as the beads flow freely in solution, attracting the nucleic acids. For this reason bead chemistries tend to have equivalent yields to columns, but are also less problematic with regards to PCR inhibitors. Following the success of silica magnetite beads, the concept of encapsulating an iron core with the adsorption chemistry of interest, in order to easily automate the procedure, has been replicated for several applications, some of which are described in this chapter.

ChargeSwitch®

An increasingly common method for nucleic acid purification exploits the native negative charge of nucleic acid molecules and employs a charge–charge interaction to bind the target molecules. The chemistry is dependent on a low pH lysis and binding environment wherein the surface ligands sequester free protons that result in a positively charged surface. The negatively charged nucleic acids therefore bind and can be purified of cellular debris and proteins by aqueous wash steps. Liberation of the nucleic acids is then readily effected by the transfer of the bound nucleic acids into a high-pH environment, where the surface loses its charge and the bound molecules are readily washed free.

A number of other kits employing this type of technology are available on the market for the isolation of genomic and plasmid DNA as well as RNA. Such kits employ a relatively mild urea-based chemical lysis and are often free of organic solvent based buffers and concentrated salts that offer significant advantages. Processes involving high quantities of salts throughout may have a detrimental effect on the signal strength observed during subsequent real-time PCR analysis of the purified nucleic acids (Freeman *et al.*, 1999). For commercial processes, the use of ethanol within the process wash buffers introduces significant cost issues into supply chain management. The hygroscopic and corrosive nature of ethanol and its flammability contribute to the increased cost of the transportation. Further, the addition of guanidinium-based salts, while excellent chaotropes and highly effective in host lysis by sequestration of water, may lead to inhibition of the PCR if these salts are carried forward (Demeke and Jenkins, 2009).

Isolation using carboxyl derivatives

A number of proprietary and non-proprietary isolation methods employ particles or planar surfaces with chemistries featuring carboxyl functional groups. The method is centred on interaction between the carboxyl group and the nucleic acid in a high salt/PEG environment (Hawkins *et al.*, 1994). The method has been successfully employed for the separation of total nucleic acids from a background of cellular debris and proteins (Horak *et al.*, 2007), as well as for the graded separation of nucleic acids by molecular weight to differentiate, for example, plasmid DNA from genomic.

For the separation of DNA from other cellular contaminants, the method employs a combination of polyethylene glycol, of molecular weight 6000–10,000, and the chloride salt of lithium, sodium, magnesium, potassium, calcium, caesium or barium. Typically, however, the method is limited to PEG-8000, at a concentration of 5–15% w/v, and sodium chloride at a concentration of 0.5–5.0 M, which is the general range required for nucleic acids to bind to carboxyl functional groups. Specific concentrations of these reagents can be optimized from this general range; for example non-specific binding of DNA can be achieved by using PEG-8000 at 10% w/v and sodium chloride at 1.25 M.

The application of this method to bulk separations is addressed through a number of specialized kits currently on the research tools market. Most commonly these feature the use of solid-phase particles with derivatized surfaces exposing the nucleic acids in the bulk medium to the carboxyl-functional group chemistry designed to capture the molecules. The use of various particle formation methods to create particles with maximized surface area leads to highly efficient methods for nucleic acid isolation and often leads to high yields. Combination with appropriate methods for the treatment and washing of these beads can result in highly effective purification of the molecule of interest. It has further been postulated that the appropriate selection of polyethylene glycol and salt concentrations can act as a selective method for nucleic acids by molecular weight in a batch process method.

Chromatographic methods, with the potential for being adapted for operation in continuous mode, have also been reported employing this chemistry for the separation of specifically sized nucleic acid strands. These methods are of particular utility in the isolation of digestion products where fine resolution of product size is essential. Affinity resins crafted with optimized porosity and bed voidage allow differentiation of nucleic acid strands to a resolution of < 10 base pairs. Column based processes, with their low intrinsic shear effects, are often the desired process route for the purification of high molecular weight nucleic acids that are difficult to isolate in-tact through bead based processes. The high level of purification achievable is a significant advantage of these systems, but yield can often be compromised. Combination of a batch method to achieve the requisite yield of nucleic acids with a chromatographic method to achieve the required purity could therefore be the desired process route.

Cellulosic beads

Recent additions to the nucleic acids purification market include those employing cellulosic surfaces for the isolation of both DNA and RNA. A prominent example is the Promega Magazorb® kit that employs a chaotropic lysis (5 M guanidinium chloride, 8 M urea) followed by PEG-salt precipitation of the nucleic acids onto paramagnetic beads comprising carboxymethyl-cellulose and diethylaminoethyl cellulose (up to 10% w/w). The precise mechanism by which nucleic acids bind to the cellulosic surface is as yet unknown, but it is postulated that binding is driven by the presence of an aqueous-aqueous two phase system brought about by the concentrated sodium chloride and PEG-4000 present in the binding buffer (Lebreton *et al.*, 2002; Zhang *et al.*, 2006; Li *et al.*, 2010). In such an environment, the highly polar nucleic acids will tend towards the polar salt rich region and the apolar cellular contaminants or overall-neutral large proteins tend towards the PEG rich region. The high salt environment of the binding buffer (1.25 M NaCl) facilitates binding of the nucleic acids to the beads as well as sequestration of apolar contaminants (10% w/v PEG-8000). Subsequent washing under lower salt conditions (0.5 M NaCl) results in the removal of proteins that have lower overall polarity than the nucleic acids that are subsequently eluted in the effectively salt free environment of the nuclease free water.

Ion exchange purification

An emerging alternative to caesium chloride ultracentrifugation techniques is ion exchange chromatography based DNA purification (Yang *et al*,. 2008) which is the basis of a number of kits currently on the market for both plasmid and genomic DNA. The significant process advantage offered by such kits is the ability to perform ion exchange purification without the need for specialist chromatography equipment, which is often expensive and occupies a large spatial footprint. However, these kits are limited to gravity flow through columns with poor diffusional efficiency leading to concentration gradients across the bed. The loose packing required for gravity flow also leads to inefficient use of the binding sites within the individual beads. Ion exchange membranes have the advantages of high flow rate and thus the potential for high-throughput applications. Such membranes have been shown to have wide utility in the capture and purification of plasmid DNA (Endres *et al.*, 2003; Teeters *et al.*, 2003; Zhang *et al.*, 2003), oligonucleotide sequences (Deshmukh *et al.*, 2000) and many other non-nucleic acid biomolecules (Roper and Lightfoot, 1993; Knudsen *et al.*, 2001; Specht *et al.*, 2004).

As with many of the techniques described here, anion exchange techniques rely upon the negatively charged nucleic acid backbone interacting with the oppositely charged anion exchange surface. In its most simple format, the anion exchange technique employs a binding environment of near neutral pH and medium ionic strength to affect binding. The high density of surface groups for binding is exploited to achieve high levels of recovery in a relatively mild environment. For bulk capture and purification of nucleic acids, without distinguishing for example plasmid from genomic DNA, or RNA from DNA, such purification is highly effective both in terms of recovery and purity. However, for more selective capture the pH of the binding environment can be manipulated to affect the charge density of the matrix and the ionic strength of the buffers tuned to yield proteins, RNA or DNA specifically. In addition to PCR (Jones and Winistorfer, 1992; Jung *et al.*, 1992), due to its highly selective nature the technique has been employed for such applications as mammalian cell transfection (Chen *et al.*, 1991), DNA sequencing (Hall *et al.*, 1989; Voss *et al.*, 1990) and transfection (Kayne *et al.*, 1988; Luytjes *et al.*, 1989).

Elution

The final step in the isolation and purification of nucleic acids is the elution of the captured nucleic acids from the medium upon which they were adsorbed. The choice of elution buffer is dependent upon the mechanism of nucleic acid binding to the isolation medium but will often be an inert buffer with low ionic strength (low-salt) and slightly alkali pH (~8). Such a buffer serves the multiple purposes of eluting the target molecules, no intrusion on the PCR process and maintaining the integrity of the nucleic acids in a standard environment. Heating the elution buffer prior to use may also be beneficial to increase yield from capture methods as it aids stripping of the nucleic acid from the binding chemistry. Eluates from these processes can then be stored for short (at 4°C) or long periods (at −20°C) before analysis by PCR. Light, in particular ultra-violet, is highly damaging to nucleic acids, so exposure must be limited to further preserve their integrity.

Conclusion

This chapter has outlined popular methods from the literature for the isolation and purification of nucleic acids from a range of sample types. The evolution of some these techniques towards easy to use kits has made them an essential aspect of the nucleic acid analyst's tool box. In selecting the appropriate method particular care must be taken in identifying the target, such as genomic versus plasmid DNA or DNA versus RNA, and

Table 7.1 Summary of the relative strengths of the described sample extraction and isolation chemistries described

				Silica		Charge switch			Ion exchange	
Criteria	CsCl	Phe/Chl	Isopropanol	Beads	Columns	beads	Carboxyl beads	Cellulosic beads	Beads	Column
Technical skill required	+++	++	++	+	+	+	+	+	++	++
Speed	+	++	++	+++	+++	+++	++	++	+++	++
Automatable	+	+	+	+++	+++	+++	++	++	+++	++
Yield	+++	+++	+++	++	++	++	++	++	++	++
Purity	+++	+++	+++	++	++	++	++	++	+++	+++

CsCl refers to caesium chloride ultracentrifugation, Phe/Chl refers to phenol–chloroform isolation.+= poor, ++=good, +++=excellent.

in a number of other considerations, such as the automatability of the technique at hand. Additionally, thought must be given to the future application of the method. If the method is destined for a QC environment, with higher throughput, and potentially lower expertise, then techniques which are easily automatable maybe more desirable, even if this requires a compromise in other areas. The techniques discussed have been summarized with some of these considerations in Table 7.1.

References

Boom R., Sol, C.J., Salimans, M.M., Jansen, C.L., Wertheim-van Dillen P.M., and van der Noordaa J. (1990). Rapid and simple method for purification of nucleic acids. J. Clin. Microbiol. *28*, 495–503.

Chen W., Andres, D., Goldstein, J., Russell, D., and Brown, M. (1991). cDNA cloning and expression of the peptide-binding beta subunit of Rat p21 farnesyltransferase, the counterpart of yeast DPR1/RAM1. Cell *66*, 327–334.

Chomczynski P., and Sacchi, N. (1987). Single-step method of RNA isolation by acid guanidinium thiocyanate-phenol-chloroform extraction. Anal. Biochem. *162*, 156–159.

Clewell D.B., and Helinski, D.R. (1970). Properties of a supercoiled deoxyribonucleic acid-protein relaxation complex and strand specificity of the relaxation event. Biochemistry *9*, 4428–4440.

Cullen, D.W., and Hirsch, P.R. (1998). Simple and rapid method for direct extraction of microbial DNA from soil for PCR. Soil Biol. Biochem. *30*, 983–993.

Demeke T., and Jenkins, G.R. (2009). Influence of DNA extraction methods, PCR inhibitors and quantification methods on real-time PCR assay of biotechnology-derived traits. Anal. Bioanal. Chem. *396*, 1977–1990.

Deshmukh R.R., Warner, T.N., Hutchison, F., Murphy, M., Leitch, W.E., De Leon P., Srivatsa, G.S., Cole, D.L., and Sanghvi, Y.S. (2000). Large-scale purification of antisense oligonucleotides by high-performance membrane adsorber chromatography. J. Chromatogr. A *890*, 179–192.

Endres H.N., Johnson, J.A., Ross, C.A., Welp, J.K., and Etzel, M.R. (2003). Evaluation of an ion exchange membrane for the purification of plasmid DNA. Biotechnol. Appl. Biochem. *128*, 132–149.

Engler H.D., Selepak, S.T., and Callaway, J. (1996). Effects of sonication of clinical specimens on detection of cytomegalovirus in shell vial cultures. Clin. Diag. Virology *5*, 67–71.

Erb, R.W., and Wagner-Döbler I. (1993). Detection of polychlorinated biphenyl genes in polluted sediments by direct DNA extraction and polymerase chain reaction. Appl. Environ. Microbiol. *59*, 4065–4073.

Freeman W.M., Walker, S.J., and Vrana, K.E. (1999). Quantitative RT-PCR: pitfalls and potential. BioTechniques *26*, 112–125.

Gross-Bellard M., Oudet, P., and Chambon, P. (1973). Isolation of high-molecular-weight DNA from mammalian cells. Eur. J. Biochem. *36*, 32–38.

Hall B., Betts, P., and Wootton, J. (1989). DNA sequence analysis of artificially evolved ebg enzyme and ebg repressor gene. Genetics *123*, 635–648.

Hawkins T.L., O'Connor-Morin T., Roy, A., and Santillan, C. (1994). DNA purification and isolation using a solid-phase. Nucleic Acids Res. *22*, 4543–4544.

Ho C.W., Tan, W.S., Kamarudin, S., Ling, T.C., and Tey, B.T. (2008). The release of hepatitis B core antigen from *Escherichia coli* by batch mode bead milling. Proc. Biochem. *43*, 206–212.

Horak D., Rittich, B., and Spanova, A. (2007). Carboxyl-functionalized magnetic microparticle carrier for isolation and identification of DNA in dairy products. J Magnetism Magentic Mat. *311*, 249–254.

Jones D., and Winistorfer, S. (1992). Recombinant circle PCR and recombinant PCR for site-specific mutagenesis without PCR product purification. Biotechniques *12*, 528–535.

Jung M., Dritschilo, A., and Kasid, U. (1992). Reliable and efficient direct sequencing of PCR double-stranded genomic DNA template. PCR Methods Appl. *1*, 171–174.

Kayne P., Kim, U., Han, M., Mullen, J., Yoshizaki, F., and Grunstein, M. (1988). Extremely conserved histone H4 N terminus is dispensible for growth but essential for repressing the silent mating loci in yeast. Cell *55*, 27–39.

Knudsen H.L., Fahmer, R.L., Xu, Y., Norling, L.A., and Blank, G.S. (2001). Membrane ion-exchange chromatography for process-scale antibody purification. J. Chromatogr. A *907*, 145–154.

Kuske, C.R., Banton, K.L., Adorada, D.L., Stark, P.C., Hill, K.K., and Jackson, P.J. (1998). Small-scale DNA sample preparation method for field PCR detection of microbial cells and spores in soil. Appl. Environ. Microbiol. *64*, 2463–2472.

Leuko S., Goh, F., Ibanez-Peral R., Burns, B.P., Walter, M.R., and Neilan, B.A. (2008). Lysis efficiency of standard DNA extraction methods for *Halococcus* species in an organic rich environment. Extremophiles: life under extreme conditions. *12*, 301–308.

Luytjes W., Krystal, M., Enami, M., Parvin, J., and Palese, P. (1989). Amplification, expression, and packaging of a foreign gene by influenza virus. Cell *59*, 1108–1113.

Maniatis, T., Fritsch, E.F., and Sambrock, J. (1982). Molecular Cloning. A Laboratory Manual (Cold Spring Harbor Laboratory Press, NY, USA), pp. 86–96.

Marra M.A., Kucaba, T.A., Hillier, L.W., and Waterston, R.H. (1999). High-throughput plasmid DNA purification for 3 cents per sample. Nucleic Acids Res. *27*, e37.

Meselson M., Stahl, F.W., and Vinograd J. (1957). Equilibrium sedimentation of macromolecules in density gradients. Proc. Natl. Acad. Sci. U.S.A. *43*, 581–583.

Miller D.N., Bryant, J.E., Madsen, E.L., and Ghiorse, W.C. (1999). Evaluation and optimization of DNA extraction and purification procedures for soil and sediment samples. Appl. Environ. Microbiol. *65*, 4715–4724.

Moré M.I., Herrick, J.B., and Silva, M.C. (1994). Quantitative cell lysis of indigenous microorganisms and rapid extraction of microbial DNA from sediment. Appl. Environ. Microbiol. *60*, 1572–1580.

Ogram, A., Sayler, G.S., and Barkay, T. (1987). The extraction and purification of microbial DNA from sediments. J. Microbiol. Methods *7*, 57–66.

Porteous L.A., Seidler, R.J., and Watrud, L.S. (1997). An improved method for purifying DNA from soil for polymerase chain reaction amplification and molecular ecology applications. Mol. Ecol. *6*, 787–791.

Radloff, R., Bauer, W., and Vinograd, J. (1967). A dye-buoyant-density method for the detection and isolation of closed circular duplex DNA in HeLa cells. Proc. Natl. Acad. Sci. U.S.A. *57*, 1514–1521.

Roper D.K., and Lightfoot, E.N. (1993). Separation of biomolecules using adsorptive membranes. J. Chromatogr. A *702*, 3–26.

Schupbach J., Tomasik, Z., Knuchel, M., Opravil, M., Gunthard, H.F., Nadal, D., and Boni, J. (2006). Optimized virus disruption improves detection of HIV-1 p24 in particles anduncovers a p24 reactivity in parients with undetectable HIV-1 RNA under long-term HAART. J. Med. Virol. *78*, 1003–1010.

Specht R., Han, B., Wickramasinghe, S.R., Carlson, J.O., Czermak, P., Wolf, A., and Reif, O.W. (2004). Densonucleosis virus purification by ion exchange membranes. Biotechnol. Bioeng. *88*, 465–473.

Teeters M.A., Conrardy, S.E., Thomas, B.L., Root, T.W., and Lightfoot, E.N. (2003). Adsorptive membrane chromatography for purification of plasmid DNA. J. Chromatogr. A *989*, 165–173.

Volossiouk T., Robb, E.J., and Nazar, R.N. (1995). Direct DNA extraction for PCR-mediated assays of soil organisms. Appl. Environ. Microbiol. *61*, 3972–3976.

Voss H., Zimmerman, J., Schwager, C., Stegemann, J., Erfle, H., Stuckey, K., and Ansorge, W. (1990). Automated fluorescent sequencing of cosmic DNA. Nucl. Acids Res. *18*, 1066.

Yang Y., Hebron, H.R., and Hang, J. (2008). High performance DNA purification using a novel ion exchange matrix. J. Biomol. Tech. *19*, 205–210.

Zeugin J.A., and Hartley, J.L. (1985). Ethanol Precipitation of DNA. Focus 7, 1–2.

Zhang S., Krivosheyeva, A., and Nochumson, S. (2003). Large-scale capture and partial purification of plasmid DNA using anion-exchange membrane capsules. Biotechnol. Appl. Biochem. *37*, 245–249.

DOI: https://doi.org/10.21775/9781912530243.07

Sample Preparation for Real-time PCR in Food Science

7

Tomáš Kuchta

Abstact
Sample preparation including DNA isolation is described as the first procedural step preceding the analysis of food by real-time PCR. Principles of, applications of and prerequisites for direct DNA isolation from food matrix are presented, providing information on chaotropic solid phase extraction, solubilization with cetyltrimethyl-ammonium bromide followed by liquid-liquid extraction, and on immunomagnetic separation. Importance of and procedures for testing the amplifiability of the isolated DNA, determination of the recovery and recovery rate of DNA from food as well as procedures improving the efficiency of DNA isolation from "difficult" food matrices are given. Various techniques for DNA quantitation, including UV spectrometry, fluorimetry and determination of amplifiable DNA by PCR, are described and their use and usefulness are discussed. The effectivity of individual approaches at the analysis of various food matrices is discussed. For the application field of rapid detection of pathogenic bacteria in food, information on cultivation enrichment, immunoseparation, centrifugation and micro-filtration is provided. The applicability of different sample preparation methods at the detection of various pathogenic bacteria in several food product types is discussed. Attention is paid also to DNA isolation from enriched samples, including description of rapid techniques for partial DNA separation, bacterial cell lysis and removal of PCR inhibitors.

Introduction

The first procedural step preceding the analysis of food by real-time PCR is DNA isolation. For the analysis of animal or plant species in food, or for quantification of microorganisms, DNA is isolated directly from the food matrix. For the detection of pathogenic microorganisms in food, enrichment has to be employed in order to achieve sufficient detection sensitivity.

Direct DNA isolation from food matrix

General considerations

Food products represent a complex matrix containing DNA in a mixture with various compounds. DNA in food is often fragmented or degraded, and the accompanying compounds may cause PCR inhibition if they manage to pass through the DNA isolation procedure. DNA isolation deserves adequate attention because it can strongly influence the results of the downstream PCR. Besides the high efficiency, methods for DNA isolation from food should not be too labourious and should not require much time. With real-time PCR, additional attention has to be paid to possible presence and co-extraction of fluorescent compounds, which may interfere with fluorescence measurement by elevating the baseline.

Two methods are the most widely used for DNA isolation from food: chaotropic solid-phase extraction (SPE) and a method of selective precipitation with cetyltrimethylammonium bromide (CTAB) and liquid-liquid extraction (CTAB-LLE). Chaotropic SPE is the faster and less labourious alternative, providing the product in a few hours. Several kits based on this principle are commercially available. The method proved to be suitable for effective DNA isolation from various food matrices. CTAB-LLE is a more labourious method, which requires two days to accomplish in its full version. Recently, its simplified, faster kit version has become available (DNeasy Mericon Food Kit; Qiagen,

Hilden, Germany), providing DNA of quality comparable to full CTAB-LLE version from many sample types including processed foods and white wine (our unpublished results). With some food matrices, CTAB-LLE is known to remove the inhibitors more effectively than chaotropic SPE, although the DNA yields may be lower (Zimmermann et al., 1998; Olexova et al., 2004). Out of other suitable DNA isolation methods, non-chaotropic SPE can be mentioned, which is available in two versions, for the food of plant and animal origin, respectively (Olexova et al., 2004). Good purification efficiency with vegetable samples has been reported also for a magnetic DNA purification system (Di Pinto et al., 2007). As a rule, several other "rapid" methods for DNA isolation are usually not effective enough to be used with complex food samples.

Chaotropic SPE

This method is based upon the formation of a complex of DNA with a chaotropic salt, which is bound to a silicagel carrier, washed by a solution of a high ionic strength, and a subsequent elution of DNA by a solution of a low ionic strength (Boom et al., 1990). Several kits based on this principle are available. These contain minicolumns of the specifically formulated silicagel and also all the necessary solutions. An example of a popular kit is NucleoSpin Food kit (Macherey-Nagel, Düren, Germany). In its case, the procedure involves the following steps:

- lysis of the homogenized sample in a solution containing chaotropic salts, denaturation agents and detergents
- removal of solid particles and of the precipitate by centrifugation
- mixing of the supernatant with the binding buffer and ethanol
- application and binding to the silicagel minicolumn
- two subsequent washings with washing buffers
- elution with a buffer of a low ionic strength

Chaotropic SPE has been shown to facilitate successful isolation of DNA from various types of food matrices, producing DNA concentrations by 10^0 x - 10^4 x higher than the minimum concentration necessary for amplification; certain problems were observed only with samples containing chocolate (Gryson et al., 2004; Olexova et al., 2004; Olexova et al., 2005; Di Pinto et al., 2007). The recovery rate determined using short linear DNA fragments and soya flour as a matrix was 8 % - 66 % (Piknova and Kuchta, 2006).

Solubilization with CTAB and liquid-liquid extraction (CTAB-LLE)

The procedure of DNA isolation using CTAB was first elaborated as a method for DNA isolation from plants (Murray and Thompson, 1980). Later it was adapted for the isolation of DNA from food of plant origin (Zimmermann et al., 1998). The method is based upon the selective precipitation of nucleic acids by the detergent CTAB in a solution of a low ionic strength (< 0,5 M NaCl). In these conditions, polysaccharides, phenolic compounds and other potentially PCR-inhibitory compounds remain soluble. The precipitate is dissolved in a solution of a higher ionic strength, and non-polar compounds are removed by the extraction to chloroform.

In practice, several protocols of the CTAB-LLE method are available. A popular version is that elaborated by the GMO laboratory of the Joint Research Centre, Ispra, Italy (Anonym, 2002). The procedure involves the following steps:

- lysis of the homogenized sample in the solution containing CTAB, proteinase and NaCl at a higher concentration
- RNA degradation
- removal of non-polar compounds by two extractions to chloroform
- DNA precipitation by CTAB in a solution with a lower NaCl concentration
- separation of the precipitate by centrifugation
- dissolution of the precipitate in a solution with a higher NaCl concentration
- removal of non-polar compounds by an extraction to chloroform
- precipitation of DNA by isopropanol and ethanol

CTAB-LLE has been shown to be successful at isolation of DNA from various types of food matrices of plant origin (Straub et al., 1999; Brunnert et al, 2001; Oraby et al., 2005), producing DNA concentrations by 10^1 x - 10^4 x higher than the minimum concentration necessary for amplification (Olexova et al., 2004). The recovery rate determined using short linear DNA fragments and soya flour as a matrix was 3 % - 54 % (Piknova and Kuchta, 2006).

Immunomagnetic separation

DNA isolation by immunomagnetic separation (IMS) is based upon the use of paramagnetic microbeads coated with an agent providing affinity to dsDNA. The process relies upon the cell lysis to release DNA with a subsequent adsorption of DNA to the surface of the microbeads. This is followed by magnetic separation of the intact DNA+microbeads complex and subsequent washing to remove any residual contaminants and potential PCR inhibitors from the isolated DNA. Finally, the complex is resuspended for use in the downstream PCR. Alternatively, DNA may be eluted from the microbeads by a short incubation at 65°C. Multiple samples may be handled in parallel and the isolation process may be automated. The method is particularly suitable for simpler matrices and/or for matrices with a lower DNA content. Several types of microbeads for IMS are commercially available. IMS has been shown to successfully isolate DNA from certain types of food matrices (Hahnen et al., 2002; Hird et al., 2003; Rizzi et al., 2003).

Testing the amplifiability of the isolated DNA

For qualitative applications, in order to avoid false negative results due to improper DNA isolation, it is necessary to confirm the amplifiability of the isolated DNA preparation at least with samples producing negative results with the specific PCR. This can be accomplished by a universal eukaryotic PCR targeting the nuclear gene encoding for 18S rRNA (Allmann et al., 1993; Brezna et al., 2006) or by a universal PCR targetting the mitochondrial gene encoding for 16S rRNA (Remler et al., 2004). At the analysis of food of animal origin, PCR widely specific for *Vertebrata* may be used, targetting the gene encoding for cytochrome b (Meyer et al., 1995) or the gene encoding for myostatin (Laube et al., 2003). At the analysis of food of plant origin, PCR widely specific for photosynthesizing plants targeting the chloroplast gene *rbcL* encoding for a large subunit of ribulosophosphatecarboxylase-oxidase (RUBISCO; Wyman et al., 2000) may be used.

Determination of the recovery and recovery rate of DNA from food

In order to determine the recovery of DNA from food, various methods of DNA quantification may be used – UV-spectrophotometry, fluorimetry of the complex of DNA with an intercalation dye, or quantification of the amplifiable DNA using real-time PCR.

UV-spectrophotometric DNA quantification is the simplest method suitable for this purpose. Nucleic acids in solution absorb ultraviolet (UV) light in the range from 210 nm to 300 nm with an absorption maximum at 260 nm. Since DNA, RNA and nucleotides have their absorption maxima at 260 nm, RNA and nucleotide contamination of DNA solutions cannot be determined by UV spectrometry. For this reason, RNA must be removed enzymatically during DNA extraction before DNA determination. For precise determination, also oligonucleotides and nucleotides derived from RNA hydrolysis should be eliminated. Otherwise, oligonucleotides and nucleotides generated by RNase treatment would lead to an overestimation of the DNA content of the sample. Moreover, the extinction of dsDNA is lower than of ssDNA and since the proportion of ssDNA in the solution is unknown, in order to avoid overestimation of the DNA content, all DNA in the test sample should be converted to its single-stranded form by using the denaturing agent NaOH. Since nucleic acids do not absorb at 320 nm, reading at 320 nm is informative for the determination of background absorption due to light scattering and UV-active compounds. It is not necessary to construct a calibration curve, provided that an appropriate molar extinction coefficient is used.The method is applicable to DNA concentrations in the range from 2 $\mu g\ ml^{-1}$ to 50 $\mu g\ ml^{-1}$. Before quantitation, suitable dilutions of the extracted DNA should be made, in order to be in the linear range of the spectrometric measurement (absorbance from 0.05 to 1). The simplest version of the method involves just the determination of absorbance at a wavelength of 260 nm and DNA concentration is calculated on the basis of a fact that the absorbance of a dsDNA solution of 5 µg ml^{-1} is approximately 0.1. In order to monitor a possible contamination by proteins, additional determination of absorbance at a wavelength of 280 nm is often carried out. The DNA preparation is taken as free of proteins, if A_{260}/A_{280} ranges from 1.8 to 2.0 (Maniatis et al., 1982). A more precise variant of this basic method is available which involves conversion of all DNA to the single-straned form using NaOH, measuring the absorbance at 260 nm and 320 nm and calculation as given below (ISO 21571, 2005):

$$c_{DNA} = F \times (A_{260} - A_{320}) \times 37$$

where c_{DNA} is the concentration in µg ml^{-1}, F is the dilution factor, A_{260} is the absorbance at 260 nm, A_{320} is the absorbance at 320 nm and 37 is the conversion factor in µg ml^{-1}.

A disadvantage of the spectrophotometric quantification method is its low sensitivity and the need for removal of RNA and nucleotides for a precise DNA determination. A better method facilitating direct DNA quantification is fluorimetry of the complex of DNA with an intercalation dye, e. g. Hoechst 33258 or PicoGreen. The quantification limit is 10 ng ml^{-1} with Hoechst 33258 and 250 pg ml^{-1} with PicoGreen. Hoechst 33258 quantitates both dsDNA and ssDNA, while the PicoGreen dye is selective for dsDNA. If less sensitive DNA quantification is sufficient, cheaper intercalation dyes, like ethidium bromide or SybrGreen I may be used (Anonym, 2005).

However, even precise DNA quantification does not always provide a relevant information on the effective, i. e. amplifiable template DNA concentration. This relates to the degree of fragmentation, and also to the presence of PCR inhibitors. The degree of fragmentation is determined using agarose gel electrophoresis (Maniatis et al., 1982). The concentration of amplifiable DNA may be determined by e.g. decimally diluting the DNA preparation and then subjecting it to PCR. In this way, the maximum dilution which is amplifiable by PCR is determined. Ideally, a universal PCR which amplifies DNA sequences present in all organisms is used, however, a specific PCR may be used as well (Olexova et al., 2004). Further information may be provided by dilution experiments if various primers targeting the same marker sequence but amplifying DNA fragments of a different size are used (Hrncirova et al., 2008).

When quantitating DNA, homogenity of the DNA preparation has to be taken into account as well. A DNA preparation which has undergone precipitation and dissolution is hardly homogenous and individual volumes pipetted out of the stock of such preparation often contain different amounts of DNA. Another practically observed phenomenon is that DNA tends to stick to tube walls and its concentration in solution, in particular in solutions of low-concentrations, tends to decrease with storage time.

At a quantitative DNA-based analysis of food, the recovery rate of DNA achieved by the isolation method needs to be determined for the given matrix. The recovery rate is the percentage of the extracted DNA out of all DNA present in the sample. It can be determined by adding a known amount of a defined exogenous DNA, carrying a specific marker sequence, to the food sample, then performing DNA isolation and, finally, performing quantitative real-time PCR targeting the marker sequence. When simulating DNA isolation from food, short linear DNA fragments are preferentially used. Recovery rates of DNA may be considerably low and may vary significantly between matrices and also between individual experiments (Piknova et al., 2004). The situation is even more complicated by the fact that recovery rates may be concentration-dependent (Piknova and Kuchta, 2006).

Procedures improving the efficiency of DNA isolation from "difficult" food matrices

From certain food matrices, it may be difficult to isolate amplifiable DNA. In particular, these include matrices with high contents of fat or contents of certain types of PCR inhibitors. Fat may interfere with partition at phase separation. On the other hand, certain inhibitors may not be separated from DNA and thus pass to the final DNA preparation. Too hard consistence of the sample may also pose a problem and needs to be dealt with by specific pre-extraction procedures.

A procedure was developed to improve bacterial DNA isolation from hard cheese. The procedure involves homogenization in a lysis buffer containing 8 mol l^{-1} urea and 1% SDS, incubation for 30 min at 45 °C, a sequence of centrifugation and washing steps to separate bacterial cells suitable for DNA isolation (Rossmanith et al, 2007). This procedure was further developed and adapted for a range of food matrices, employing a different non-ionic detergent, Lutensol (Mayrl et al., 2009).

A pre-extraction procedure was developed for DNA isolation from olive oil, involving a series of centrifugations and freeze-thaw cycles using liquid nitrogen. The procedure, when followed by CTAB-LLE provided DNA suitable for downstream analysis by PCR (Busconi et al., 2003). Testolin and Lain (2005) successfully isolated amplifiable DNA from olive oil by chaotropic SPE, finding unnecessary any pre-treatment such as pre-extraction with hexane.
With chocolate-containing samples, true removal of PCR inhibitors is sometimes not successful and it is necessary to include facilitators in the PCR mixture.

DNA extraction from enriched or concentrated food samples

General considerations

An important application field of real-time PCR is the rapid detection of pathogenic bacteria in food. In this application, PCR is preceded by target cell multiplication (by cultivation enrichment) and/or concentration (by immunoseparation, centrifugation and/or microfiltration). These processes are followed by DNA extraction. It should be noted that involvement of cultivation enrichment increases the sensitivity of the detection, but causes the lost of information on the quantity of the analyte in the sample

Cultivation enrichment

During cultivation enrichment, bacterial cells present in the food sample are allowed to multiply in a culture medium. The simplest and most often used approach is a single-step enrichment in a non-selective or resuscitation medium. This enrichment should, however, last for at least 16 h or overnight, because shorter enrichment does not provide enough target bacterial cells for the downstream detection. The reason is the fact, that lag phases for bacterial cultures of cells from food are considerably long, because the cells may be stressed or may multiply slower when present in an excess of other, often faster-multiplying bacterial cells (D'Aoust et al., 1990). Nevertheless, the single-step enrichment may produce false positive results caused by the presence of dead bacterial cells in the sample, because the original sample is, as a standard, diluted only 10x (Drahovska et al., 2001). Another drawback of the low dilution rate is the danger of false negative results caused by food-borne PCR inhibitors (Salomonsson et al., 2005).

These problems may be solved if a twostep enrichment is used. In this case, dead bacterial cells as well as foodborne PCR inhibitors would be diluted and would not affect the results. The second enrichment may be carried out in the same medium as the first one (Salomonsson et al., 2005). However, additional improvement in the sensitivity of the entire method may be achieved if the second enrichment medium is selective, because then the competing microflora is suppressed (Krascsenicsova et al., 2006; Oravcova et al., 2008). The second, selective enrichment step does not require too much additional labour and, if its duration is limited, the entire method including the cell lysis and PCR, may produce results on the next day after the sample collection. Certain attention has to be paid to the composition of the second enrichment medium concerning PCR inhibitors, if only a simple DNA preparation is intended to be used prior to PCR. Some media contain such PCR inhibitors (Lantz et al., 1998) and to avoid problems, special PCR-compatible selective media have been developed (Knutsson et al., 2002).

The main advantage of the cultivation enrichment is that the numbers of target bacterial cells increase while those of dead cells remain on the original level. These procedures are also less laborious compared to the others, requiring less hands-on time (Scheu et al., 1998; Rijpens and Herman, 2002).

Immunoseparation methods

Immunoseparation systems employ antibodies bound to microbeads to separate specific bacterial cells from the homogenized food sample. To separate the microbeads with bound bacterial cells from the rest of the sample, magnetic force or low-speed centrifugation can be used. In the former case, in immunomagnetic separation (IMS; Cudjoe et al., 1995; Uyttendaele et al., 1999), paramagnetic beads are used. Working with manual separators, IMS is a considerably laborious technique. This drawback may be overcome by the use of an automatic immunomagnetic separator (Lynch et al., 2004).

Cheaper plastic microbeads can be used when separation by low-speed speed centrifugation (~100 x *g*) is utilized. This approach is suitable for upscaling, which is necessary when low numbers of target bacterial cells are present in the food sample. Once the bacterial cells are bound to the microbeads and separated, they are washed and lysed. Because food matrix, solid particles and fat, may interfere with the separation process, filtration techniques should be used prior to separation (Kaclikova et al., 2001).

The main advantage of immunoseparation is its speed. The drawbacks are labour intensity and price. The questionable selectivity should not be a problem if sufficient recovery rate is achieved.

Centrifugation_methods

Centrifugation facilitates the separation of bacterial cells from the food matrix on the basis of the difference in the density of the particles. In the simplest case, the sample is homogenized in an appropriate solution and the suspension is centrifuged at a high speed (~10000 x *g*) yielding a sediment of bacterial cells. In practical conditions, it is usually necessary to use a preliminary centrifugation, repeated centrifugation with washing of the sediment and other steps to remove PCR inhibitors (Herman et al., 1995; Stevens and Jaykus, 2004).

Another approach is the use of flotation, when the food sample is homogenized and centrifuged in a discontinuous density gradient with densities of individual layers set specifically with respect to the buoyant density of the cells of the given species. In contrast to the simple sedimentation centrifugation, here the matrix particles sediment in the lower phase, bacterial cells are concentrated in the interphase and certain components of the food matrix are present in the upper phase or remain on the surface of the multilayer system. Bacterial cells are taken up from the interphase of the gradient and are subsequently washed and lysed (Lindqvist et al., 1999; Wolffs et al., 2004).

Microfiltration methods

Microfiltration facilitates the separation of bacterial cells from the food matrix on the basis of the difference in the size of the particles. The sample is homogenized in an appropriate solution, and the suspension is centrifuged or filtered through a coarse filter to remove larger particles originating from the food matrix. Then, the suspension is passed through a hydrophilic membrane microfilter (e.g. 0.45 μm pore size) and microbial cells are retained. The cells are subsequently eluted from the filter, washed and lysed. Unfortunately, performance of this separation method with food matrices is usually problematic mainly because of clotting of the filters. Finding efficient sequence of filters of different pore sizes is very difficult and always specific for the given food matrix. For this reason, usefulness of microfiltration methods for separation of microbial cells from food samples is restricted to liquid matrices (Minarovicova et al., 2010).

DNA isolation from enriched samples

As a template for PCR, DNA preparation of different purity may be used – purified DNA, partially purified cell lysate, crude cell lysate or even directly the bacterial cells. From the biochemical point of view, highly purified DNA as produced by chaotropic SPE is optimal. However, from the point of view of the requirement for time and labour, this approach is unacceptable for practical use.

Partial purification of the cell lysate to remove PCR inhibitors is a widely used alternative, which is rapid in terms of taking 15 – 30 min. It is moderately labourious, since usually only a few working steps are necessary –

centrifugation, adding the reagent, incubation, centrifugation. Systems for partial cell lysate purification are commercially available, e. g. Chelex 100 (Bio-Rad, Hercules, California, USA) or PrepMan Ultra (Applied Biosystems, Foster City, California, USA). These lysate purification systems usually provide amplifiable DNA preparation, however, their efficiency may not be high enough with certain food matrices and dilution of the DNA solution prepared may be necessary.

Even faster and less laborious is to prepare DNA by boiling bacterial cells in a buffer for 10 – 15 min (Abolmaaty et al., 1998). This technique is satisfactory to produce amplifiable DNA from cultures after a two-step enrichment, where food-borne inhibitors had been diluted. However, it is recommended that cells are washed by centrifugation before the lysis to remove inhibitors originating in the enrichment medium. The simplest, fastest as well as cheapest approach is to add bacterial cells directly to PCR (Krascenicsova et al., 2007). However, the danger of PCR inhibition is too high for the practical use of this approach in food analysis.

References

Abolmaaty, A., El-Shemy, M.G., Khallaf, M.F., Levin, R.E. (1998) Effect of lysing methods and their variables on the yield of *Escherichia coli* O157:H7 DNA and its PCR amplification. J. Microbiol. Meth. *34*, 133-141.

Allmann, M., Candrian, U., Höfelein, C., Lüthy, J. (1993) Polymerase chain reaction (PCR). A possible alternative to immunochemical methods assuring safety and quality of food. Z. Lebensm. Unters. Forsch. *196*, 248251.

Anonym (2002) Training course on the analysis of food samples for the presence of genetically modified organisms. User manual. (Ispra: Joint Research Centre).

Anonym (2005) Overview of common fluorescent dyes used in nucleic acid quantitation. Application note. (Hayward: Phenix Research Products).

Boom, R., Sol, C.J., Salimans, M.M., Jansen, C.L., Wertheim-Van Dillen, P.M., van der Noordaa, J. (1990) Rapid and simple method for purification of nucleic acids. J. Clin. Microbiol. *28*, 495-503.

Brezna, B., Hudecova, L., Kuchta, T. (2006) Detection of pea in food by realtime polymerase chain reaction (PCR). Eur. Food Res. Technol. *222*, 600-603.

Brunnert, H.J., Spener, F., Börchers, T. (2001) PCR-ELISA for the CaMV-35S promoter as a screening method for genetically modified Roundup Ready soybeans. Eur. Food Res. Technol. *213*, 366-371.

Busconi, M., Foroni, C., Corradi, M., Bongiorni, C., Cattapan, F., Fogher, C. (2003) DNA extraction from olive oil and its use in the identification of the production cultivar. Food Chem. *83*, 127-134.

Cudjoe, K.S., Hagtvedt, T., Dainty, R. (1995) Immunomagnetic separation of *Salmonella* from foods and their detection using immunomagnetic particle (IMP)-ELISA. Int. J. Food Microbiol. *27*, 11-25.

D'Aoust, J.Y., Sewell, A., Jean, A. (1990) Limited sensitivity of short (6 h) selective enrichment for detection of foodborne *Salmonella*. J. Food Prot. *53*, 562-565.

Di Pinto, A., Forte, V.T., Guastadisegni, M.C., Martino, C., Schena, F.P., Tantillo, G. (2007) A comparison of DNA extraction methods for food analysis. Food Control *18*, 76-80.

Drahovska, H., Turna, J., Piknova, L., Kuchta, T., Szitasova, I., Skarkova, A., Sasik, M. (2001) Detection of *Salmonella* by polymerase chain reaction targeted to *fimC* gene. Biologia (Bratislava) *56*, 611-616.

Gryson, N., Messens, K., Dewettinck, K. (2004) Evaluation and optimisation of five different extraction methods for soya DNA in chocolate and biscuits. J. Sci. Food Agric. *84*, 1357-1363.

Hahnen, S., Offermann, S., Miedl, B., Rüger, B., Peterhänsel, C. (2002) Automated DNA preparation from maize tissues and food samples suitable for real-time PCR detection of native genes. Eur. Food Res. Technol. *215*, 443-446.

Herman, L.M.F., De Block, J.H.G.E., Moerman, R.J.B. (1995) Direct detection of *Listeria monocytogenes* in 25 milliliters of raw milk by a two-step PCR with nested primers. Appl. Environ. Microbiol. *61*, 817-819.

Hird, H., Lloyd, J., Goodier, R., Brown, J., Reece, P. (2003) Detection of peanut using real-time polymerase chain reaction. Eur. Food Res. Technol. *217*, 265-268.

Hrncirova, Z., Bergerova, E., Siekel, P. (2008) Effects of technological treatment on DNA degradation in selected food matrices of plant origin. J. Food Nutr. Res. *47*, 23-28.

ISO 21571. (2005) Foodstuffs - Methods of analysis for the detection of genetically modified organisms and derived products - Nucleic acid extraction. (Geneva: International Organization for Standardization).

Kaclikova, E., Kuchta, T., Kay, H., Gray, D. (2001) Separation of *Listeria* from cheese and enrichment media using antibody-coated microbeads and centrifugation. J. Microbiol. Meth. *46*, 6367.

Knutsson, R., Fontanesi, M., Grage, H., Radstrom, P. (2002) Development of a PCR-compatible enrichment medium for *Yersinia enterocolitica*: amplification precision and dynamic detection range during cultivation. International J. Food Microbiol. *72*, 185-201.

Krascsenicsova, K., Kaclikova, E., Kuchta, T. (2006) Growth of *Salmonella enterica* in model mixed cultures during a two-step enrichment. New Microbiologica *29*, 261-267.

Krascsenicsova, K., Kaclikova, E., Kuchta, T. (2007) Comparison of blocked thermostable DNA polymerases with different activation periods at the direct detection of enterobacterial cells by real-time polymerase chain reaction. J. Rapid Meth. Automat. Microbiol. *15*, 240-243.

Laube, I., Spiegelberg, A., Butschke, A., Zagon, J., Schauzu, M., Kroh, L., Broll, H. (2003) Methods for the detection of beef and pork in foods using real-time polymerase chain reaction. Int. J. Food Sci. Technol. *38*, 111118.

Lantz, P.G., Knutsson, R., Blixt, Y., Al-Soud, W.A., Borch, E., Radstrom, P. (1998) Detection of pathogenic *Yersinia enterocolitica* in enrichment media and pork by a multiplex PCR: a study of sample preparation and PCR-inhibitory components. Int. J. Food Microbiol. *45*, 93-105.

Lindqvist, R., Norling, B., Thisted-Lambertz, S. (1997) A rapid sample preparation method for PCR detection of food pathogens based on buoyant density centrifugation. Lett. Appl. Microbiol. *24*, 306-310.

Lynch, M.J.B., Leon-Velarde, C.G., McEwen, S., Odumeru, J.A. (2004) Evaluation of an automated immunomagnetic separation method for the rapid detection of *Salmonella* species in poultry environmental samples. J. Microbiol. Meth. *58*, 285-288.

Maniatis, T., Fritsch, E.F., Sambrook, J. (1982) Molecular cloning. (New York: Cold Spring Harbor Laboratory).

Mayrl, E., Roeder, B., Mester, P., Wagner, M., Rossmanith, P. (2009) Broad range evaluation of the matrix solubilization (matrix lysis) strategy for direct enumeration of foodborne pathogens by nucleic acids technologies. J. Food Prot. *72*, 1225-1233.

Meyer, R., Hofelein, C., Luthy, J., Candrian, U. (1995) Polymerase chain reaction - Restriction fragment length polymorphism analysis: A simple method for species identification in food. J. AOAC Int. *78*, 1542-1551.

Minarovicova, J., Kaclikova, E., Kuchta, T. (2010) A method for the detection of *Cryptosporidium parvum* in apple juice based on microfiltration and real-time polymerase chain reaction. J. Food Nutr. Res. *49*, 160-164.

Murray, M.G., Thompson, W.F. (1980) Rapid isolation of high molecular weight plant DNA. Nucl. Acids Res. *8*, 4321-4325.

Olexova, L., Dovicovicova, L., Kuchta, T. (2004) Comparison of three types of methods for the isolation of DNA from flours, biscuits and instant paps. Eur. Food Res. Technol. *218*, 390-393.

Olexova, L., Kuchta, T., Pangallo, D. (2005) Determinazione di tracce di nocciole in prodotti dolciari. Ingr. Alim. *4*, 16-18.

Oraby, H.A.S., Hassan, A.A., Mossallam, A.A.A. (2005) Screening food products for the presence of CaMV 35S promoter and NOS3 terminator. J. Sci. Food Agric. *85*, 1974-1980.

Oravcova, K., Trncikova, T., Kuchta, T., Kaclikova, E. (2008) Limitation in the detection of *Listeria monocytogenes* in food in the presence of competing *Listeria innocua*. J. Appl. Microbiol. *104*, 429437.

Piknova, L. - Kuchta, T. - Drahovska, H. - Pangallo, D. (2004) Determinazione del tasso di recupero del DNA dagli alimenti. Ind. Alim. *43*, 1129-1132.

Piknova, L., Kuchta, T. (2006) Concentration dependence of recovery rates of exogenous DNA fragments from soya flour. Eur. Food Res. Technol. *223*, 143-145.

Remler, P., Mülleder, U., Ruppitsch, W., Rassi, E. (2004) Entwicklung von Methoden zum qualitativen und quantitativen Nachweis und zur Unterscheidung von Tier- und Fischmehl in

Futtermitteln durch Nachweis von DNA mit PCR. Endbericht. (Graz: Technische Universität).

Rijpens, N., Herman, L. (2002) Molecular methods for identification and detection of bacterial food pathogens. J. AOAC Int. *85*, 984-995.

Rizzi, A., Panebianco, L., Giaccu, D., Sorlini, C., Daffonchio, D. (2003) Stability and recovery of maize DNA during food processing. Ital. J. Food Sci. *15*, 499-510.

Rossmanith, P., Süß, B., Wagner, M, Hein, I. (2007) Development of matrix lysis for concentration of gram positive bacteria from food and blood. J. Microbiol. Meth. *69*, 504-511.

Salomonsson, A.C., Aspan, A., Johansson, S., Heino, A., Haggblom, P. (2005) *Salmonella* detection by polymerase chain reaction after pre-enrichment of feed samples. J. Rapid Meth. Automat. Microbiol. *13*, 96110.

Scheu, P.M., Berghof, K., Stahl, U. (1998) Detection of pathogenic and spoilage microorganisms in food with the polymerase chain reaction. Food Microbiol. *15*, 13-31.

Stevens, K.A., Jaykus, L.A. (2004) Direct detection of bacterial pathognes in representative dairy products using a combined bacterial concentration-PCR approach. J. Appl. Microbiol. *97*, 1115-1122.

Straub, J.A., Hertel, C., Hammes, W.P. (1999) The fate of recombinant DNA in thermally treated fermented sausages. Eur. Food Res. Technol. *210*, 62-67.

Testolin, R., Lain, O. (2005) DNA extraction from olive oil and PCR amplification of microsatellite markers. J. Food Sci. *70*, C108-C112.

Uyttendaele, M., van Boxstael, S., Debevere, J. (1999) PCR assay for detection of the *E. coli* O157:H7 *eae*-gene and effect of the sample preparation method on PCR detection of heat-killed *E. coli* O157:H7 in ground beef. Int. J. Food Microbiol. *52*, 85-95.

Wolffs, P., Knutsson, R., Norling, B., Radstrom, P. (2004) Rapid quantification of *Yersinia enterocolitica* in pork samples by a novel sample preparation method, flotation, prior to real-time PCR. J. Clin. Microbiol. *42*, 1042-1047.

Wyman, M., Davies, J.T., Crawford, D.W., Purdie, D.A. (2000) Molecular and physiological responses of two classes of marine chromophytic phytoplankton (Diatoms and prymnesiophytes) during the development of nutrient-stimulated blooms. Appl. Environ. Microbiol. *66*, 2349-2357.

Zimmermann, A., Luthy, J., Pauli, U. (1998) Quantitative and qualitative evaluation of nine different extraction methods for nucleic acids on soya bean samples. Z. Lebensm. Unters. Forsch. *207*, 81-90.

DOI: https://doi.org/10.21775/9781912530243.08

Internal and Other Controls for Real-time PCR Validation

8

Martin A. Lee, Dario L. Leslie and David J. Squirrell

Abstract

A range of factors can cause false-negative results in real-time PCR through effects on one or more of the reaction components. Consequently applications requiring a high level of confidence need to be designed to control for the occurrence of false-negatives. Whilst an external, or batch, control is often used, the ideal control is an internal one included in the reaction cocktail in a multiplex assay. Here we discuss the application and development of molecular mimics as controls in real-time PCR and explain concepts and experimental considerations to aid in the optimization of controlled multiplexed assays.

Introduction

In real-time PCR assays there is inherent scope for variability in the outcome because the reagent mixes are relatively complex, with a dozen or more components, and they are often made up in very small volumes, limiting the accuracy of their composition. In addition, the PCR process is exponential, so variations in starting conditions tend to be amplified, affecting the course followed by individual reactions. The course may include assay failure if problems with the sample, the reagents, or the instrumentation have occurred.

In many applications, including infectious disease testing in clinical and veterinary diagnostics, food quality control and environmental analysis, the majority of PCR results are genuinely negative, but the consequences of false-negatives may be serious. There is therefore a pressing need to control for the occurrence of false-negatives.

For most molecular tests the use of reference material in a batch test is the only control that can be implemented. However, PCR and other nucleic acid amplification techniques provide not only extremely sensitive detection, but also have the advantage that they can be multiplexed to include an internal control (IC). An IC is a second target molecule that can be amplified with, but distinguished from, other products in the same tube. Ideally this should be able to control for almost all of the reagents in a reaction cocktail and for variations in instrument operation. The exception the IC cannot control for is a specific probe for another target. The IC can be produced as a synthetic molecule that gets co-amplified with the target species, using the same set of amplimers, but comprising an internal sequence differing from the target to allow it to be bound by a distinguishable probe. In this, the preferred embodiment, the IC acts as a 'molecular mimic'.

Before real-time PCR, IC approaches used amplicons with different molecular masses and post-amplification analysis by a separation method. For example, the molecular mimic could contain both priming sites, but an internal sequence changed by insertions or deletions (Ursi *et al.*, 1992). The products of amplification could be analysed by the use of agarose gel electrophoresis with ethidium bromide staining. There are a large number of such examples reported in the literature. The main drawbacks of the approach were the time-consuming nature of the analysis method, the tendency of smaller amplicons to be amplified more efficiently, and unpredictable generation of specific and control product hetero-duplexes confusing the data interpretation (Henley *et al.*, 1996).

Because of its sensitivity, PCR is routinely used in applications where there may be only a few molecules of the target species present in the sample. Particularly where the test is for an RNA species, loss of the target during sample preparation and PCR set-up can lead to inaccurately low or false-negative outcomes. To cover this eventuality, the IC may additionally be used as a 'process control' and introduced with the sample to indicate loss of target molecules during sample storage or sample processing. This can ensure that methods to overcome false-negatives caused by degradation of the target have succeeded. These methods include: the addition of carrier RNA or DNA; the use of an RNAse inhibitor; the use of reagents treated and tested for removal of RNase and DNase activity; and always wearing disposable gloves when handling any components so as to avoid transfer of RNAse or DNAse activity from the skin.

False-positives are also a problem in nucleic acid amplification based tests, but are not addressable through the use of an IC. Measures for controlling these are covered in the endnote to this chapter.

Real-time PCR approaches with fluorescence reporting are particularly useful for high-confidence applications because the assays can be carried out in the same tube, in multiplex, with measurements *in situ* so the tube never has to be reopened. This closed tube format allows an IC to control for negatives whilst significantly reducing opportunities for cross-contamination by amplification products and thereby reducing the risk of false-positives. Most commercial fluorimeter instruments can co-amplify and detect at least two targets in a quantitative manner. However, increasing the level of multiplexing adds significantly to the complexity of designing and optimizing the assay. Multiplexing of real-time assays is discussed in detail in Lee *et al.* (2009) and reference should be made to that chapter for optimizing multiplex assays.

In this chapter we present various strategies for producing internal controls for different fluorescent reporting arrangements and discuss some experimental considerations for their optimization. The experimental approaches described here are based largely on our own experiences and we make some suggestions for the development and implementation of controlled assay systems. Whilst users will find that for many applications the use of a homologous or competitive control is not necessary, the information will be of equal use to those developing heterologous controls and synthetic standards alike. Competitive homologous internal controls are the focus of this chapter. In the literature, the term internal control is often used incorrectly so we begin with some definitions.

Nomenclature

The use of a second amplicon can serve to control, normalize or standardize the result. This chapter only covers the first application, the use of molecular mimics to check reaction outcomes, but the strategies described are equally valid in the development of internal and external mimics for other applications, such as their use as external or internal standards. The nomenclature used to describe the types of second amplicon is summarized in Table 2.1. It is easy to see how an internal control using homologous amplimers will add more

Table 2.1 Definitions and types of standard, control and reference nucleic acids.

Type	Second amplicon type: Standard	Control	Reference	
	A standard is a nucleic acid preparation that has a set or known concentration. Unknown samples may be compared to one or more standards and an absolute value for the unknown can be determined by numerical interpolation/extrapolation from experimentally determined values	A control is an amplicon added to the PCR and allowed to amplify to verify the integrity of one or more reagent(s) in the cocktail	A nucleic acid target that is used to compare the sample for either a qualitative or relative quantitative analysis.	
External: amplified in a different well	The standards are amplified separately from the target species in different wells on the same instrument*	"Batch control" This involves the use of a control in a separate well on the instrument. This approach cannot control for false negatives in different samples. It controls only for successful thermal cycling and the integrity of a common core reaction mix*	A related nucleic acid that is used for a qualitative comparison of amplification.	
Internal: amplified in the same well	Homologous		Exogenous	This is an internal reference that is added to the sample and is co-amplified with the target nucleic acid using a second set of amplimers
	A competitive standard that has the same amplimers as the specific target	A competitive control that is included in the same well of the instrument that has the same amplimers as the specific assay and controls for the core reagents and primers, and the thermal cycling		
	Heterologous		Endogenous	A 'housekeeping gene'. This is usually a gene that is naturally present in the sample at constant levels such that the up or down regulation of the target species can be relatively quantified using a second set of amplimers.
	A non-competitive standard with different amplimers to those of the specific target	A control that is included in the same well of the instrument but has different amplimers to that of the specific assay that controls for both the thermal cycling and the core reagents		

*External standards and controls may also be homologous, or heterologous, to avoid the risk of cross-contamination.

confidence to a result by confirming that conditions in the PCR are adequate for allowing amplification to occur. However, this controls only for the amplification and, where a nucleic acid reporter probe is also involved, this has to be controlled for as a separate factor.

Strategies for development of internal controls and their analysis

The strategy for detecting an internal control depends upon the instrument's capabilities and the fluorescent reporting chemistry. The simplest approach is to use a sequence-specific probe for the IC with a reporting dye of different emission wavelength to that of the probe for the specific target. This requires that the emission wavelengths of the dyes are either spectrally separated enough to not overlap, or that the emissions can be separated by the instrument through the means of a dye de-convolution algorithm. For example on the ABI 7700, a 5′ nuclease assay may use FAM as the specific reporter probe and a dye such as VIC, JOE or HEX as the reporter for the control probe. The cycle-by-cycle increase in the emissions of the dyes reports amplification of the respective species. This strategy may be applied to virtually all fluorescent chemistries and instruments.

The signal from the IC does not necessarily need to be quantitative since it serves only to report the integrity of the PCR reagents and the amplification conditions. The ability to detect the species at the end of thermal cycling should suffice for most applications. For assays based on hybridization run on an instrument that can determine melting point (T_m), discrimination between target and control species can be obtained by designing the control amplicon and/or probe to have a T_m different to that of the specific nucleic acid target. This may be achieved using DNA-binding dyes for amplicons with different melting points (Lee *et al.*, 1999), or labelled probes for amplicons with different internal sequences. When using specific probes melting point can be used in tandem with the use of different emission wavelengths. The merits of sequence-specific probes over DNA binding dyes have been discussed in Chapter 1. Whilst the use of DNA binding dyes as probes has some disadvantages in terms of specificity relative to sequence-specific probes, the correlation between amplification of the IC and the specific target is enhanced as they are detected through by the same signalling reagent. Equally, the target sequence with a polymorphism introduced into it can used in the design of the IC so that a single hybridization probe can be used for both IC and target amplicons with T_m used to differentiate the binding partner. This represents one of the few approaches where a sequence-specific reporter probe can be controlled for as well as the amplification. In this embodiment the function of the probe function can be better controlled than in approaches with independent probes for target and IC. Using dual hybridization probes in either quantification and/or melting point analysis it is possible to use one common probe, either the donor or the acceptor, to provide a better approach for control of sequence-specific probes. The methods for generating these mimics are now briefly described.

Synthesizing and optimizing molecular mimics for use as internal controls

There are several recombinant DNA techniques that can be used to generate a homologous internal control mimic. In each case the objective is to create a synthetic target that can co-amplify with the specific amplicon and that has a 'motif' which may be easily distinguished from that of the specific amplicon. This distinction can be achieved by the use of dyes with different emission wavelengths and/or a shift in the melting point as described above. Several methods have been reported. These include: insertion of the specific target into a vector and insertion/deletion of a sequence between the amplimer regions (Ursi *et al.*, 1992; Zimmerman *et al.*, 1996; Brightwell *et al.*, 1998), the amplification of a generic target from a related sequence (deWit *et al.*, 1993), the use of overlapping and tailed PCRs (Müller *et al.*, 1998; Sachadyn *et al.*, 1998), site-directed mutagenesis (Nash *et al.*, 1995), random misprimed products (Mergenthaler *et al.*, 2004) and polymerase extension of oligonucleotides and subsequent PCR (Rosenstraus *et al.*, 1998). In addition a complete synthesis of the gene is possible for smaller amplicons (Zimmerman *et al.*, 2000) and some commercial suppliers will now make the gene and supply it in a vector ready for use. Since the ability to discriminate amplicons by size is not a requirement, the amplicon may be similar, or the same in size, and thus achieve a similar amplification efficiency. Gene synthesis is now available commercially and so the entire sequence may be requested in a choice of vectors. The only constraints are cost and time. However, two main techniques that we use to generate mimics are:

- Direct insertion of amplimer and probe sequences into cloning vectors using recombinant techniques. This can be done by amplifying amplimer/probe sites, or the use of oligonucleotide linkers. In either case the use of restriction overhangs will assist in the correct directional insertion of the fragments.
- Generation of targets using a PCR approach of amplimers tailed with specific amplimer sequences to amplify alternative probe regions. The products contain the specific amplimer sequences incorporated into the PCR product. This product may be subsequently cloned into the desired vector (Fig. 2.1). The use of the additional overhangs with unique restriction sites will assist in directional cloning that may be required for some applications (e.g. sense-specific RT-PCR).

For reverse transcriptase PCR (RT-PCR) a mimic may be made using the methods described and subsequent insertion of the product into a vector. *In-vitro* transcription using the T7/T3 promoters allows the RNA molecule to be then synthesized in quantity. The use of modified nucleotides that make the transcripts resistant to RNase(s) could form one strategy that would allow for a more stable RNA control (Capodici *et al.*, 2002). Alternatively, the sequence could be cloned into an RNA virus such as the bacteriophage MS2. Armoured RNA™ is a product available from Ambion INC (Pasloske *et al.*, 1998). This is a method for packaging recombinant RNA into pseudoviral particles using a plasmid system whereby the MS2 bacteriophage coat protein is located downstream of the promoter and the target gene of interest. RNA encoding the coat protein and the target sequence is transcribed under the control of the inducer. Coat protein is translated and subsequently binds to recombinant RNA to form the product. The packaged molecule is resistant to RNase digestion yet easily extracted for amplification. The process has been shown to stabilize the RNA compared to *in vitro* transcripts made from plasmid and is an ideal method for RNA IC production. The authors would expect that any recombinant RNA virus containing the control sequence would improve the stability of the nucleic acid over naked RNA and an alternative protected RNA control system has been developed based on cowpea mosaic virus (King *et al.*, 2007). It should be noted that for providing a robust and inexpensive exogenous IC for RT-PCR extraction or amplification control native MS2 sequences have proved to be useful (Dreier *et al.*, 2005).

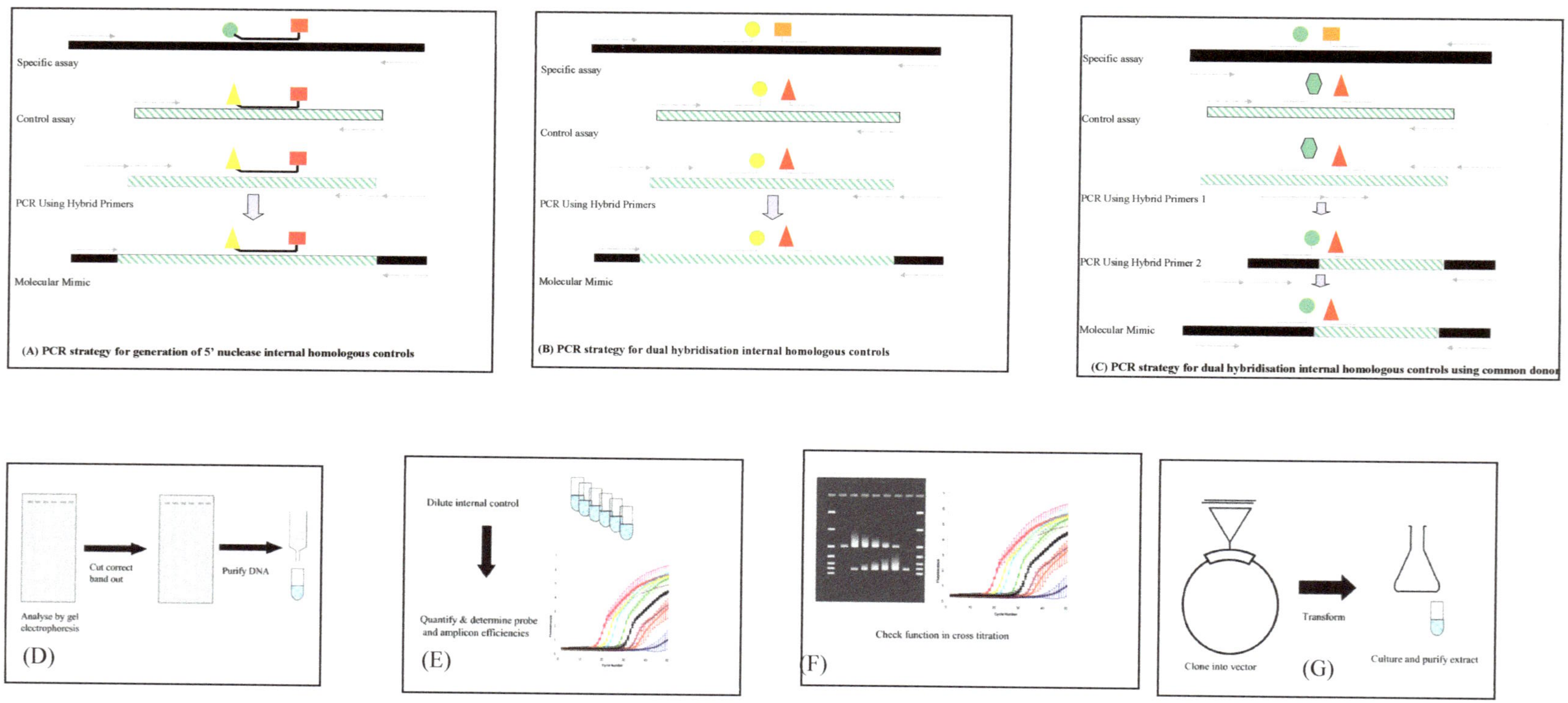

Figure 2.1 Schematic showing strategies for rapidly generating and evaluating competitive internal homologous controls by PCR. (A) For assays such as the 5′ nuclease where there is only one probe. (B) For dual hybridization assays using two additional probes. (C) Dual hybridization probes using one common probe for both the control and specific assay. (D) Amplify the IC using PCR and extract the correct band from the gel. (E) Serially dilute the product and determine the amplification and probe efficiencies (Fig. 2.2). (F) Cross-titre the IC with specific target to assess the efficacy of the IC in multiplex PCR before cloning into a suitable vector. (G) Transformation, culture, extraction and use.

Experimental considerations

In designing an experiment the sequence of the specific target and the assay are important considerations. (Reference should be made to Chapter 1 covering the utility, design and sequence dependency of different reporting chemistries.) The design of the assay is best facilitated by software tools that can automatically select amplimers and probes and predict (score) amplimer–amplimer and amplimer–probe interactions for the multiplexed assay. Here we highlight some of the considerations that we consider important in achieving a successful IC implementation.

Control sequence

The choice of control sequence is the most important factor since the other considerations are mostly dependent on this. With nucleic acid probes one suggested approach is to utilize a sequence that has been previously reported/optimized, or that is commercially available for the same assay type. For example, the ABI human β-actin sequence for a 5′ nuclease assay is useful because the probe is efficient and available as a kit. Ideally, the probe sequence should not be of the same origin (or endogenous to the sample) as the target for the specific test. For example, it would not be appropriate to use the β-actin sequence for a human test. Providing the probe sequence has been shown to work, any other sequence would be suitable, but preferably the GC content for the sequence should be similar to that of the specific target.

Amplicon efficiency

Whilst the differences in amplification efficiencies of the target and IC may be compensated for by changing the concentration of the initial target, they should ideally be similar and can be determined empirically. This is most important if the mimic is to be used for competitive quantification since the amplification efficiency can vary significantly between amplicons (Zimmerman *et al.*, 1996). It is useful to include native sequence either side of the amplimer when designing the mimic to ensure that the effects of neighbouring sequence is maintained. The amplification efficiency is best determined using a DNA binding dye such as SYBR®Green-1 combined with an efficient hot-start to amplify a broad range of concentrations across the dynamic range of the assay. Whilst using a probe directly to do this is satisfactory, checking the amplification efficiency without the probe allows it to be measured directly and is useful in understanding the efficiency of the probe when it is later evaluated. It will avoid dismissing a satisfactory reaction for the sake of iterating the probe design and/or as a result of a failure in probe synthesis. The efficiency (E) can be calculated by using the slope value for the derived standard curve using (Fig. 2.2);

$$E = 10^{-1/\text{slope}}$$

The efficiency of different amplicons for the same (or similar) dilution series can then be directly determined.

Probe efficiency

The efficiency of a sequence-specific probe is important if the signal from the IC is to be commensurate with that from the target of the assay. Probe efficiency can be affected by a number of factors that are discussed in Chapter 1. The probe sequence to be chosen

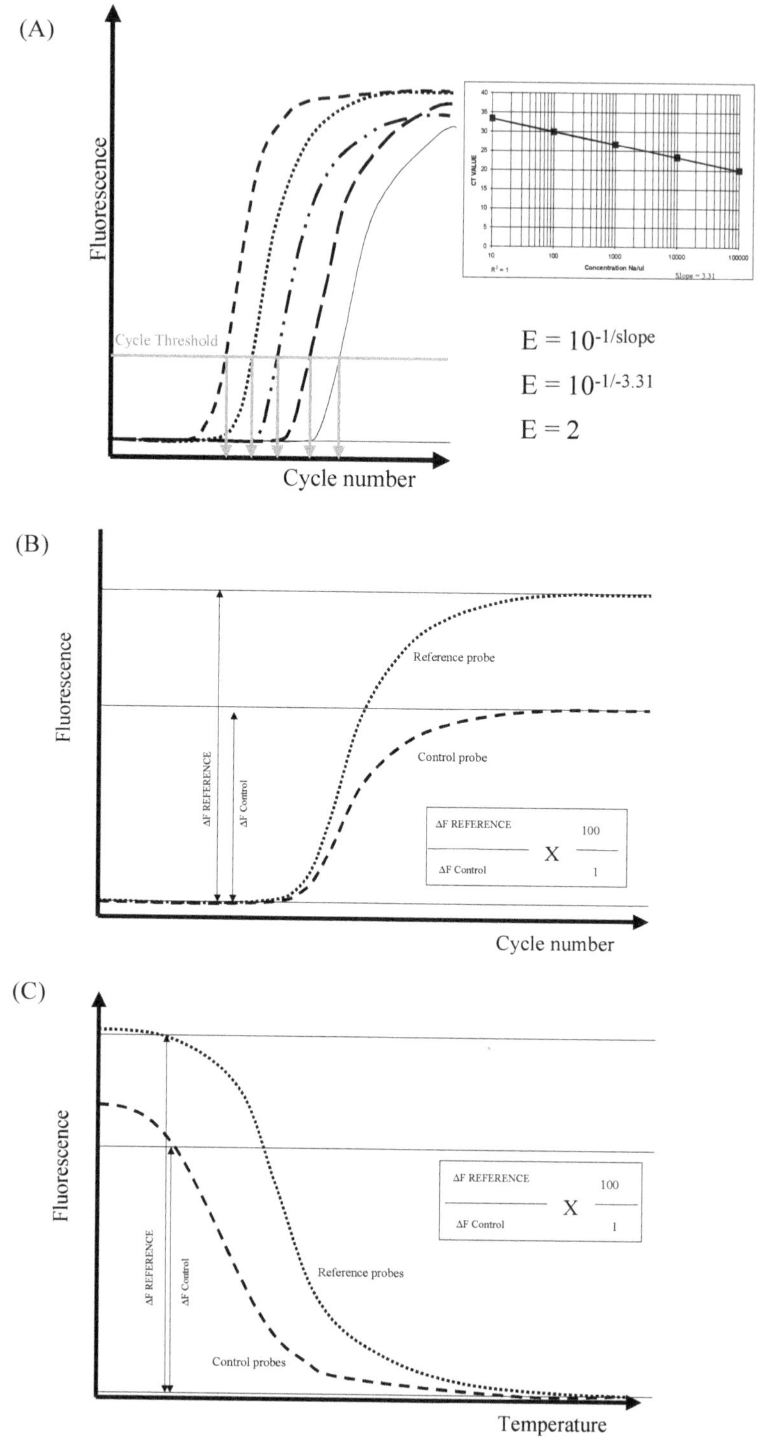

Figure 2.2 Determining and comparing amplification and probe efficiencies of different amplicons. (A) Determining amplification efficiency (E). Ideally, using a generic method such as a DNA binding dye, e.g. SYBR® Green I, the cycle threshold (CT) values are determined for a dilution series of target material over a range of concentrations and the plot of fluorescence intensity against cycle number generates the amplification curves. The derived CT values can be used to generate a standard curve. The slope of the curve can then be used to calculate E for the range tested. This value can be used to compare E with other amplicons using

should be checked for complementarities to other probes, amplimers and amplicons in the reaction. This is most important when the probe melting point approach is used since the signal from a control probe that is too large may swamp a low signal obtained in positives. When assessing potential probes for the mimic sequence it is useful to score the probe efficiency relative to a reference assay. For example, the percentage of signal–noise ratio for both 5′ nuclease assay specific and control probes can be compared to that of another assay such as the ABI Human β-actin probe. For this to be done the reaction for either probe must be taken to completion (plateau). These data may be expressed as:

$$\frac{\text{Fluorescence signal value specific assay/control assay at (cycle 50 – cycle 1)}}{\text{Fluorescence signal value } \beta\text{-actin (cycle 50 – cycle 1)}} \times \frac{100}{1}$$

This provides a comparative means for matching candidate control sequences to given assays. Likewise, the same may be done for comparing dual hybridization probes using:

$$\frac{\text{Fluorescence signal value specific assay/control assay at } (95^{\circ}\text{C}–60^{\circ}\text{C}^{1})}{\text{Fluorescence signal value } \beta\text{-globin } (95^{\circ}\text{C}–60^{\circ}\text{C}^{1})} \times \frac{100}{1}$$

This approach may be carried out using any reference probe and is equally applicable to other assay systems to estimate the relative probe efficiency (see Fig. 2.2). The quotient used for comparison should be based on the values derived from the maximum signal (when the target is in excess of the probe) subtracted from the minimum signal (the background) for each probe. The minimum signal is usually at the start of the amplification or when the probe is dissociated from the target at elevated temperatures.

Control vector

The choice of control vector is important since the vector may affect the stability and amplification efficiency. Plasmid constructs are commonly used and the efficiency of amplification may drop dramatically at lower target numbers (< 100 copies per reaction). Linearizing the plasmid by digestion at a unique restriction site in a vector can improve amplification efficiency. The efficiency of closed plasmids is often seen to drop off dramatically at ~ 100 to 1000 copies. The use of other vectors such as bacteriophage λ vectors and M13 (ssDNA rescue by co-infection with helper phage) may also be useful approaches for generating large amounts of linear IC molecules. PCR product itself is fine for use as a control although a vector represents the best method of safe storage of the desired sequence. We recommend the initial amplification of control DNA by PCR since the products may be used to evaluate potential mimics prior to the cloning of the molecule

---

the same or similar dilution series. (B) Determining probe efficiency. Using the 5′ nuclease assay and other assays by comparing the total change in fluorescence (as a function of a reference probe/assay) when the reaction is allowed to go to completion. (C) Using probes based on hybridization by comparing (as a function of a reference probe/assay) the change in fluorescence when the probe is hybridized (at a specific temperature) to when it is unhybridized (at high temperatures).

into a vector. However, the use of an *in-vivo* amplification method for the final synthesis of DNA to be used for controls is superior in that individual clones may be readily sequenced to check their correct identity and this process avoids the spurious artefacts that can be generated in PCR. These artefacts can be minimized during the evaluation stages by cutting the predicted PCR band out of an agarose gel and using any one of a number of methods for cleaning up the product. These include column purification and/or dialysis, before dilution and evaluation as suitable mimics.

Number of molecules

When using a competitive control the number of molecules to be added into the reaction must be determined empirically so that the required sensitivity of the specific assay is not compromised. The control must be amplified effectively in the absence of specific target. With increasing numbers of specific target molecules in the reaction the control amplification may be outcompeted. We have observed that it is possible to stop this if the T_m of the specific amplicon is higher than that of the control. Reducing the denaturing temperature of the thermal cycle (which is possible on some thermal cyclers) to below that of the specific target, but higher than that of the control, suppresses specific amplification thus allowing the IC to amplify. However, this is not necessary since the amplification of the control only serves to validate the reagents in the absence of the specific target. A cross-titration of specific target to control should be carried out to determine the correct amount of control molecules to include in a reaction. In practice, this amount is routinely determined to be between 10 and 100 copies per reaction if the specific assay is to maintain a sensitivity of ten or more copies of specific target. Accurate determination of the IC mimic concentration is therefore important. The use of fluorescent dyes and standards is inherently more accurate than ultra-violet optical density quantification at 260 nm. An alternative method for statistically determining the correct concentration of a stock of internal controls using PCR is provided by Rosenstraus *et al.* (1998). This method describes using a dilution series of the control preparation and performing multiple amplifications. At a given dilution the amplification will become stochastic owing to the presence of low or no copies of the template in the reaction in accordance with Poisson's law. The relationship between the average number of internal control copies in a dilution (C) and the probability (P_n) that no molecules exist in a sample of this dilution is given by:

$$C = -\ln(P_n)$$

where P_n is determined by counting the number of negative replicates used to calculate C.

Target nucleic acid

The control must be RNA if it is to be used for controlling the RT step. If a specific primer initiated RT step is used then utilizing the same primers in the mimics will control for both the RT and the PCR and negates the requirement for a second mimic for the PCR step in combined RT-PCR. Depending on the application, controls at both the RNA and DNA levels may be required in a two-step RT-PCR. The RT efficiency of the mimic and the specific target must be similar if the mimic is to be used as a standard in any application. When the RT-PCR is designed to be sense specific, directional cloning and the choice of vector orientation is important.

Dye/channel assignment

The most efficient dye and/or detector should be used for the specific target assay. This is of importance where a single excitation source is used in a universal acceptor arrangement (see Chapter 1), in which case it is usually the shorter wavelength dye that will be most efficiently excited. When a universal donor arrangement is used, it will be the acceptor to which energy is most efficiently transferred from the donor. In either multiplex arrangement, the optimum channel will depend on the instrument's specification and the dyes used.

Synthesis

If the IC is to be a synthetic oligonucleotide then the authors would recommend using a different supplier to the company that will be making your amplimers. The risk of cross-contamination by the IC of an amplimer is such as to have been reported numerous times in PCR discussion forums..

Summary and future improvements

The further development of instruments and methods of detection will increase the ability to multiplex PCR to higher levels in a quantitative manner. This introduces new possibilities for controlling for and/or normalizing sample extraction, and/or RT-PCR steps. This will be achieved through a combination of competitive (homologous) and non-competitive (heterologous) internal controls for each process. It is not always possible to control for these processes with one control since each may be subjected to different inhibitory factors. The ability to do this will improve the usefulness of the PCR for both research and diagnostic applications. The ability to control for all such processes will be of most use in emerging applications where remote testing in the non-laboratory environment necessitates the use of internal validation.

End note: causes of false-positives and measures to reduce their occurrence

False-positives are a significant problem in real-time PCR and in other analysis methods that depend upon nucleic acid amplification. They are often caused by cross-contamination from positive samples or reaction products, and sometimes from unwitting contamination of PCR reagents.

Preventative methods against cross-contamination are essential for eliminating the occurrence of false-positives. These are most effective when used in combination and include: good laboratory practice (see Espy *et al.*, 2006) in terms of segregating clean and contaminated materials and work areas, especially separation of PCR preparation and analysis areas; the use of closed tube assay formats to contain reaction products; using cabinets with UV treatment for PCR set-up; UV laboratory air scrubbers for removing aerosolized nucleic acids; and using uracil N-glycosylase carry over prevention chemistry (Longo *et al.*, 1990).

As mentioned above concerning synthesis of the IC, synthesis of assay target nucleic acids in the same facility as used to produce the amplimers should be strictly avoided. Contamination from workers in production plants making sampling kits, reagents and consumables has been known to cause false-positives in PCR testing of human samples. PCR users should also be aware that reaction components such as polymerase enzymes and hotstart antibodies will have been derived from natural sources that may contain, for example, *E. coli* or mouse DNA as contaminants.

In isothermal methods that are not dependent upon thermal cycling, amplification can commence as soon as reaction mixtures are brought together. This means that, for example, the pipette tips used to load the reaction vessels may well contain residues in which amplified products and by-products can be generated and thereby pose a contamination risk.

In batch assays, therefore, a no template control should routinely be incorporated in every run to control for false-positives.

References

Brightwell, G., Pearce, M., and Leslie, D. (1998). Development of internal controls for PCR detection of *Bacillus anthracis*. Mol. Cell. Probes *12*, 367–377.

Capodici, J., Kariko, K., and Weissman, D. (2002). Inhibition of HIV-1 infection by small interfering RNA-mediated RNA interference. J. Immunol. *169*, 5196–5201.

Dreier, J., Störmer, M., and Kleesiek, K. (2005). Use of bacteriophage MS2 as an internal control in viral reverse transcription-PCR assays. J. Clin. Micro. *43*, 4551–4557.

Espy, M.J., Uhl, J.R., Sloan, L.M., Buckwalter, S.P., Jones, M.F., Vetter, E.A., Yao, J.D.C., Wengenack, N.L., Rosenblatt, J.E., Cockerill, I.I.I., F.R., *et al.* (2006). Real-time PCR in clinical microbiology: applications for routine laboratory testing. Clin. Microbiol. Rev. *19*, 165–256.

Henley, W.N., Schuebel, K.E., and Nielsen, D.A. (1996). Limitations imposed by heteroduplex formation on quantitative RT-PCR. Biochem. Biophys. Res. Comm. *226*, 113–117.

King, D.P., Montague, N., Ebert, K., Reid, S.M., Dukes, J.P., Schädlich, L., Belsham, G.J., and Lomonossoff, P. (2007). Development of a novel recombinant encapsidated RNA particle: evaluation as an internal control for diagnostic RT-PCR. J. Virol. Methods *146*, 218–225.

Lee, M.A., Brightwell, G., Bird, H., Leslie, D., and Hamilton, A. (1999). Fluorescent detection techniques for real-time multiplex strand specific detection of *Bacillus anthracis* using rapid PCR. J. Applied. Microbiol. *87*, 218–223.

Lee, M.A., Leslie, D., and Squirrell, D.J. (2009). Internal and external controls for reagent validation. In Real-time PCR: Current Technology and Applications, Logan, J., Edwards, K., and Saunders, N., eds. (Caister Academic Press, Norfolk, UK), pp. 95–104.

Longo, M.L., Berninger, M.S., and Harley, J.L. (1990). The use of uracil glycosylase to control carry-over contamination in polymerase chain reactions. Gene *93*, 125–128.

Mergenthaler, E., Fodor, M., and Sűle, S. (2004). A new method to develop internal controls for the determination of phytoplasma concentration. Acta. Phyto. Ent. Hung. *39*, 1–7.

Müller, F.-M.C., Schnitzler, N., Coot, O., Kockelkorn, P., Haase, G., and Li, Z. (1998). The rationale and method for constructing internal control DNA used in Pertussis polymerase chain reaction. Diag. Microbiol. Infect. Dis. *31*, 517–523.

Mullis, K.B., and Faloona, F.A. (1987). Specific synthesis of DNA *in vitro* via a polymerase catalyzed chain reaction. Meth. Enzymol. *155*, 335–351.

Nash, K.A., Klein, J.S., and Inderlied, C.B. (1995). Internal controls as performance monitors and quantitative standards in the detection by polymerase chain reaction of herpes simplex virus and cytomegalovirus in clinical specimens. Mol. Cell. Probes *9*, 347–356.

Pasloske, B.L., Walkerpeach, C.R., Obermoeller, R.D., Winkler, M., and DuBois, D.B. (1998). Armoured RNA technology for production of ribonuclease-resistant viral RNA controls and standards. J. Clin. Micro. *36*, 3590–3594.

Rosenstraus, M., Wang, Z., ChangS.-Y., DeBonville, D., and Spadoro, J.P. (1998). An internal control for routine diagnostic PCR: design properties, and effect on clinical performance. J. Clin. Microbiol. *36*, 191–197.

Rossen, L., Norskov, P., Holmstrom, K., and Rasmussen, O.F. (1992). Inhibition of PCR by components of food samples, microbial diagnostic assays and DNA extraction solutions. International J. Food. Microbiol. *17*, 37–45.

Sachadyn, P., and Kur, J. (1998). The construction and use of a PCR internal control. Mol. Cell. Probes *12*, 259–262.

Saiki, R.K., Scharf, S., Faloona, F.A., Mullis, K.B., Horn, G.T., Erlich, H.A., and Arnheim, N. (1985). Enzymatic amplification of beta-globin genomic sequences and restriction site analysis for diagnosis of sickle cell anemia. Science *230*, 1350–1354.

Ursi, J.-P., Ursi, D., Ieven, M., and Pattyn, S.R. (1992). Utility of an internal control for the polymerase chain reaction. Application to detection of Mycoplasma pneumoniae in clinical specimens. APMIS *100*, 635–639.

Wilson, I.G. (1997). Inhibition and facilitation of nucleic acid amplification. Appl.Environ. Microbiol. *63*, 3741–51.

deWitt, D., Wooton, M., Allan, B., and Steyn, L. (1993). Simple method for production of internal control DNA for Mycobacterium tuberculosis polymerase chain reaction assays. J. Clin. Micro. *31*, 2204–2207.

Zimmerman, K., and Mannhalter, J.W. (1996). Technical aspects of quantitative competitive PCR. Biotechniques *21*, 268–279.

Zimmerman, K., Rieger, M., Groß, P., Turecek, P.L., and Schwarz, H.P. (2000). Sensitive single-stage PCR using custom synthesised internal controls. Biotechniques *28*, 694–702.

DOI: https://doi.org/10.21775/9781912530243.09

Oligonucleotide Primers and Probes: Use of Chemical Modifications to Increase or Decrease the Specificity of qPCR

9

Scott D. Rose, Richard Owczarzy, Joseph R. Dobosy and Mark A. Behlke

Abstract

Although the vast majority of primers and probes employed in qPCR applications today are synthesized using unmodified DNA bases, selective use of chemically modified bases and non-base modifying groups can prevent primer dimer artefacts, improve specificity, and allow for selective amplification of sequences that differ by as little as a single base. A wide variety of chemical modifications have been characterized for use in qPCR. As a general class, the modifications that are in greatest use today increase the binding affinity of the oligonucleotides (i.e. increase the melting temperature, T_m). T_m-enhancing modifications allows both primers and probes to be shorter, improving the differential T_m ($\Delta T_m = T_{m\ match} - T_{m\ mismatch}$) between perfect match and mismatch hybridization. These modifications have widespread application in allele-specific PCR and in the detection of single nucleotide polymorphisms (SNPs). Conversely, a second class of base modifications are in common use that decrease specificity and improve duplex formation in the presence of base mismatches. Although these modifications lower T_m, they have less of an impact on primer stability than do actual mismatched bases. Universal bases permit use of primers and probes in polymorphic loci when it is desirable to detect all sequence variants and minimize mismatch discrimination.

Introduction

The primers and probes used in real-time quantitative PCR (qPCR) are synthetic oligonucleotides and can be manufactured using natural DNA bases or can comprise RNA bases, artificially modified bases, or a variety of pendant chemical groups not found in nature. While most needs are well met using unmodified oligodeoxynucleotides, the use of chemical modifications can improve qPCR performance in a number of areas. This chapter will cover three specific areas of interest, with a primary focus on increasing specificity and allelic discrimination methods:

1. increasing specificity, with a focus on SNP discrimination;
2. preventing primer-dimer or other PCR artefacts;
3. decreasing specificity using universal bases.

Increasing specificity

Specificity relates to the ability of the qPCR reaction to generate a positive signal from a desired nucleic acid sequence while simultaneously not generating a signal from any similar yet distinct sequences also present in the target nucleic acid sample. In some instances this involves discrimination of related sequences within a gene family where multiple base differences are present. In its most extreme form, this involves discrimination of sequences that differ by only a single base, i.e. a SNP. In broadest terms, qPCR approaches to SNP discrimination can be categorized as those methods which employ biased amplification, where one allele is selectively amplified over other alleles, and those methods which do not employ biased amplification, where all related sequences are amplified equally and a detection probe is employed to distinguish base identity.

SNP discrimination using unbiased amplification

PCR-based methods that discriminate between related sequences using unbiased amplification employ primers placed in regions of the sequence that are identical between alleles, positioning the SNP between the two primers. Sequence identity is usually interrogated by differential hybridization with a third probe oligonucleotide. Success is entirely dependent upon having a sufficiently high ΔT_m between the 'match' and 'mismatch' targets so that the 'match probe' hybridizes to the target site during amplification and the 'mismatch probe' does not. A variety of different assay formats exist that employ this strategy, most of which show improved performance using chemically modified oligonucleotides.

Linear dual-labelled probes

One of the more popular formats for quantitative sequence detection is the 5′-nuclease assay, which was first described by Gelfand and colleagues (Cetus Corporation) in 1991 (Holland *et al.*, 1991). This assay employs a fluorescence-quenched dual-labelled linear probe that binds to the target nucleic acid during the annealing phase of each cycle and is degraded during the extension phase of PCR. The probe has a reporter fluorophore on one end and a quencher on the other end, which absorbs fluorescence emission from the reporter dye (the intact probe is 'dark'). The reaction is cycled using a two-step programme with incubation at a single temperature during which time both the primer annealing and polymerase extension steps occur. The T_m of a sequence is the temperature at which 50% of the possible duplexes are formed and 50% remains in the single-stranded state. Generally, the probe is designed to have a T_m 5°C or more above the annealing temperature so that it is almost completely hybridized to the target template strand during DNA synthesis. During elongation, the DNA polymerase encounters the probe which is quickly degraded by the enzyme's 5′–3′ exonuclease activity. Degradation physically separates the reporter fluorophore from the quencher, resulting in the appearance of a detectable signal (the degraded probe is 'bright'). This assay format is often referred to as the TaqMan® assay.

The 5′-nuclease assay was initially developed to measure gene expression levels; it was quickly adapted for use in allelic discrimination. This method allows for a convenient, homogenous closed-tube assay where the same primers are employed to amplify both alleles. Two probes are present in the reaction that are labelled using detectably distinct fluorophores, one specific for allele 'A' and the other specific for allele 'B'. The presence of one or both alleles in a sample can be detected in real time by monitoring rise of signal in two separate fluorescence channels (Livak, 1999). The probe oligonucleotides usually need to be 20 bases or more in length to be completely hybridized to the target nucleic acid at the 60°C annealing/extension temperature (precise length will of course vary with sequence). Unfortunately, single base mismatches in probe sequences this long are often insufficient to produce a large enough ΔT_m between match and mismatch probes to obtain clean separation of signal between alleles. This problem is easily solved by using probes that are made with T_m-enhancing modifications. Modifications that increase T_m permit use of shorter probes at the same reaction temperature. A single base mismatch has a greater impact on duplex stability in short sequences than in long sequences, so the ΔT_m between match and mismatch improves as probe length decreases. Many factors contribute to the T_m and ΔT_m and it is important to be able to accurately predict these values when designing assays.

The ability of a probe oligonucleotide to distinguish between two species solely on the basis of hybridization is a thermodynamic problem that is influenced by the sequence context, the precise match versus mismatch base pair being studied, the length of the probe, and physical parameters of the hybridization reaction, including temperature, probe concentration, and buffer composition. As a general rule of thumb, a good place to start is to design the 'match probe' to have a T_m around 3–5°C above the reaction temperature and the 'mismatch probe' to have a T_m around 5–7°C below the reaction temperature, giving a ΔT_m of 10°C between the match and mismatch hybridization events. For reactions run at a standard temperature of 60°C, this means selecting sequences with T_m values around 64–65°C and 54–55°C for the match and mismatch probes respectively. Obviously having the ΔT_m as large as possible is desirable; however, it is generally not necessary to exceed 10°C and sometimes in actual practice probes with a lower ΔT_m may still perform well. While empiric testing of different probe designs is sometimes necessary, the use of computer assisted design tools that employ the latest thermodynamic parameters can often be used to predict probe sequences that discriminate well without further optimization, or at least greatly accelerates the process of finding good probes.

Typically, reactions are performed at around 60°C with 250 nM probe in a buffer that contains ~50 mM monovalent cation (sodium or potassium), ~3 mM divalent cation (magnesium), and 0.8 mM deoxyribonucleotide triphosphates (dNTPs). It is important to include the total dNTP concentration in thermodynamic calculations since dNTPs bind Mg^{2+} ions and reduce the concentration of free Mg^{2+} present in the reaction. Buffer composition can vary and unfortunately many vendors do not disclose the exact salt content of their PCR reagents, which complicates probe design. Thermodynamic parameters have been determined that allow for the accurate prediction of the T_m of short DNA oligonucleotides (SantaLucia *et al.*, 1996, 1998). In addition, the stabilizing effects of both monovalent and divalent cations are well characterized (Owczarzy *et al.*, 2004, 2008a), and the relative contribution of some fluorescent dyes and quencher have been measured (Moreira *et al.*, 2005). Sufficient detail is known about the thermodynamic effects of DNA base mismatches to model hybridization events of the kind needed to design SNP probes (SantaLucia and Hicks, 2004). A variety of algorithms and sequence analysis tools are available as part of commercial software packages or are freely accessible on the internet and that can help with design of both primers and probes for qPCR applications, including SNP detection (Owczarzy *et al.*, 2008b).

Early approaches to increase T_m and improve SNP discrimination through use of chemical modification focused on alterations of the sugar and base groups. Although 2′-*O*-methyl RNA (2′OMe RNA) is often thought of as being a T_m-increasing modification, it actually has increased affinity when binding to an RNA target while the T_m effects with a DNA target are variable. Use of this modification was not found to improve probe performance. In fact, extensive modification with 2′OMe RNA can impair nuclease cleavage of the probe and can be detrimental (Kuimelis *et al.*, 1997). Propynyl pyrimidines are a class of modified base where a propyne group is attached at the C5 position of the pyrimidine ring. This modification was first used to increase binding affinity of antisense oligonucleotides (Wagner *et al.*, 1993), but *in vivo* toxicity limited utility of this application. The propynyl pyrimidines, however, are quite effective in increasing T_m and improve performance of probes used in the 5′-nuclease assay in qPCR (Kuimelis *et al.*, 1997); each base substitution leads to a 0.5–1.5°C increase in T_m. Nearest neighbour parameters have not been reported for these bases; thus T_m predictions cannot be accurately made and empirical testing usually must be done when designing probes of this type. Unfortunately, these modifications are only available for pyrimidines, limiting benefit to pyrimidine-rich probe sequences. To be maximally useful, T_m-enhancing base modifications should be available for all four nucleotides. Structures of the various modifications discussed in this chapter are shown in Fig. 8.1.

Locked-nucleic acids (LNAs) are a T_m-enhancing modification simultaneously developed by Jesper Wengel and his group at the University of Southern Denmark and by Takeshi Imanishi and his group at the University of Osaka. LNAs have a methylene bridge that connects the 2′-O with the 4′-C of the ribose. This bicyclic structure locks the ribose in the 3′-*endo* conformation and increases T_m by 2–6°C per substitution (Kvaerno and Wengel, 1999; Obika *et al.*, 1997; Singh *et al.*, 1998; Singh and Wengel, 1998). LNA nucleotides are commercially available for all four nucleotides (guanine, adenosine, 5-methylcytosine, and thymidine) and have found widespread use in many applications, including antisense, RNA interference, and PCR. Substitution of LNA nucleotides for DNA in probes allows use of much shorter sequences at the standard 60°C reaction temperature and these shorter probes show a significant improvement in the ΔT_m between match and mismatch hybridization compared with parent unmodified (longer) sequences (Letertre *et al.*, 2003; Costa *et al.*, 2004; Johnson *et al.*, 2004; Ugozzoli *et al.*, 2004). The early reports from 2003–2004 which first demonstrated the utility of the LNA modification in dual-labelled hydrolysis probes substituted LNA nucleotides for DNA bases using a variety of different modification patterns with a primary focus on simply improving T_m and reducing probe length. This design strategy was effective; however, no systematic attempt was made to optimize the specific pattern of LNA modifications around the SNP site to maximize the ΔT_m between match and mismatch. In fact, the SNP site was often avoided for modification because it was uncertain if the increased binding affinity of the LNA nucleotide would improve or hurt SNP discrimination if placed directly at the site of mismatch. This question was recently addressed by Owczarzy and colleagues in 2006 (Integrated DNA Technologies), who carefully studied the effect that placement of LNA nucleotides at and around the SNP site had on T_m. Melting temperatures were experimentally measured for a large series of modified and unmodified oligonucleotides hybridized to perfect match or single base mismatch complements. The test set included all possible base variations at the SNP site and the adjacent bases, comparing the effects that LNA substitution had at these sites (You *et al.*, 2006). It was found that simply placing three consecutive LNA nucleotides spanning the SNP site (with the SNP positioned centrally) gave the maximum ΔT_m for all base combinations except for one, where the mismatch is a G:T pair and the guanine base is in the probe. In the single G:T exception, LNA nucleotides actually stabilized the

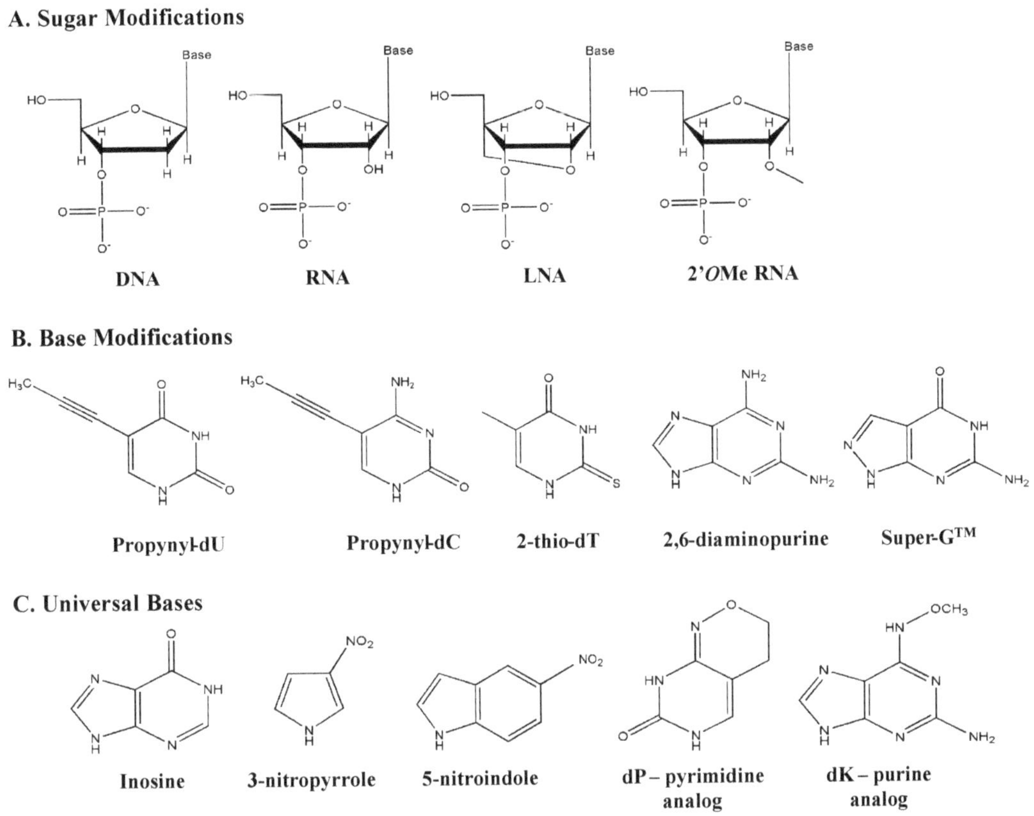

Figure 8.1 Oligonucleotide modifications. (A) Modifications to the ribose. DNA, RNA, LNA (locked nucleic acids), and 2′-*O*-methyl (2′OMe) are shown. (B) Base modifications. Propynyl-dU, propynyl-dC, 2-thiothymidine, 2,6-diaminopurine (2,6-DAP, or 2-amino-dA), and Super-G™ are shown. (C) Universal bases. Inosine, 3-nitropyrrole, 5-nitroindole (5-NI), dP, and dK are shown.

mismatch relative to the unmodified sequences, making SNP discrimination worse. LNA modifications can still be used to improve probes having a G:T mismatch by positioning the LNA nucleotides in flanking regions of the probe and reducing probe length. Alternatively, the probe can be designed to the other strand, in which case the mismatch is a T:G pair (with the thymidine base in the probe), and the standard 'LNA triplet' design strategy can be used.

When using the 'LNA triplet' modification pattern described above, it is important to balance the actual T_m of the modified probe so that 'match' hybridization stays around 64–65°C (i.e. 4–5°C above reaction temperature) by trimming bases from each end of the parent unmodified sequence to compensate for the increased binding affinity of the LNA nucleotides. Additional LNA nucleotides can be substituted in the flanking regions of the probe to allow further shortening of the probe sequence, which will continue to improve mismatch discrimination. Ideally, this process would be done using design software that calculates the T_m of different modification patterns and automatically optimizes design. Assistance with estimating the T_m of LNA-modified oligos and design of SNP-discriminating probes can be found on the Exiqon website at: http://lnatools.com/. The interactive tools on this site employ thermodynamic calculations that use experimentally determined parameters and unpublished proprietary rule sets for probe design. Unfortunately, enough inaccuracy exists in these calculations that empiric testing of different probe designs is often required. Experimentally determined nearest neighbour (NN) parameters were published for isolated LNA nucleotides in LNA/DNA hybrids by Celadon Laboratories in 2004 (McTigue *et al.*, 2004). Additional parameters are needed to enable predictions to be made with sufficient accuracy for the T_m and ΔT_m of LNA containing sequences to design probes for use in SNP assays, including parameters for sequential LNA nucleotides and LNA/DNA mismatches. These missing parameters were recently obtained from studies performed at Integrated DNA Technologies (IDT) and were published in 2011 (Owczarzy *et al.*, 2011); the resulting new algorithms enable the accurate calculation of the T_m of complex LNA-modified oligonucleotides of this type and can be accessed at: http://www.idtdna.com/analyzer/Applications/OligoAnalyzer/ or http://biophysics.idtdna.com.

A SNP in the human *SMAD7* gene (NM_005904) was employed here as a model site to demonstrate these principles. This SNP comprises a C/T variation (rs4939827) and the C-SNP is associated with increased risk of colorectal carcinoma (Tenesa *et al.*, 2008). Genomic sequence at this SNP site is shown in Fig. 8.2A. Probes were designed using existing DNA NN parameters and the new LNA match/mismatch parameters described above, keeping the T_m for the perfect match around 65°C. Calculations were made for a reaction mix having a probe concentration of 250 nM, 50 mM KCl, 3.0 mM $MgCl_2$, and 0.8 mM dNTPs (equivalent to a final free Mg^{++} concentration of 2.2 mM). Unmodified probes were designed as a control and modified probes were designed using the LNA 'triplet pattern' at the SNP site. Additional LNA residues were placed in flanking sequence as needed to allow the probes to be shortened from 20–22 bases (for the unmodified probes) to 14 bases; final LNA content ranged from three to four bases. Sequences of the probes designed are shown in Fig. 8.2B. Although not necessary for the sequences studied here, note that probes ending in a 5′-G base should have an additional non-complementary base inserted between the G and the reporter dye, to reduce the impact of G-quenching (which is particularly problematic for fluorescein) (Kaji *et al.*, 2009). Sequences of the SMAD7 probes are shown in Table 8.1. T_m values for the mismatch alleles were calculated using the new thermodynamic parameters and are also shown in Table 8.1. The predicted T_m values for the mismatch targets ranged from 50.9–55.5°C for the LNA-modified probes and resulted in a ΔT_m (match minus mismatch) of 10.0–15.7°C. In contrast, the unmodified probes showed predicted T_m's for the mismatch target ranging from 59.6–64.6°C, with a ΔT_m for match minus mismatch of only 1.5–5.6°C. Accuracy of the new T_m algorithms employed was tested by performing melting curve analysis on the SMAD7 probes. Experimental T_m values were obtained for both match and mismatch targets.

Melting experiments were conducted on a Beckman DU650 spectrophotometer (Beckman-Coulter) with a Micro T_m Analysis accessory, a Beckman High Performance Peltier Controller (to regulate the temperature), and 1 cm path-length cuvettes. UV-absorbance values at 268 nm wavelength were recorded using a PC interfaced to the spectrophotometer and temperature was increased at 0.1 degree increments. Pre-annealed duplexes were melted at 2 μM concentration in 10 mM Tris pH 8.3 (at 25°C), 50 mM KCl, and 2.2 mM $MgCl_2$ (equivalent to a PCR reaction buffer containing 3.0 mM $MgCl_2$ with 0.8 mM dNTPs). The melting profiles were analysed using methods that have been previously described (Owczarzy, 2005). Results are shown in Table 8.1. Experimental melting data were obtained using 2 μM duplex, which will raise T_m by 2–4°C compared

A) ***SMAD7*** **SNP locus, rs4939827**

```
GGGGCTGTCTCTAATCCACCATGCTCACAGCCTCATCCAAAAGAGGAAA(C/T)AG
GACCCCAGAGCTCCCTCAGACTCCTCAGGAAACACAGACAATGCTGGGGTTTAGAG

rs4939827 For
GGCTGTCTCTAATCCACCAT

rs4939827 Rev
CCCAGCATTGTCTGTGTTTC
```

B) Probe Alignments

```
5'        FAM-AGGAAACAGGACCC-FQ            3'
5'    FAM-AAAGAGGAAACAGGACCCCA-FQ          3'

5' CATCCAAAAGAGGAAACAGGACCCCAGAGCTCCCT 3'
3' GTAGGTTTTCTCCTTTGTCCTGGGGTCTCGAGGGA 5'

3'          MGB-CTTTGTCCTGGGGT-FAM         5'


5'        HEX-AGGAAATAGGACCC-FQ            3'
5'     HEX-AAGAGGAAATAGGACCCCAGAG-FQ       3'

5' CATCCAAAAGAGGAAATAGGACCCCAGAGCTCCCT 3'
3' GTAGGTTTTCTCCTTTATCCTGGGGTCTCGAGGGA 5'

3'          MGB-CTTTATCCTGGGGT-VIC         5'
```

Figure 8.2 Assays for the C/T SNP rs4939827 in the human *SMAD7* gene. (A) Genomic sequence of the region surrounding the C/T SNP site is shown. For and Rev primer sequences are shown and their binding sites are underlined. (B) Both strands of the genomic sequence of the 'C' and 'T' alleles are shown aligned with the different probe sequences that were tested. The SNP site is indicated in larger font. DNA bases are uppercase. LNA nucleotides are underlined. Reporter dyes employed (FAM, HEX, VIC) are indicated. FQ indicates the Iowa Black FQ dark quencher. MGB indicates a minor groove binder modification group linked to an Eclipse dark quencher.

Table 8.1 Predicted versus measured T_m values for SMAD7 probes

	Predicted T_m (°C) (0.25μM)			Measured T_m (°C) (2.0μM)		
Probe sequence	Match	Mismatch	ΔT_m	Match	Mismatch	ΔT_m
5′ FAM-AGGAAACAGGACCC-FQ 3′	66.6	50.9	15.7	68.3	54.1	14.2
5′ FAM-AAAGAGGAAACAGGACCCCA-FQ 3′	65.2	59.6	5.6	65.5	56.4	9.1
3′-MGB-CTTTGTCCTGGGGT-FAM 5′	66.0	NA	NA	65.7	56.5	9.2
5′-HEX-AGGAAATAGGACCC-FQ 3′	65.5	55.5	10.0	67.6	59.1	8.5
5′-HEX-AAGAGGAAATAGGACCCCAGAG-FQ 3′	66.1	64.6	1.5	65.1	63.0	2.1
3′-MGB-CTTTATCCTGGGGT-FAM 5′	66.0	NA	NA	67.4	57.7	9.7

Probes specific for the 'C' allele were labelled with FAM at the 5′-end. Probes specific for the 'T' allele were labelled with HEX or VIC at the 5′-end, as indicated. The SNP site is indicated in larger, bold font. LNA nucleotides are underlined. FQ indicates the Iowa Black FQ dark quencher. MGB indicates a minor groove binder modification group linked to an Eclipse dark quencher. The MGB probes were designed to the opposite strand and are accordingly shown in reverse orientation. Note that predicted T_m values generated using a probe concentration of 0.25μM were used for probe design; measured T_m values are done at 2μM duplex concentration to accommodate needs for spectrometric methods. NA=mismatch predictions not available.

with the 0.25 μM duplex concentration used in qPCR (0.25 μM was used for the original T_m predictions when designing the probes). Measured T_m values were found to all be within a few degrees of the expected values for three of the four probes, well within expected experimental error for these methods. The unmodified 'C'-SNP probe, however, showed a lower than predicted T_m against the mismatched target, resulting in a ΔT_m of over 9°C.

The SMAD7 probes and primers shown in Fig. 8.2 and Table 8.1 were functionally tested in real-time qPCR. Reactions were run on a BioRad CFX384 in 384 well plates using 10 μl reaction volumes. Reactions consisted of either 1X Taqman Genotyping Master Mix (Life Technologies) or a homemade Immolase Master Mix (0.4 U Immolase DNA polymerase (Meridian Bioscience, Cincinnati, Ohio), 1 μl of 10× polymerase buffer (160mM $(NH_4)_2SO_4$, 670 mM Tris-HCl pH 8.3, 0.1% Tween-20), 3 mM $MgCl_2$, 800 μM dNTPs), 250 nM of each probe, 500 nM of the forward and reverse primers, and 2 ng human genomic DNA. Cycling conditions included an initial 10 minutes soak at 95°C, followed by 40 cycles of 15 seconds at 95°C and 60 seconds at 60°C. Human genomic DNAs that were homozygous C/C or T/T or heterozygous C/T at SNP locus rs4939827 were obtained from the Coriell Institute for Medical Research Cell Repository (http://ccr.coriell.org/, GM07019 (T/T), GM07029 (C/T), GM07034 (C/C)). 'C'-SNP results were collected in the FAM fluorescence channel and 'T'-SNP results were collected in the HEX fluorescence channel. Amplification plots are shown in Fig. 8.3. The LNA-modified probes (Fig. 8.3a) resulted in unambiguous base

A. LNA Modified Probes

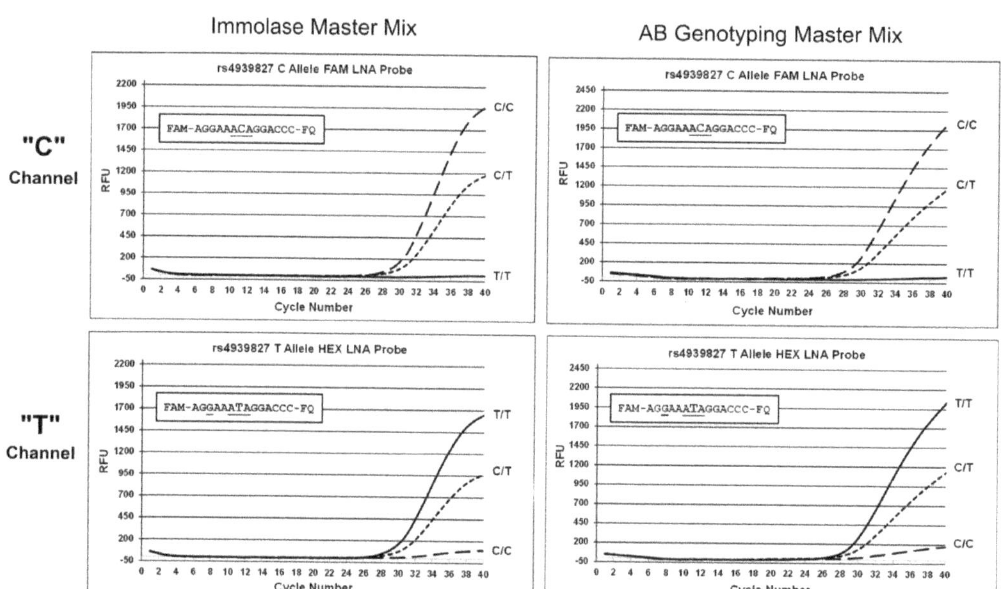

Figure 8.3 5′-Nuclease assay amplification plots for different probe designs at a SNP locus. The 5′-nuclease qPCR assay was performed using 2ng of human genomic DNA homozygous or heterozygous at the SMAD7 rs4939827 locus using probes of different design and two different buffer systems, the Immolase Master Mix or the Applied Biosystems (AB) Genotyping Master Mix. Reactions were run in multiplex format with probes specific for both alleles present in the reaction and were monitored in separate fluorescent channels. Probe

B. Std Unmodified Probes

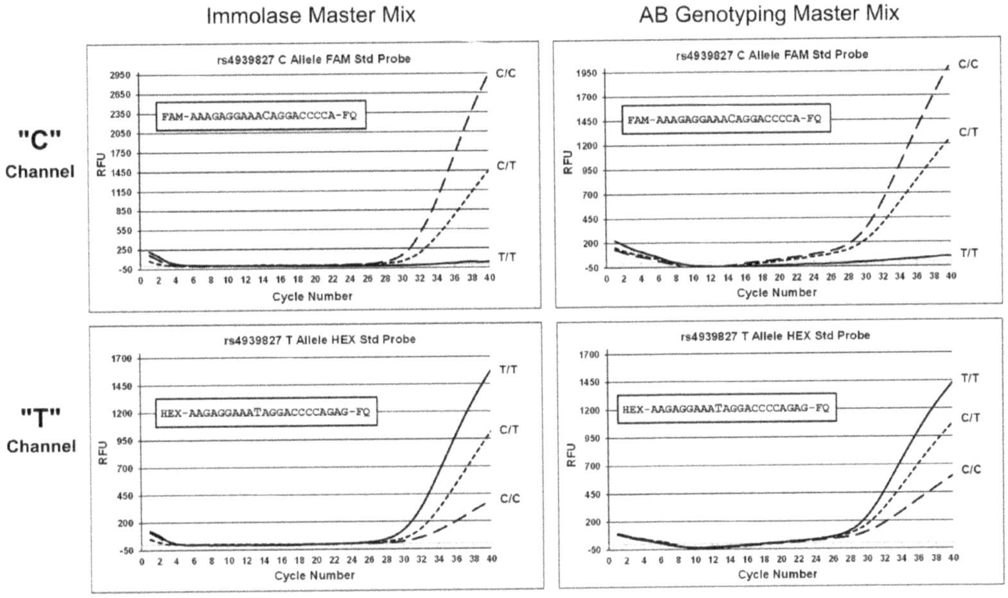

C. MGB Modified Probes

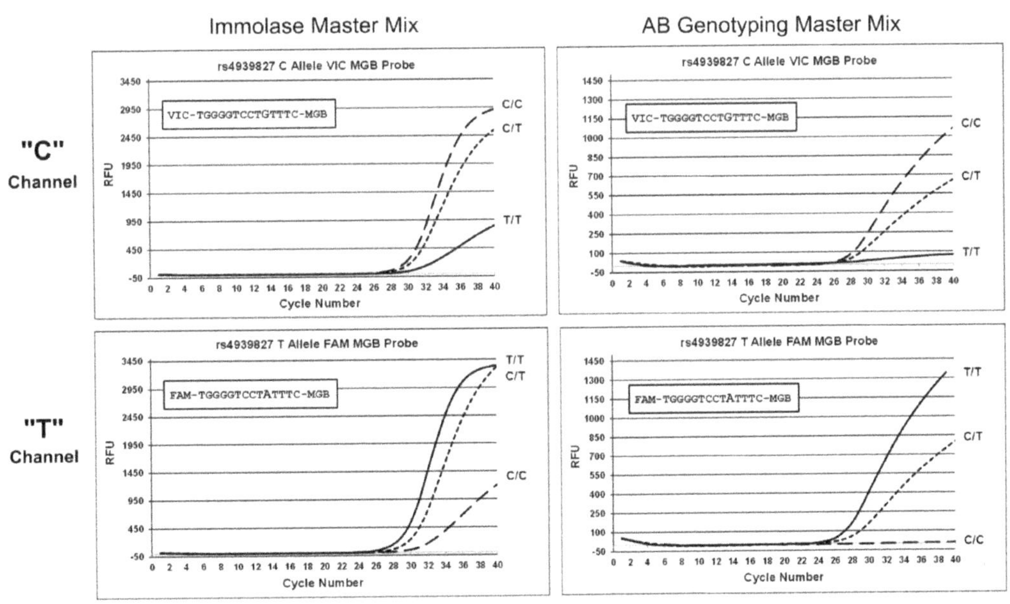

sequences are inset at the upper left corner of the amplification plots. Reporter dyes FAM, HEX, and VIC are indicated. FQ=Iowa Black FQ dark quencher. MGB=minor groove binder with Eclipse dark quencher. LNA nucleotides are underlined and the position of the SNP base is indicated in larger font. (A) Reactions using LNA-modified dual-labelled probes. (B) Reactions using unmodified DNA dual-labelled probes. (C) Reactions using MGB-modified dual-labelled probes.

discrimination, with the 'C'-probes only detecting amplification with C/C and C/T DNAs and the 'T'-probes only detecting amplification with T/T and C/T DNAs. Similar results were obtained in both of the master mix buffer systems. Importantly, these results were obtained without the need for any empiric optimization and probes were simply designed using the method described above.

The unmodified DNA dual-labelled probes were also tested and results are shown in Fig. 8.3b. The unmodified 'C'-SNP probe gave good discrimination but the unmodified 'T'-SNP probe gave poor discrimination. These results are consistent with the experimentally measured T_m and ΔT_m values obtained for match and mismatch pairs for the unmodified probes (Table 8.1). Interestingly, predictions made using nearest neighbour parameters suggested that both assays would show poor SNP discrimination, whereas in actual practice one worked well and the other failed. Nevertheless, both probes must perform well together, and genotypes could not be reliably called using the unmodified probe assay. Thus, use of ΔT_m-enhancing LNA nucleotides improved the ability of qPCR to distinguish between alleles at this locus, converting a failed assay into a successful assay.

Three additional modified bases were reported by Epoch Biosciences that can be used to improve performance of dual-labelled probes (and primers) in qPCR. The Super A™ and Super T™ bases represent modifications of adenosine and thymidine that increase stability when base paired with unmodified A or T. Substitution of these bases in a probe sequence will increase T_m, particularly for AT-rich sequences (Afonina *et al.*, 2002a; Dempcy *et al.*, 2003). Nearest neighbour parameters have not been reported for these bases, however, so T_m predictions cannot be accurately made to assist with design. The third base, Super G™ (8-aza-7-deazaguanine, or PPG) does not increase T_m but does disrupt the Hoogsteen-bonding potential for the N-7 position of the guanine ring (Kutyavin *et al.*, 2002). Hoogsteen-bonding is required for formation of G-tetrads (or G-quadruplexes), a structure that is naturally found in telomeres and can also cause unwanted aggregation of synthetic oligonucleotides (Simonsson, 2001; Reddy and Hardin, 2003). Substituting Super G™ in G-runs (sequences having four or more sequential guanine residues) can prevent G-tetrad formation and eliminate the various complications that arise from tetrad formation.

Another strategy to increase T_m, shorten probe length, and improve specificity is use of non-nucleoside T_m-enhancing modifying groups. Compounds have been known for many years that will bind DNA duplexes with high affinity, such as intercalators and minor groove binders (Bailly and Henichart, 1991). A novel compound of this class was developed by Epoch Biosciences in 1996 that binds the minor groove of a DNA duplex and significantly stabilizes hybridization of a short DNA oligonucleotide to its complementary sequence. This group is a complex molecule, 1,2-dihydro-(3*H*)-pyrrolo[3,2-*e*]indole-7-carboxylate ($CDPI_3$), which is more conveniently called 'MGB' (Kutyavin *et al.*, 1997). The thermodynamic properties of the MGB vary with sequence, length, and chemistry. When attached to the end of a short A:T 8mer (poly-dT_8:poly-dA_8), the MGB leads to greater than a 40°C increase in T_m. The increase in binding affinity is greater for AT-rich sequences than for GC-rich sequences. T_m changes for phosphorothioate DNA have similar magnitude as unmodified DNA; however, the impact is much less when the MGB is attached to a 2′OMe-modified oligonucleotide sequence, possibly because geometry of the minor groove is different in RNA (or RNA:DNA heteroduplexes) than DNA.

The first application proposed for this new chemistry was modification of the 5′-end of PCR primers to enable use of very short oligos (eight to ten bases long) to prime DNA synthesis (Afonina *et al.*, 1997). It was proposed that this would have particular utility in designing primers which were specific for short blocks of highly conserved sequence (for example, in amplifying a broad range of viral serotypes, it is necessary to place primers in very limited areas where sequence is sufficiently conserved between species). Importantly, the MGB group lies in the minor groove of the DNA helix and occupies space corresponding to around six bases, so if this group is positioned at the 5′-end of a primer it will not overlap or interfere with the 3′-terminus of an oligonucleotide that is even as short as eight to ten bases. It was quickly appreciated that the MGB group would also have significant utility if used for dual-labeled probes in the 5′-nuclease assay. In this format, the MGB group is placed at the 3′-end of the probe with a dark quencher positioned between the MGB and the 3′-DNA residue and a reporter fluorophore is placed at the 5′-end. Use of the MGB was demonstrated to permit shortening some probe sequences from 27 bases to as little as 12 bases in length while keeping a constant 65°C T_m, thanks to a 5–20°C contribution from the MGB (Kutyavin *et al.*, 2000, 2003). While probes of this design work well for general gene quantification purposes, the real value becomes apparent when performing allelic discrimination assays. It is thought that the group binds the 3′-end of the probe/target duplex, and, if a SNP site is positioned within the terminal roughly eight bases from the 3′-end (within or near the MGB binding domain), the resulting ΔT_m for match versus mismatch is often in the 10–20°C range, making for very robust SNP assays.

Use of the MGB modification at the 3′-end of a probe for 5′-hydrolysis format qPCR assays was exclusively licensed from Epoch Bioscience by Applied Biosystems and is the basis for the current line of TaqMan® probes. Probes of this design are in widespread use today for all qPCR applications; this format is most often used for low to mid-volume assay throughput needs (De la Vega *et al.*, 2005). The PrimerExpress™ assay design software from Applied Biosystems will provide an estimate of T_m for MGB-modified probes. However, no estimates are given for the ΔT_m when paired with mismatched targets. Applied Biosystems (now Life Technologies) also offers a probe design service via their website and has predesigned or validated probes available for many targets. Predesigned MGB-TaqMan assays are available for most genes and many SNPs from Applied Biosystems and are easily searched/ordered using tools on their website. However, these probes are only provided as fully formulated 'assays', where all probes and the primers come pre-mixed in a single tube, making it difficult to study the performance of individual probes. Further, the sequences of the probes are not provided when ordered via this route. Therefore, instead of ordering pre-designed assays, MGB-Taqman probes for the SMAD7 SNP locus were designed using the PrimerExpress™ software and these sequences were obtained from Applied Biosystems as a custom synthesis. Experimental T_m values were determined for the MGB probes using identical methods and buffers as was previously used for the LNA probes. Results are reported in Table 8.1. The ΔT_m comparing match versus mismatch hybridization was 9–10°C, a value grossly similar to that obtained using the LNA modified probes. The SMAD7 MGB-probes were used in allelic discrimination qPCR assays, using the same methods as outlined previously for the unmodified and LNA-modified probes. The results are shown in Fig. 8.3c. Performance of the MGB probes was excellent and complete discrimination of the C/C and T/T genotypes was achieved when using Applied Biosytems genotyping mastermix. However, performance of the MGB probes was very different when reactions were done using a different buffer (the Immolase Master Mix). Anneal/extension temperatures were varied with little benefit (not shown).

It is not clear why the 'C' allele probe showed good functional discrimination while the 'T' allele probe did not, even though both showed similar ΔT_m values (in a buffer having similar ionic composition to the Immolase Master Mix). These results illustrate the important point that buffer composition is a critical variable when performing demanding applications such as SNP discrimination in qPCR. Some commercial reagents contain proprietary components and do not reveal ionic composition, making it difficult to predict performance. Empiric comparison of different reaction mixes may be needed to optimize performance.

The MGB group can also be added to the 5′-end of the probe. In this configuration, 5′-nuclease activity of the polymerase is blocked and the probe instead functions in a hybridization format with fluorescence signal resulting from the reduced quenching that occurs upon hybridization (Afonina *et al.*, 2002a,b). In single stranded form, the end-to-end distance in the random coil conformation is short and the reporter dye is effectively quenched. When hybridized to target, the duplex exhibits more rigid-rod conformation than single strands and the end-to-end length increases, reducing quenching. Fluorescence emission therefore increases from the reporter dye in the absence of probe degradation (Livak *et al.*, 1995).

Other chemical groups can be used to increase T_m and improve probe performance. Behr and colleagues from the University of Strasbourg developed methods to attach cationic spermine groups to oligonucleotides using a novel spermine phosphoramidite. Multiple spermine groups can be joined in serial fashion, creating a polycation. Attachment of this positively charged polymer to one of the ends of an oligonucleotide helps overcome the electrostatic repulsion between two annealing anionic DNA strands, increasing binding affinity (i.e. increasing T_m) (Noir *et al.*, 2008). Methods employing this technology in molecular biology assays have been commercialized by PolyPlus-Transfection, Inc. This modification has been called 'ZNA', for 'zip nucleic acids', when used in synthetic oligonucleotides. Like the effects described previously for the MGB group, addition of a 5′-ZNA modification permits short primers to be used in PCR (Moreau *et al.*, 2009). Similarly, attachment to the 3′-end of a dual-labelled probe is compatible with use in the 5′-nuclease assay and increases probe T_m, enabling use of shorter probes with better performance in allelic discrimination assays than unmodified probes (Paris *et al.*, 2010). Unlike the MGB group, which is usually positioned at the very 3′-end, ZNA probes position the quencher at the 3′-end and typically four spermine groups are placed between the quencher and the 3′-most base of the nucleic acid sequence. This poly-spermine tail flips over and binds the DNA nucleotides near the 3′-end of the probe, which fortuitously also results in the quencher being positioned closer to the reporter fluorophore (at the 5′-end). Forcing a shorter dye-quencher distance increases quenching, lowering background fluorescence and further improves probe performance.

Another variant of a linear hybridization probe for SNP discrimination was described by Yamane (Wakunaga Pharmaceutical Co., Japan) and were called 'MagiProbes' (Yamane, 2002). In this case, a reporter fluorophore and a pyrene quencher were positioned centrally in the probe near the SNP site in adjacent or nearly adjacent positions. When hybridized to a perfect match target, the pyrene group intercalates into the DNA duplex resulting in loss of quencher function and fluorescence emission seen from the reporter. Conversely, when hybridized to a mismatch target, the pyrene group remains external to the duplex and continues to quench the reporter.

Hairpin dual-labelled probes

Hairpin fluorescent dual-labelled probes, called 'Molecular Beacons', were first described by Tyagi and Kramer (Public Health Research Institute, New York, NY, USA) in 1996 (Tyagi and Kramer, 1996). These probes have a reporter fluorophore on one end and a quencher on the other end; a target hybridizing domain occupies the central loop portion of the probe sequence and two short sequences are placed on either end which are self-complementary (and non-complementary to target). In the absence of target, the Molecular Beacon forms a hairpin, which forces the reporter and quencher into close proximity. As a result, quenching efficiency is very high and probes of this design have a very good signal-to-noise ratio. Further, hairpin formation competes with target binding and improves the thermodynamics of match/mismatch discrimination, increasing the relative ΔT_m. With careful adjustment of the hairpin stem length, good SNP discrimination can sometimes be achieved without additional chemical modification (Tyagi *et al.*, 1998; Bonnet *et al.*, 1999; Tsourkas *et al.*, 2003b). As with linear probes, LNA nucleotides can be included within the target binding domain which will allow for use of shorter probe sequences, further improving the ΔT_m. Non-base modifiers (like MGB or ZNA), however, cannot easily be used in Molecular Beacons as these groups would primarily increase stability of the stem and would be detrimental to target binding. Synthesizing probes with 2′-O-Me-RNA in place of DNA increases nuclease stability and permits use of Molecular Beacons in live cells to directly visualize mRNA by fluorescence microscopy (Bratu *et al.*, 2003); note that shorter stem domains must be used with this chemistry as the T_m of the stem is much higher for 2′O-Me–2′O-Me duplexes (Tsourkas *et al.*, 2003a). The use of a 2′-O-Me sugar backbone, however, does not confer any benefit for use in qPCR applications. A version of Molecular Beacons called 'Scorpions' was described by Brown and colleagues (University of Southampton, UK) where the probe is tethered to a PCR primer and configured so that target binding can occur at specific points in the reaction cycle (Thelwell *et al.*, 2000). This design showed good sensitivity and specificity compared with traditional probes and in particular worked well with fast cycling times.

Enzymatic interrogation of SNP identity

The SNP discrimination methods discussed above all rely upon differential hybridization of a probe oligonucleotide with different target nucleic acids to distinguish base identity at the site of interest. It is also possible to perform SNP discrimination using probes with high affinity that bind well to both match and mismatch targets that include modification(s) which permit the probes to interact with an enzyme that is capable of discriminating between paired and unpaired bases at the SNP site. The best example of this class of SNP assay uses RNase H enzymes to selectively cleave probes. RNases H are a family of enzymes that specifically cleave at RNA bases in an RNA:DNA heteroduplex. The enzymes do not cleave single stranded RNA or DNA. Two major classes of RNase H enzymes exist, RNase H1 and RNase H2 (Ohtani *et al.*, 1999). RNase H1 enzymes require a stretch of at least four consecutive RNA bases to show maximal enzyme activity and cleave between RNA residues (Hogrefe *et al.*, 1990). In contrast, RNase H2 enzymes will cleave a substrate having only a single RNA residue in an otherwise DNA sequence and cleavage occurs at the 5′-side of the RNA base at what is actually a DNA linkage (Eder and Walder, 1991; Eder *et al.*, 1993). Cleavage patterns for RNase H1 and RNase H2 enzymes are illustrated in Fig. 8.4. RNase H enzymes have been isolated from a variety of prokaryotic species that are thermostable

(Kanaya, 2001), some of which can be used at elevated temperatures in molecular biology assays or even be included in PCR reactions.

Peter Duck (Meiogenics, MD, USA) and Joseph Walder (University of Iowa and Integrated DNA Technologies) described 20 years ago a method to use chimeric RNA:DNA oligonucleotides as probes to detect complementary sequences with isothermal signal amplification. Probes were made with DNA bases at both ends and contained a stretch of at least four RNA residues in the central region. When hybridized to a DNA target, RNase H1 will cleave within the RNA domain, cutting the probe in half. The shortened probe dissociates from the target, permitting a new probe molecule to hybridize and repeat the cycle (isothermal cycling); cleavage products accumulate and can be detected by a variety of methods (Duck *et al.*, 1990; Walder and Walder, 1995). Harvey and Han (Eximus Biotech, Maryland) adapted this method to a fluorescence format by adding a dye and a quencher to the chimeric probe molecule on opposite sides of the cleavable RNA domain, so that the probe is dark in the intact state and bright after RNase H cleavage (Harvey *et al.*, 2004). Cleavage by RNase H1 is less efficient if a mismatch is present in the RNA cleavage domain between the probe and the DNA target. This method can be linked to PCR by using a thermostable RNase H1 enzyme and the resulting assay can be used to discriminate SNPs (Harvey *et al.*, 2008). In this case, a commercially available RNase H1 enzyme from a *Thermophilus* sp. was employed (Epicenter, Madison, WI).

While this method works, more than one cleavable linkage exists within the 4-RNA central domain and it is not entirely clear where it is best to position the SNP when designing the assay. If an RNase H2 enzyme is employed, the same assay strategy can be used with a probe that has only a single RNA base (i.e. a single cleavable linkage). Hou and colleagues (Shanghai Jiaotong University, China) characterized the ability of the RNase H2 enzyme from *Chlamydia pneumoniae* to discriminate mismatches when a probe containing a single RNA residue was hybridized to targets having mismatches at different positions and

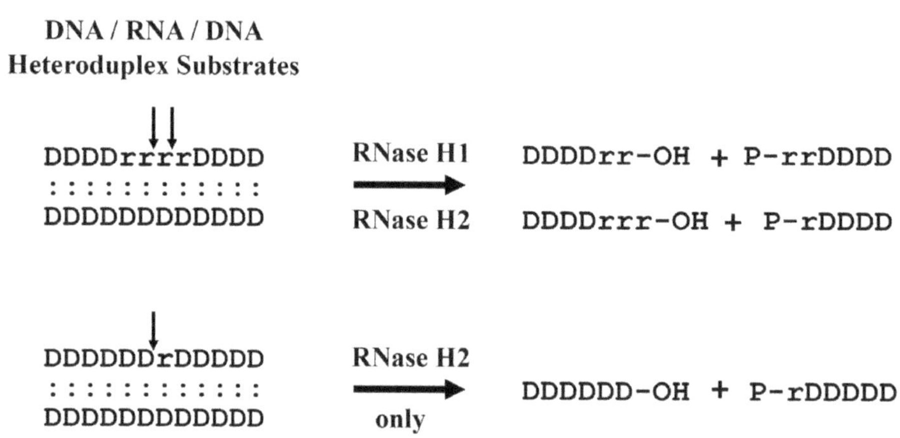

Figure 8.4 RNase H cleavage patterns. Two classes of RNA:DNA heteroduplex substrates are shown. The top row shows a substrate having four consecutive RNA bases and the bottom row shows a substrate having a single RNA base. Arrows indicate expected sites for RNase H cleavage for each substrate. Expected products following cleavage by RNase H1 and/or RNase H2 are shown on the right. D=DNA, r=RNA, OH=3′-hydroxyl, P=5′-phosphate.

found that discrimination was best when the SNP was placed at the RNA base; the bases on either side of the RNA base (at the '+1' and '–1' positions) also showed high selectivity (Hou *et al.*, 2007a). They adapted this assay for use with Molecular Beacons, positioning the RNA base centrally within the target binding loop domain of the probe. Since the *C.p.* RNase H2 enzyme is not thermostable, the Molecular Beacon hybridization/cleavage steps were run post-amplification as an end-point assay (Hou *et al.*, 2007b,c). The same group later improved this method by using a thermostable RNase H2 enzyme from *Thermus thermophilus*, which permitted a homogeneous assay format where the probe and cleaving enzyme were present and functioned in real time during PCR (Liu *et al.*, 2010).

SNP discrimination using biased amplification

PCR-based methods that discriminate between related sequences using biased amplification lead to differential accumulation of product with one allele being preferentially amplified over the other allele. If the different alleles are present at roughly similar levels in the test nucleic acid sample, SNP discrimination methods that use biased or unbiased amplification can perform equally well. However, if one allele is under-represented in the population, unbiased amplification is often only capable of detecting the more prevalent species. Detection of a rare allele in the presence of a high background of wild-type sequence requires use of a method that employs biased amplification, especially if that species is present at levels of 10^{-2} to 10^{-6} or lower. Unlike unbiased amplification, biased amplification normally requires that one of the primers overlaps the SNP site. The system is designed so that the match template supports amplification while the mismatch template does not. Sequence identity is determined by the presence or absence of amplification. Probe-based methods discussed previously usually rely upon differential hybridization between match versus mismatch targets (i.e. thermodynamic discrimination). For biased amplification, primers typically bind to both alleles and discrimination is based upon enzymatic events, e.g. whether or not the oligonucleotide will prime DNA synthesis or not. Most of the methods to perform biased amplification with enzymatic interrogation employ chemically modified oligos, and the chemical modifications enable the process.

Allele-specific PCR

The simplest system to perform allele-specific PCR (ASPCR) employs primers where one primer overlaps the SNP, which is positioned as the 3′-end base of the primer. Differences in the priming efficiency on match versus mismatch templates determine the level of selectivity. In practice, actual discrimination can vary from very low to fairly high and depends largely on the sequence context and precise bases present as the mismatch pair at the SNP site. Also, different polymerases show different degrees of selectivity using the same primers and template (Mendelman *et al.*, 1990). A survey was performed at IDT where we examined the relative discrimination seen in a synthetic amplicon using allele-specific PCR. Four unmodified primers were made having the four possible bases at the 3′-position; four templates were made having the four possible bases at the SNP site. All 16 pairwise combinations were tested in a SYBR® Green format, and reactions were tested individually (not in a multiplex format). The ΔCq (the cycle number at which signal was first detected for the mismatch minus the cycle number at which signal was first detected for the match) ranged from 1.2 to 10.7 cycles, with an average ΔCq of 5.4 for the 12 mismatch pairings (Dobosy *et al.*, 2011). In other words, sometimes this method works well and sometimes it does not. Careful optimization of primer design, reaction temperature, buffer composition, and enzyme choice can improve results. Traditionally, these reactions are performed individually in separate tubes, as there is no simple way to distinguish between the two different allele products when using unmodified primers. Using modified dye labelled primers (such as the hairpin fluorescence-quenched Amplifluor™ system), it is possible to perform assays for both alleles in a single tube. Interestingly, results are often significantly improved in multiplex format and it appears that kinetic competition between alleles increases specificity and reliability (Myakishev *et al.*, 2001).

Use of chemically modified bases can further improve the specificity of ASPCR. In particular, placing an LNA nucleotide at the 3′-end of the allele-specific primer will decrease priming on mismatched templates without compromising efficiency of PCR on matched templates. Usually only a single LNA modification is used on each primer, which is placed at the 3′-end with the LNA nucleotide pairing with the SNP (Latorra *et al.*, 2003; Nakitandwe *et al.*, 2007). This strategy works best when using a DNA polymerase that does not have 3′–5′ exonuclease activity. If a high fidelity DNA polymerase is used that has proofreading function (3′–5 exonuclease activity), then the enzyme can remove the terminal 3′-base from the primer (error correction) and amplification will proceed whether a mismatch is present or not. In this scenario, it is preferred to position the LNA nucleotide at the 'N-2' position, the base adjacent to the 3′-end (Di Giusto and King, 2004). In this design, the LNA modification confers structure to the primer that makes this last linkage relatively nuclease resistant so that the 3′-base is retained in the primer and can contribute to specificity as intended.

The 'amplification refractory mutations system' (ARMS) and 'mismatch amplification mutation assay' (MAMA) are methods that introduce an artificial second mismatch into allele-specific PCR primers as a way to increase selectivity (Newton *et al.*, 1989; Cha *et al.*, 1992). As with standard ASPCR, the SNP under interrogation is positioned at the 3′end of the primer. A second mismatch is introduced in the primer directly adjacent to the 3′-base, resulting in the 'match' primer having a single mismatch with target (the artificial mismatch) and the 'mismatch' primer having two mismatches with the target (the artificial mismatch plus the SNP). At this position, the single internal mismatch is permissive and the 'match' primer does not show any significant decrease in priming efficiency. However, the double mismatch is very inefficient when priming on the mismatch target and ΔCqs of ten or more cycles can often be achieved, even for base mismatches that are unfavourable for standard ASPCR. The primers used in the original versions of these assays are unmodified. In many cases, the specificity of ARMS or MAMA primers can be further enhanced by placing an LNA modification at the SNP site located at 3′ end.

Biased amplification using blocked primers

Another approach to perform biased amplification employs primers that are blocked at the 3′-end which cannot prime DNA synthesis until the block is removed. The blocked 3′-OH group can be removed by cleavage using either an endonuclease (internal cleavage) or a 3′-exonuclease; cleavage activates the primer, allowing PCR to proceed. To function as a SNP assay, the 3′-deblocking step must be selective and only occur when a perfect match is present between the primer and the target at the cleavage site. Specificity of the cleavage process is critical to success.

One successful version of this assay strategy was developed by Liu and Sommer (City of Hope National Medical Center, Duarte, CA, USA) and is called pyrophosphorolysis-activated polymerization, or PAP (Liu and Sommer, 2000, 2002). Like traditional ASPCR primers, PAP primers have the SNP site positioned at the 3′-end of the primer sequence; however, PAP primers are also blocked at the 3′-end. The 3′-end block is removed by the polymerase enzyme via pyrophosphorolysis, which is essentially a reverse polymerization reaction that occurs in the presence of sufficiently high levels of inorganic pyrophosphate (Deutscher and Kornberg, 1969). High levels of inorganic pyrophosphate inhibit DNA polymerases, so careful buffer composition optimization must be done. PAP reactions also employ lower levels of deoxynucleotide triphosphates than are typically employed in PCR (e.g. 50 μM of each dNTP with 300 mM sodium pyrophosphate). Pyrophosphorolysis occurs only when the terminal base of the primer is base-paired with a complementary base on the target; a mismatch base will not function as a substrate for this enzymatic reaction. A variant of the PAP method uses both blocked forward and blocked reverse primers, called Bi-PAP (bidirectional pyrophosphorolysis-activated polymerization). The SNP is positioned at the 3′-end of both primers. As a result, the primers overlap by one base, but this short overlap is not sufficient to allow primer-dimer amplification (the primers are blocked – after deblocking, the primers no longer overlap). For amplification to occur, successful PAP must occur on each primer independently, greatly increasing specificity of the final assay. Bi-PAP has been shown to be capable of rare allele detection at levels as low as 10^{-7}–10^{-8} (Shi *et al.*, 2007). A wide variety of 3′-blocking strategies have been tested and the best results are obtained when the 3′-base is a dideoxynucleotide (ddN). Widespread use of PAP therefore requires the availability of primers with a 3′-terminal dideoxy base for each of the four bases; ddA and ddC synthesis reagents are available as commercial solid supports, so synthesis of primers ending in an A or C base is straightforward. Unfortunately, ddT and ddG supports are not available and primers ending in a T or G base have to be made using expensive reverse synthesis or by enzymatic methods, limiting the general utility of this approach.

Biased amplification using RNase H and blocked-cleavable primers

Another method in this class of assays employs blocked primers that have a single RNA base positioned near the 3′-end of the primer. Following hybridization to a complementary sequence, the terminal blocking group is removed by action of a thermostable RNase H2 enzyme (in contrast with PAP, this time the 'deblocking enzyme' acts via internal endonuclease cleavage). RNA-containing blocked-cleavable primers can be used in PCR and provide improved specificity when compared with traditional priming methods, resulting in lower background, reduced mispriming, and elimination of primer-dimer formation (Walder *et al.*, 2009; Dobosy *et al.*, 2011). PCR done using blocked-cleavable primers has been called 'rhPCR'. Many of the features of RNase H enzymes were reviewed in the previous section. It is important to note that the cleavage reaction requires that the primer must be duplexed with a DNA target. RNase H2 will not degrade single-stranded nucleic acids; it requires an RNA–DNA heteroduplex substrate. When a blocked primer of this design hybridizes to the target, the substrate has a single RNA base embedded in an otherwise DNA sequence. RNase H2 cleaves at the 5′-side of the RNA base at the DNA linkage, leaving a 3′-hydroxyl at this end and a cleavage product which is an activated, functional primer. The second cleavage product contains a 5′-phosphate, the single RNA base, a few DNA bases, and the 3′-blocker; this is an inactive fragment that is short, dissociates from the target, and is 'disposable'. Importantly, the cleavage reaction is sensitive to correct base pairing such that primer deblocking is inhibited by the presence of a base mismatch in the vicinity of the cleavage site. The basic scheme for performing rhPCR with blocked-cleavable primers and RNase H2 is outlined in Fig. 8.5.

Ideally the RNase H2 enzyme employed in rhPCR will be very thermostable. A number of suitable enzymes have been isolated from archaeal sources, including *Pyrococcus furiosus* (Sato *et al.*, 2003), *Methanococcus jannaschii* (Lai *et al.*, 2003), *Thermococcus kodakaraensis* (Haruki *et al.*, 1998; Mukaiyama *et al.*, 2004), and *Pyrococcus abyssi* (Walder *et al.*, 2009; Le Laz *et al.*, 2010; Dobosy *et al.*, 2011). These enzymes require divalent cations for catalysis and can employ either Mg^{++} ions (usually in the 2–5 mM range) or Mn^{++} ions (usually in the 0.3–1.5 mM range). Turnover rate is fast enough that primer annealing and cleavage can be done in real time during PCR using the same primer anneal/extend times that are usually employed (30–60 seconds dwell time) with no loss of priming efficiency. Cycling should be performed as a two-step PCR (e.g. 60°C–95°C steps), using the same temperature for both primer annealing and extension reactions. A three-step PCR (e.g. 60°C–72°C–95°C steps) will be inefficient as primer annealing and deblocking cannot occur during the 72°C extension phase.

For a primer/target heteroduplex to be an efficient substrate for RNase H2 cleavage, a minimum amount of nucleic acid must be present in duplex form both 5′ and 3′ to the RNA base. These requirements have been studied for *Pyrococcus abyssi* RNase H2 and 8–10 DNA bases are needed 5′-to the cleavable cleavage site and four or more DNA bases are needed 3′-to the cleavage site for maximal activity. A variety of methods can be used to block the 3′-end of the primer, including dideoxy nucleotides or non-base modifiers placed attached to the 3′-end of the terminal DNA residue. A C3 spacer group (1,3-propanediol) is inexpensive and works well as a blocking group (Walder *et al.*, 2009; Dobosy *et al.*, 2011). Using these design parameters, a typical rhPCR primer will have a 'DNA primer domain' towards the 5′-end (20–25 bases), a single RNA base, four DNA bases on the 3′-side of the RNA, and a terminal 3′-C3 spacer (or other blocking group). This basic 'first generation' rhPCR design primer will prevent primer-dimer formation and improve specificity. In addition to primer hybridization, mismatches at or near the RNA cleavage site will reduce the rate of cleavage and prevent efficient PCR. It is not necessary that the cleavage event be 'all or none' at the

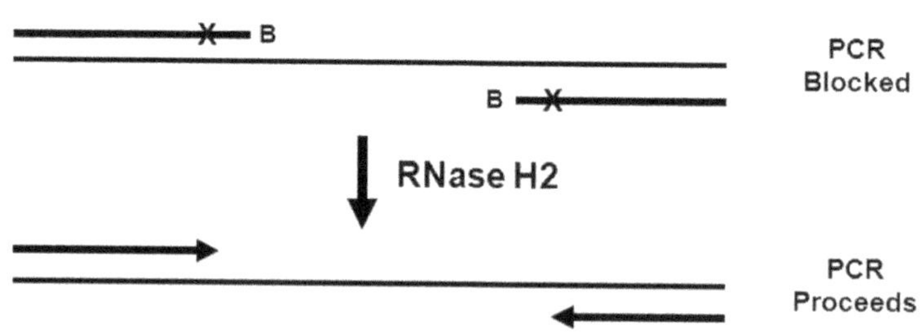

Figure 8.5 rhPCR using blocked-cleavable primers. Blocked primers are aligned against template on the top. X=scissile linkage, which in rhPCR is a single RNA base. B=blocking group, most commonly a C3-spacer (1,3-propanediol). Following cleavage, the RNA base and 3′-nucleic acids are removed and fully active primers remain having an unblocked terminal 3′-hydroxyl group (arrowheads).

RNA base to have a dramatic effect on PCR. If a mismatch reduces primer deblocking from 100% to 50% efficiency, then half of potential PCR priming events never happen and PCR amplification reduces from the standard exponential 2^n (where n = cycle number) to 1.5^n. Over 30 cycles, this represents a 5600-fold reduction in amplification.

The relative ability of a base mismatch to reduce cleavage has been studied for the RNase H2 enzymes from *Thermococcus kodakaraensis* (Liu *et al.*, 2010) and *Pyrococcus abyssi* (Walder *et al.*, 2009; Dobosy *et al.*, 2011). The best sites to place a mismatch (i.e. a SNP) are at positions complementary to the RNA base (position '0') or complementary to the DNA base which is immediately 5′-to the RNA base (position '–1'). RNase H2 cleavage occurs at the linkage between these two bases. From an assay standpoint, however, it is preferable to position the SNP at the RNA base. Since cleavage occurs on the 5′-side of the RNA base, if an unwanted cleavage event does occur in the presence of a mismatch, then the resulting primer will have this base excised and the first base of primer extension by the DNA polymerase will be encoded by the SNP. This ensures that the daughter product will faithfully reproduce the actual sequence present in the target nucleic acid. If the mismatch was positioned at the DNA base (position '–1'), cleavage still occurs immediately 5′-to the RNA base and then the SNP base encoded by the primer remains in the primer and is locked into the extension product, converting the PCR product to the primer sequence.

When the 'first generation' rhPCR primers are used in SNP assays, the observed ΔCqs (mismatch – match) typically range from 5–15 cycles depending on the sequence context and precise base mismatches involved. While it is easy to distinguish the correct genotype present from these results, this level of discrimination is lower than would be expected given the known selectivity of the RNase H2 cleavage event. The basis for this 'lower than expected selectivity' was investigated. Reaction products were cloned from rhPCR done using mismatched templates and sequence was determined. It was unexpectedly found that the amplicons had converted to the primer sequence, which should not have been possible if primer cleavage was only occurring at the 5′-side of the RNA base (the SNP was positioned at this site) (Walder *et al.*, 2009). It appears that a very low frequency (less than 1 in 1000) of cleavage events occur at a non-canonical cleavage position, such that the enzyme cuts on the 3′-side of the RNA base. Cutting at this position results in the primer-encoded SNP base being incorporated in the amplicon, which converts the PCR product to the primer sequence. In subsequent cycles, this product will amplify at 100% efficiency as it is now perfect match to the primer. Thus the frequency of non-canonical cleavage ultimately determines the specificity of rhPCR.

Various patterns of chemical modification were tested in the rhPCR primers to decrease the frequency of non-canonical cleavage on the 3′-side of the RNA base. Substituting non-base groups in the short DNA domain on the 3′-side of the RNA base creates substrates that are less easily cleaved by RNase H2 (such primers require 10- to 100-fold higher enzyme concentrations to achieve reaction efficiency similar to what is seen using a 'first generation' primer); however, the frequency of non-canonical cleavage is reduced and ΔCq values of 10–20+ cycles are easily achieved using this 'second generation' primer design. Currently the favoured 'second generation' primer design incorporates a DNA base, two C3-spacers, and a final DNA base on the 3′-side of the RNA residue. No 3′-blocking group is required with this primer design as the C3-spacers reduce or totally eliminate the ability of the terminal DNA base to prime DNA synthesis. Even if priming did occur, the spacers prevent the primer from functioning as a template for DNA synthesis so second strand synthesis would terminate at the primer, further blocking PCR. This 'second generation' primer design should be employed for allelic discrimination assays; however, the 'first generation' primer design is still preferred for other applications since lower concentrations of enzyme are needed to perform each reaction.

Examples of SNP discrimination using biased amplification

A series of SNP discrimination assays were performed using several of the methods included in the 'biased amplification' section using the same SNP in the *SMAD7* gene previously used in probe studies in this chapter (NM_005904, C/T variation, rs4939827). This site is a relatively challenging SNP assay for these kinds of methods, since one of the mismatch pairings is T:G. The T:G/G:T mismatch pairs are thermodynamically fairly stable and as a result are difficult to distinguish. Primers used included: standard ASPCR, 3′-LNA ASPCR, ARMS, 3′-LNA ARMS, and both the first generation and second generation rhPCR blocked-cleavable primers. Primers are shown in Fig. 8.6 aligned on the target SNP site.

The SNP discrimination assays were run in singleplex format on a Roche LightCycler® 480 in 384 well plates using 10 µl reaction volumes. Reactions consisted of 5 µl of 2′ BIO-RAD iQ™ SYBR® Green Supermix (BIO-RAD, Hercules, CA), 200 nM each of the forward and reverse primers, and 2 ng human genomic DNA. Cycling conditions included an initial 5 minutes incubation at 95°C, followed by 80 cycles of 10 seconds at 95°C and 30 seconds at 60°C (note – reactions were run 80 cycles only to demonstrate the selectivity of the blocked primers; a standard 40–45 cycle run would be used in a typical assay). Human genomic DNAs that were homozygous C/C or T/T at SNP locus rs4939827 were obtained from the Coriell Institute for Medical Research Cell Repository (http://ccr.coriell.org/, GM18562 and GM18537). The quantification cycle number (Cq) was determined using the absolute quantification/2nd derivative method. All reactions were run in triplicate and data reported in Fig. 8.6 represents an average of the three data points. The rhPCR reactions included recombinant *Pyrococcus abyssi* RNase H2 in the reaction mix (2.6 mU per 10 µl reaction for the first generation primers and 50 mU per 10 µl reaction for the second generation rC-containing primer and 200 mU for the rU containing primer) (Walder *et al.*, 2009).

At the locus, unmodified ASPCR primers were ineffective and showed < 1 cycle different in Cq values seen between match and mismatch targets. Adding a single LNA nucleotide to the 3′-end of this primer design boosted the ΔCq to six to nine cycles. Unmodified ARMS primers performed even better, showing a ΔCq of 9–11 cycles. Adding a 3′-LNA nucleotide to the ARMS primers increased discrimination to 10–13 cycles; however, a three cycle delay in the match Cq was observed for the 'T' allele. The first generation blocked-cleavable primers (rhPCR) gave a ΔCq of around 11 cycles and the second generation version of this assay gave the best results in this survey with ΔCq values of 28–30 cycles. It is important to note that these assays were performed in singleplex format. The ASPCR approach often shows improved specificity when run in a multiplex format (which requires modification to add the appropriate dyes/quenchers to the primers to permit multiplex fluorescence detection) where kinetic competition between primers for each allele occurs. Nevertheless it is clear that such biased amplification methods can work well, even at difficult SNP sites like the rs4939827 locus used in the present example.

Assay		Primer Seq	Cq match	Cq mismatch	ΔCq
Control	5′	CAGCCTCATCCAAAAGAGGAAA	26.4	--	--
ASPCR	5′	CAGCCTCATCCAAAAGAGGAAAC	26.7	27.4	0.7
3′-LNA	5′	CAGCCTCATCCAAAAGAGGAAAC	25.9	32.3	6.4
ARMS	5′	CAGCCTCATCCAAAAGAGGAATC	26.5	35.6	9.1
ARMS+LNA	5′	CAGCCTCATCCAAAAGAGGAATC	27.7	41.3	13.6
rhPCR, Gen1	5′	CAGCCTCATCCAAAAGAGGAAAcAGGAx	27.2	38.4	11.2
rhPCR, Gen2	5′	CAGCCTCATCCAAAAGAGGAAAcAxxA	27.2	55.7	28.5
	5′	GCTCACAGCCTCATCCAAAAGAGGAAACAGGACCCC 3′			
	3′	CGAGTGTCGGAGTAGGTTTTCTCCTTTGTCCTGGGG 5′			
Control	5′	CAGCCTCATCCAAAAGAGGAAA	26.4	--	--
ASPCR	5′	CAGCCTCATCCAAAAGAGGAAAT	26.6	27.0	0.4
3′-LNA	5′	CAGCCTCATCCAAAAGAGGAAAT	25.9	35.1	9.2
ARMS	5′	CAGCCTCATCCAAAAGAGGAATT	26.7	37.5	10.8
ARMS+LNA	5′	CAGCCTCATCCAAAAGAGGAATT	30.6	40.6	10.0
rhPCR, Gen1	5′	CAGCCTCATCCAAAAGAGGAAAuAGGAx	27.2	38.1	10.9
rhPCR, Gen2	5′	CAGCCTCATCCAAAAGAGGAAAuAxxA	28.6	59.3	30.7
	5′	GCTCACAGCCTCATCCAAAAGAGGAAATAGGACCCC 3′			
	3′	CGAGTGTCGGAGTAGGTTTTCTCCTTTATCCTGGGG 5′			

Figure 8.6 Biased amplification assays at a C/T SNP in the human *SMAD7* gene. Genomic sequence of the region surrounding the C/T SNP site is shown. SNP discrimination is based upon amplification from the For primer sequences, which are shown aligned with the genomic target. Assays tested include unmodified allele-specific PCR primers (ASPCR), 3′-LNA (variant of ASPCR), ARMS (note second mutation added at the next to last base from the 3′-end, an A–T-change), ARMS with 3′-LNA, first generation rhPCR blocked-cleavable primer with RNase H2 cleavage, and second generation rhPCR blocked-cleavable primer with RNase H2 cleavage. A control primer was also used which did not include the SNP base under interrogation. DNA bases are uppercase. LNA nucleotides are underlined. RNA bases are lowercase. 'X' is a C3 spacer (1,3-propanediol group). Reactions were performed in singleplex format using SYBR® Green detection using 2ng input homozygous C/C or T/T DNA. Cq values obtained for the match and mismatch templates are shown as well as the D*Cq* (mismatch *Cq* minus match *Cq*).

Biased amplification using template-blocking agents

Unlike the methods to perform biased amplification previously discussed, this version uses primers that do not overlap the SNP site and are non-discriminatory. A high affinity oligonucleotide or peptide nucleic acid (PNA) is made that is complementary to the target and overlaps the SNP site, which is positioned between the primers. Like a hybridization probe, the blocking agent must bind to the match target but not bind to the mismatch target under the reaction conditions employed. Unlike fluorescent probes, here the binding must be stable enough to block processivity of the DNA polymerase, interrupting amplification. Both PNA oligomers and LNA oligonucleotides have been shown to work in this application (Shinozaki *et al.*, 2007; Tatsumi *et al.*, 2008). It is essential that either the blocking agent is nuclease resistant or that the DNA polymerase employed lacks a 5′–3′ exonuclease activity. It can be challenging to find oligomers that bind with sufficient affinity to block the polymerase yet do not bind in the presence of a single mismatch. As a result, this method has not found widespread use.

Preventing primer–dimer formation

In spite of efforts to design primers that do not have significant overlap with themselves or each other, sometimes it is necessary for biological reasons to position primers at locations in a sequence which are not ideal and permits primer-dimer formation. 'Hot start' PCR methods have decreased some of the problems seen with primer-dimer formation; however, this does not solve all problems and sometimes it is necessary to use chemically modified primers that have reduced this risk. The blocked-cleavable rhPCR primers discussed above offer one method to prevent primer-dimer formation and will not be discussed further in this section.

Another method to minimize primer-dimer formation was described by Howard Gamper and colleagues from Epoch Pharmaceuticals (Kutyavin *et al.*, 1996; Gamper *et al.*, 2006) and relies upon the base modifications 2,6-diaminopurine (2,6-DAP, or 2-amino-dA) and 2-thiothymine (2-thio-dT). The modified base 2,6-DAP is an adenosine analogue and will pair with thymine bases in the normal fashion. Similarly 2-thio-dT is a thymine analogue which forms a stable base pair with native adenosine. However, 2-thio-dT will not pair with 2,6-diaminopurine. Thus primers made using these modified bases will hybridize to their intended targets and can support PCR with equal efficiency as unmodified primers; however, the modified primers cannot easily hybridize to each other as the 2,6-DAP and 2-thio-dT bases (A and T replacements) will not base pair. These modifications are fairly expensive and this method has not found widespread use, even though it is biochemically quite elegant and works well in practice.

A different approach was described by Gerald Zon, Richard Hogrefe, and colleagues from TriLink BioTechnologies (Lebedev *et al.*, 2008). Instead of using a base modification, the internucleotide linkage was modified at the 3′-most and/or next to last linkage with a 4-oxo-1-pentyl (OXP) phosphotriester (PTE) group. This modification impairs the ability of the oligonucleotides to prime DNA synthesis; however, the modifying group is heat labile and rapidly dissociates from the primers during the initial 95°C heat denaturation step, after which the primers revert to unmodified form and function normally. This prevents primer-dimer formation during the initial warming phase of PCR and is a supplement to standard hot-start methods. In particular, this kind of primer modification should be very useful in one-tube RT-PCR chemistries where the primers are present with an active polymerase and traditional 'hot start' is not employed. This method is inexpensive and primers with this chemistry can be obtained directly from TriLink BioTechnologies (San Diego, CA).

Decreasing specificity using universal bases

The previous sections of this chapter focused on methods to increase specificity. Sometimes it is also necessary to decrease specificity. For example, the SNPs that occur in outbred populations can interfere with assays intended to amplify all variants. This problem can be particularly difficult with infectious disease targets, where different strains or serotypes within the same bacterial species or the same virus may have widely varying sequences. This greatly complicates attempts to make a single assay capable of detecting all important relatives. One approach to deal with this problem is to introduce mixed base sites, where synthetic oligonucleotides are made having a mixture of bases inserted at a specific position so that within the population of oligos some subpopulation will always have a perfect match to the target. This strategy works well if the amount of variability is limited. However, as target variability increases, then the number of different sequence variants of the primers or probes that exist in the mixed population can become very high with the result that the actual molar concentration of perfect match primers is too low to support PCR without having to employ unacceptably high concentrations of oligonucleotide.

A solution to reduce complexity of the oligonucleotides that will bind multiple targets is to incorporate modified bases which can pair with more than one of the natural bases, or at least are minimally destabilizing when paired with different natural bases. An excellent review of this topic was published by Loakes in 2001 (Loakes, 2001). The first 'inert' base used for this purpose was inosine, a naturally occurring hypoxanthine variant. Inosine is commonly used today, but is an inferior 'universal base' because it has very different effects on T_m depending on which base it is mispaired with (Martin *et al.*, 1985). Inosine most stably pairs with cytosine and behaves as a guanine analogue that simply does not have as strong a mismatch thermodynamic cost when paired with other bases as does guanine. The approximate rank order of stability for inosine base paring is C > A > G > T (Kawase *et al.*, 1986). Unlike other universal bases that may show more uniform thermodynamic effects when paired with any native base, inosine has the advantage that it can be inserted in primers and supports template function. When present in a nucleic acid which is used as a template in DNA synthesis, inosine directs incorporation of cytosine, consistent with it being considered a 'G analogue'.

The first artificial base that seemed to have true 'universal base' properties was 3-nitropyrrole; however, this modification was fairly destabilizing when used in primers and was quickly replace with 5-nitroindole (5-NI) (Nichols *et al.*, 1994; Loakes *et al.*, 1995). 5-NI destabilizes short duplexes by 2–5°C per insertion depending on location but shows reasonably little difference between templates having different bases mispaired with it. 5-NI is a good choice to use in hybridization probes as a universal base; however it does not function well in primers. Most polymerases do not recognize 5-NI as a base and chain termination usually occurs when 5-NI is encountered in a template and thus 5-NI containing oligonucleotides do not support PCR.

Additional compounds have been identified that behave as universal purine or pyrimidine bases. The synthetic base analogue 6H,8H-3,4-dihydropyrimido[4,5-c][1,2]oxazin-7-one (called 'p') is a pyrimidine replacement that will pair with the purine bases A or G. The synthetic base analogue N6-methoxy-2,6-diaminopurine (called 'k') is a purine analogue that will pair with the pyrimidine bases C or T (Bergstrom *et al.*, 1997). These bases are particularly useful as they support template function when used in a primer. The 'p' base directs incorporation of dA (60%) and dG (40%) and the 'k' base directs incorporation of dT (87%) and dC (13%) (Hill *et al.*, 1998). A variety of other base analogues have been synthesized and described in the literature; however, the set described here are commercially available and can be obtained from a variety of sources.

Conclusions

A wide variety of chemical modification can be introduced into synthetic oligonucleotides that alter their properties as PCR primers or probes. Modifications can be made to any part of the nucleic acid molecule, including the nucleobases, the internucleotide linkages, and the ribose ring. The modifications can increase binding affinity (T_m) and are commonly used to shorten primers and probes, improving specificity when this is desired. Conversely, modified bases are available which decrease specificity and are permissive for mismatch pairing. Modifications can also be used to prevent formation of primer-dimers by interfering with the ability of the oligonucleotides to prime DNA synthesis or interact with each other. All of the modifications discussed in this chapter are easily obtained from commercial sources. New modifiers and applications for their use are likely to continue to be developed in the future.

Acknowledgements

The authors thank all members of the research staff at Integrated DNA Technologies for their contributions over the past decade to our understanding of the fine details of qPCR and the impact that chemical modifications have on assay performance. In particular, we thank Derek Thomas, Ashley Dvorak, Andrey Tautarov, and Michael Christodoulou for assistance with the present manuscript. All original work discussed herein was supported by funding from Integrated DNA Technologies, Inc.

References

Afonina, I., Zivarts, M., Kutyavin, I., Lukhtanov, E., Gamper, H., and Meyer, R.B. (1997). Efficient priming of PCR with short oligonucleotides conjugated to a minor groove binder. Nucleic Acids Res. *25*, 2657–2660.

Afonina, I., Belousov, Y.S., Metcalf, M., Mills, M., Mills, A., Sanders, S., Kutyavin, I., Walburger, D.K., Gorn, V.V., Shishkina, I., *et al.* (2002a). Single nucleotide polymorphism detection with MGB Eclipse assays. J. Clin. Ligand Assay *25*, 268–275.

Afonina, I.A., Reed, M.W., Lusby, E., Shishkina, I.G., and Belousov, Y.S. (2002b). Minor groove binder-conjugated DNA probes for quantitative DNA detection by hybridization-triggered fluorescence. Biotechniques *32*, 940–944, 946–949.

Bailly, C., and Henichart, J.P. (1991). DNA recognition by intercalator-minor-groove binder hybrid molecules. Bioconjug. Chem. *2*, 379–393.

Bergstrom, D.E., Zhang, P., and Johnson, W.T. (1997). Comparison of the base pairing properties of a series of nitroazole nucleobase analogs in the oligodeoxyribonucleotide sequence 5′-d(CGCXAATTYGCG)-3′. Nucleic Acids Res. *25*, 1935–1942.

Bonnet, G., Tyagi, S., Libchaber, A., and Kramer, F.R. (1999). Thermodynamic basis of the enhanced specificity of structured DNA probes. Proc. Natl. Acad. Sci. U.S.A. *96*, 6171–6176.

Bratu, D.P., Cha, B.J., Mhlanga, M.M., Kramer, F.R., and Tyagi, S. (2003). Visualizing the distribution and transport of mRNAs in living cells. Proc. Natl. Acad. Sci. U.S.A. *100*, 13308–13313.

Cha, R.S., Zarbl, H., Keohavong, P., and Thilly, W.G. (1992). Mismatch amplification mutation assay (MAMA): application to the c-H-ras gene. PCR Methods Appl. *2*, 14–20.

Costa, J.M., Ernault, P., Olivi, M., Gaillon, T., and Arar, K. (2004). Chimeric LNA/DNA probes as a detection system for real-time PCR. Clin. Biochem. *37*, 930–932.

De la Vega, F.M., Lazaruk, K.D., Rhodes, M.D., and Wenz, M.H. (2005). Assessment of two flexible and compatible SNP genotyping platforms: TaqMan SNP Genotyping Assays and the SNPlex Genotyping System. Mutat. Res. *573*, 111–135.

Dempcy, R.O., Podyminogin, M.A., and Reed, M.W. (2003). Synthesis, characterization, and application of substituted pyrazolopyrimidine nucleosides. Curr. Protoc. Nucleic Acid Chem. *Chapter 1*, Unit 1.8.

Deutscher, M.P., and Kornberg, A. (1969). Enzymatic synthesis of deoxyribonucleic acid. 28. The pyrophosphate exchange and pyrophosphorolysis reactions of deoxyribonucleic acid polymerase. J. Biol. Chem. *244*, 3019–3028.

Di Giusto, D.A., and King, G.C. (2004). Strong positional preference in the interaction of LNA oligonucleotides with DNA polymerase and proofreading exonuclease activities: implications for genotyping assays. Nucleic Acids Res. *32*, e32.

Dobosy, J.R., Rose, S.D., Beltz, K.R., Rupp, S.M., Powers, K.M., Behlke, M.A., and Walder, J.A. (2011). RNase H-dependent PCR (rhPCR): improved specificity and single nucleotide polymorphism detection using blocked cleavable primers. BMC Biotechnol. *11*, e80.

Duck, P., Alvarado-Urbina, G., Burdick, B., and Collier, B. (1990). Probe amplifier system based on chimeric cycling oligonucleotides. Biotechniques 9, 142–148.

Eder, P.S., and Walder, J.A. (1991). Ribonuclease H from K562 human erythroleukemia cells. Purification, characterization, and substrate specificity. J. Biol. Chem. *266*, 6472–6479.

Eder, P.S., Walder, R.Y., and Walder, J.A. (1993). Substrate specificity of human RNase H1 and its role in excision repair of ribose residues misincorporated in DNA. Biochimie 75, 123–126.

Gamper, H.B., Jr., Arar, K., Gewirtz, A., and Hou, Y.M. (2006). Unrestricted hybridization of oligonucleotides to structure-free DNA. Biochemistry 45, 6978–6986.

Haruki, M., Hayashi, K., Kochi, T., Muroya, A., Koga, Y., Morikawa, M., Imanaka, T., and Kanaya, S. (1998). Gene cloning and characterization of recombinant RNase HII from a hyperthermophilic archaeon. J. Bacteriol. *180*, 6207–6214.

Harvey, J.J., Lee, S.P., Chan, E.K., Kim, J.H., Hwang, E.S., Cha, C.Y., Knutson, J.R., and Han, M.K. (2004). Characterization and applications of CataCleave probe in real-time detection assays. Anal. Biochem. *333*, 246–255.

Harvey, J.J., Brant, S.R., Knutson, J.R., and Han, M.K. (2008). SNP analysis using catacleave probes. J. Clin. Lab. Anal. *22*, 192–203.

Hill, F., Loakes, D., and Brown, D.M. (1998). Polymerase recognition of synthetic oligodeoxyribonucleotides incorporating degenerate pyrimidine and purine bases. Proc. Natl. Acad. Sci. U.S.A. 95, 4258–4263.

Hogrefe, H.H., Hogrefe, R.I., Walder, R.Y., and Walder, J.A. (1990). Kinetic analysis of *Escherichia coli* RNase H using DNA-RNA-DNA/DNA substrates. J. Biol. Chem. 265, 5561–5566.

Holland, P.M., Abramson, R.D., Watson, R., and Gelfand, D.H. (1991). Detection of specific polymerase chain reaction product by utilizing the 5′–3′ exonuclease activity of *Thermus aquaticus* DNA polymerase. Proc. Natl. Acad. Sci. U.S.A. *88*, 7276–7280.

Hou, J., Liu, X., Pei, D., and Liu, J. (2007a). RNase HII from Chlamydia pneumoniae discriminates mismatches incorporation into DNA-rN1-DNA/DNA duplexes. Biochem. Biophys. Res. Commun. *356*, 988–992.

Hou, J., Liu, X., Wang, J., Liu, J., and Duan, T. (2007b). Molecular beacons for isothermal fluorescence enhancement by the cleavage of RNase HII from Chlamydia pneumoniae. Anal. Biochem. *371*, 162–166.

Hou, J., Liu, X., Zheng, Y., and Liu, J. (2007c). A method for HLA genotyping using the specific cleavage of DNA-rN(1)-DNA/DNA with RNase HII from *Chlamydia pneumoniae*. Oligonucleotides *17*, 433–444.

Johnson, M.P., Haupt, L.M., and Griffiths, L.R. (2004). Locked nucleic acid (LNA) single nucleotide polymorphism (SNP) genotype analysis and validation using real-time PCR. Nucleic Acids Res. *32*, e55.

Kaji, T., Ito, S., Iwai, S., and Miyasaka, H. (2009). Nanosecond to submillisecond dynamics in dye-labeled single-stranded DNA, as revealed by ensemble measurements and photon statistics at single-molecule level. J. Phys. Chem. B *113*, 13917–13925.

Kanaya, S. (2001). Prokaryotic type 2 RNases H. Methods Enzymol. *341*, 377–394.

Kawase, Y., Iwai, S., Inoue, H., Miura, K., and Ohtsuka, E. (1986). Studies on nucleic acid interactions. Stabilities, I. of mini-duplexes (dG2A4XA4G2-dC2T4YT4C2) and self-complementary d(GGGAAXYTTCCC) containing deoxyinosine and other mismatched bases. Nucleic Acids Res. *14*, 7727–7736.

Kuimelis, R.G., Livak, K.J., Mullah, B., and Andrus, A. (1997). Structural analogues of TaqMan probes for real-time quantitative PCR. Nucleic Acids Symp. Ser., 255–256.

Kutyavin, I.V., Rhinehart, R.L., Lukhtanov, E.A., Gorn, V.V., Meyer, R.B., Jr., and Gamper, H.B., Jr. (1996). Oligonucleotides containing 2-aminoadenine and 2-thiothymine act as selectively binding complementary agents. Biochemistry 35, 11170–11176.

Kutyavin, I.V., Lukhtanov, E.A., Gamper, H.B., and Meyer, R.B. (1997). Oligonucleotides with conjugated dihydropyrroloindole tripeptides: base composition and backbone effects on hybridization. Nucleic Acids Res. 25, 3718–3723.

Kutyavin, I.V., Afonina, I.A., Mills, A., Gorn, V.V., Lukhtanov, E.A., Belousov, E.S., Singer, M.J., Walburger, D.K., Lokhov, S.G., Gall, A.A., *et al.* (2000). 3′-minor groove binder-DNA probes increase sequence specificity at PCR extension temperatures. Nucleic Acids Res. *28*, 655–661.

Kutyavin, I.V., Lokhov, S.G., Afonina, I.A., Dempcy, R., Gall, A.A., Gorn, V.V., Lukhtanov, E., Metcalf, M., Mills, A., Reed, M.W., *et al.* (2002). Reduced aggregation and improved specificity of G-rich oligodeoxyribonucleotides containing pyrazolo[3,4-d]pyrimidine guanine bases. Nucleic Acids Res. *30*, 4952–4959.

Kutyavin, I., Lokhov, S., Lukhtanov, E., and Reed, M.W. (2003). Chemistry of minor groove binder–oligonucleotide conjugates. Curr. Protoc. Nucleic Acid Chem. *Chapter 8*, Unit 8.4.

Kvaerno, L., and Wengel, J. (1999). Investigation of restricted backbone conformations as an explanation for the exceptional thermal stabilities of duplexes involving LNA (Locked Nucleic Acid):[dagger] synthesis and evaluation of abasic LNA. Chemical Communications, 657–658.

Lai, B., Li, Y., Cao, A., and Lai, L. (2003). Metal ion binding and enzymatic mechanism of Methanococcus jannaschii RNase HII. Biochemistry *42*, 785–791.

Latorra, D., Campbell, K., Wolter, A., and Hurley, J.M. (2003). Enhanced allele-specific PCR discrimination in SNP genotyping using 3′ locked nucleic acid (LNA) primers. Hum. Mutat. *22*, 79–85.

Le Laz, S., Le Goaziou, A., and Henneke, G. (2010). Structure-specific nuclease activities of Pyrococcus abyssi RNase HII. J. Bacteriol. *192*, 3689–3698.

Lebedev, A.V., Paul, N., Yee, J., Timoshchuk, V.A., Shum, J., Miyagi, K., Kellum, J., Hogrefe, R.I., and Zon, G. (2008). Hot start PCR with heat-activatable primers: a novel approach for improved PCR performance. Nucleic Acids Res. *36*, e131.

Letertre, C., Perelle, S., Dilasser, F., Arar, K., and Fach, P. (2003). Evaluation of the performance of LNA and MGB probes in 5′-nuclease PCR assays. Mol. Cell. Probes *17*, 307–311.

Liu, Q., and Sommer, S.S. (2000). Pyrophosphorolysis-activated polymerization (PAP): application to allele-specific amplification. Biotechniques *29*, 1072–1076, 1078, 1080 passim.

Liu, Q., and Sommer, S.S. (2002). Pyrophosphorolysis-activatable oligonucleotides may facilitate detection of rare alleles, mutation scanning and analysis of chromatin structures. Nucleic Acids Res. *30*, 598–604.

Liu, X.P., Hou, J.L., and Liu, J.H. (2010). A novel single nucleotide polymorphism detection of a double-stranded DNA target by a ribonucleotide-carrying molecular beacon and thermostable RNase HII. Anal. Biochem. *398*, 83–92.

Livak, K.J. (1999). Allelic discrimination using fluorogenic probes and the 5′ nuclease assay. Genet. Anal. *14*, 143–149.

Livak, K.J., Flood, S.J., Marmaro, J., Giusti, W., and Deetz, K. (1995). Oligonucleotides with fluorescent dyes at opposite ends provide a quenched probe system useful for detecting PCR product and nucleic acid hybridization. PCR Methods Appl. *4*, 357–362.

Loakes, D. (2001). Survey and summary: The applications of universal DNA base analogues. Nucleic Acids Res. *29*, 2437–2447.

Loakes, D., Brown, D.M., Linde, S., and Hill, F. (1995). 3-Nitropyrrole and 5-nitroindole as universal bases in primers for DNA sequencing and PCR. Nucleic Acids Res. *23*, 2361–2366.

McTigue, P.M., Peterson, R.J., and Kahn, J.D. (2004). Sequence-dependent thermodynamic parameters for locked nucleic acid (LNA)-DNA duplex formation. Biochemistry *43*, 5388–5405.

Martin, F.H., Castro, M.M., Aboul-ela, F., and Tinoco, I., Jr. (1985). Base pairing involving deoxyinosine: implications for probe design. Nucleic Acids Res. *13*, 8927–8938.

Mendelman, L.V., Petruska, J., and Goodman, M.F. (1990). Base mispair extension kinetics. Comparison of DNA polymerase alpha and reverse transcriptase. J. Biol. Chem. *265*, 2338–2346.

Moreau, V., Voirin, E., Paris, C., Kotera, M., Nothisen, M., Remy, J.S., Behr, J.P., Erbacher, P., and Lenne-Samuel, N. (2009). Zip Nucleic Acids: new high affinity oligonucleotides as potent primers for PCR and reverse transcription. Nucleic Acids Res. *37*, e130.

Moreira, B.G., You, Y., Behlke, M.A., and Owczarzy, R. (2005). Effects of fluorescent dyes, quenchers, and dangling ends on DNA duplex stability. Biochem. Biophys. Res. Commun. *327*, 473–484.

Mukaiyama, A., Takano, K., Haruki, M., Morikawa, M., and Kanaya, S. (2004). Kinetically robust monomeric protein from a hyperthermophile. Biochemistry *43*, 13859–13866.

Myakishev, M.V., Khripin, Y., Hu, S., and Hamer, D.H. (2001). High-throughput SNP genotyping by allele-specific PCR with universal energy-transfer-labeled primers. Genome Res. *11*, 163–169.

Nakitandwe, J., Trognitz, F., and Trognitz, B. (2007). Reliable allele detection using SNP-based PCR primers containing Locked Nucleic Acid: application in genetic mapping. Plant Methods *3*, 2.

Newton, C.R., Graham, A., Heptinstall, L.E., Powell, S.J., Summers, C., Kalsheker, N., Smith, J.C., and Markham, A.F. (1989). Analysis of any point mutation in DNA. The amplification refractory mutation system (ARMS). Nucleic Acids Res. *17*, 2503–2516.

Nichols, R., Andrews, P.C., Zhang, P., and Bergstrom, D.E. (1994). A universal nucleoside for use at ambiguous sites in DNA primers. Nature *369*, 492–493.

Noir, R., Kotera, M., Pons, B., Remy, J.S., and Behr, J.P. (2008). Oligonucleotide–oligospermine conjugates (zip nucleic acids): a convenient means of finely tuning hybridization temperatures. J. Am. Chem. Soc. *130*, 13500–13505.

Obika, S., Nanbu, D., Hari, Y., Morio, K.-i., In, Y., Ishida, T., and Imanishi, T. (1997). Synthesis of 2′-*O*,4′-*C*-methyleneuridien and -cytidine. Novel bicyclic nucleotsides having a fixed C3,-*endo* sugar puckering. Tetrahedron Lett. *38*, 8735–8738.

Ohtani, N., Haruki, M., Morikawa, M., and Kanaya, S. (1999). Molecular diversities of RNases H. J. Biosci. Bioeng. *88*, 12–19.

Owczarzy, R. (2005). Melting temperatures of nucleic acids: discrepancies in analysis. Biophys. Chem. *117*, 207–215.

Owczarzy, R., You, Y., Moreira, B.G., Manthey, J.A., Huang, L., Behlke, M.A., and Walder, J.A. (2004). Effects of sodium ions on DNA duplex oligomers: improved predictions of melting temperatures. Biochemistry *43*, 3537–3554.

Owczarzy, R., Moreira, B.G., You, Y., Behlke, M.A., and Walder, J.A. (2008a). Predicting stability of DNA duplexes in solutions containing magnesium and monovalent cations. Biochemistry *47*, 5336–5353.

Owczarzy, R., Tataurov, A.V., Wu, Y., Manthey, J.A., McQuisten, K.A., Almabrazi, H.G., Pedersen, K.F., Lin, Y., Garretson, J., McEntaggart, N.O., *et al.* (2008b). IDT SciTools: a suite for analysis and design of nucleic acid oligomers. Nucleic Acids Res. *36*, W163–169.

Owczarzy, R., You, Y., Groth, C.L., and Tataurov, A.V. (2011). Stability and mismatch discrimination of locked nucleic acid-DNA duplexes. Biochemistry *50*, 9352–9367.

Paris, C., Moreau, V., Deglane, G., Voirin, E., Erbacher, P., and Lenne-Samuel, N. (2010). Zip nucleic acids are potent hydrolysis probes for quantitative PCR. Nucleic Acids Res. *38*, e95.

Reddy, M.S., and Hardin, S.H. (2003). Features in short guanine-rich sequences that stimulate DNA polymerization *in vitro*. Biochemistry *42*, 350–362.

SantaLucia, J., Jr. (1998). A unified view of polymer, dumbbell, and oligonucleotide DNA nearest-neighbor thermodynamics. Proc. Natl. Acad. Sci. U.S.A. *95*, 1460–1465.

SantaLucia, J., Jr., and Hicks, D. (2004). The thermodynamics of DNA structural motifs. Annu. Rev. Biophys. Biomol. Struct. *33*, 415–440.

SantaLucia, J., Jr., Allawi, H.T., and Seneviratne, P.A. (1996). Improved nearest-neighbor parameters for predicting DNA duplex stability. Biochemistry *35*, 3555–3562.

Sato, A., Kanai, A., Itaya, M., and Tomita, M. (2003). Cooperative regulation for Okazaki fragment processing by RNase HII and FEN-1 purified from a hyperthermophilic archaeon, Pyrococcus furiosus. Biochem. Biophys. Res. Commun. *309*, 247–252.

Shi, J., Liu, Q., and Sommer, S.S. (2007). Detection of ultrarare somatic mutation in the human TP53 gene by bidirectional pyrophosphorolysis-activated polymerization allele-specific amplification. Hum. Mutat. *28*, 131–136.

Shinozaki, M., O'Day, S.J., Kitago, M., Amersi, F., Kuo, C., Kim, J., Wang, H.J., and Hoon, D.S. (2007). Utility of circulating B-RAF DNA mutation in serum for monitoring melanoma patients receiving biochemotherapy. Clin. Cancer Res. *13*, 2068–2074.

Simonsson, T. (2001). G-quadruplex DNA structures – variations on a theme. Biol. Chem. *382*, 621–628.

Singh, S.K., and Wengel, J. (1998). Universality of LNA-mediated high-affinity nucleic acid recognition. Chem. Commun., 1247–1248.

Singh, S.K., Koshkin, A. A., Nielsen, P., and Wengel, J. (1998). LNA (locked nucleic acids): synthesis and high-affinity nucleic acid recognition. Chem. Commun., 455–456.

Tatsumi, K., Mitani, Y., Watanabe, J., Takakura, H., Hoshi, K., Kawai, Y., Kikuchi, T., Kogo, Y., Oguchi-Katayama, A., Tomaru, Y., *et al.* (2008). Rapid screening assay for KRAS mutations by the modified smart amplification process. J. Mol. Diagn. *10*, 520–526.

Tenesa, A., Farrington, S.M., Prendergast, J.G., Porteous, M.E., Walker, M., Haq, N., Barnetson, R.A., Theodoratou, E., Cetnarskyj, R., Cartwright, N., *et al.* (2008). Genome-wide association scan identifies a colorectal cancer susceptibility locus on 11q23 and replicates risk loci at 8q24 and 18q21. Nat. Genet. *40*, 631–637.

Thelwell, N., Millington, S., Solinas, A., Booth, J., and Brown, T. (2000). Mode of action and application of Scorpion primers to mutation detection. Nucleic Acids Res. *28*, 3752–3761.

Tsourkas, A., Behlke, M.A., and Bao, G. (2003a). Hybridization of 2′-O-methyl and 2′-deoxy molecular beacons to RNA and DNA targets. Nucleic Acids Res. *31*, 5168–5174.

Tsourkas, A., Behlke, M.A., Rose, S.D., and Bao, G. (2003b). Hybridization kinetics and thermodynamics of molecular beacons. Nucleic Acids Res. *31*, 1319–1330.

Tyagi, S., and Kramer, F.R. (1996). Molecular beacons: probes that fluoresce upon hybridization. Nat. Biotechnol. *14*, 303–308.

Tyagi, S., Bratu, D.P., and Kramer, F.R. (1998). Multicolor molecular beacons for allele discrimination. Nat. Biotechnol. *16*, 49–53.

Ugozzoli, L.A., Latorra, D., Puckett, R., Arar, K., and Hamby, K. (2004). Real-time genotyping with oligonucleotide probes containing locked nucleic acids. Anal. Biochem. *324*, 143–152.

Wagner, R.W., Matteucci, M.D., Lewis, J.G., Gutierrez, A.J., Moulds, C., and Froehler, B.C. (1993). Antisense gene inhibition by oligonucleotides containing C-5 propyne pyrimidines. Science *260*, 1510–1513.

Walder, J.A., and Walder, R.Y. (1995). Nucleic acid hybridization and amplification method for detection of specific sequences in which a complementary labeled nucleic acid probe is cleaved. US Patent 5,403,711.

Walder, J.A., Behlke, M.A., Rose, S.D., and Dobosy, J.R. (2009). RNase H-based assays utilizing modified RNA monomers. US Patent Application 2009/0325169.

Yamane, A. (2002). MagiProbe: a novel fluorescence quenching-based oligonucleotide probe carrying a fluorophore and an intercalator. Nucleic Acids Res. *30*, e97.

You, Y., Moreira, B.G., Behlke, M.A., and Owczarzy, R. (2006). Design of LNA probes that improve mismatch discrimination. Nucleic Acids Res. *34*, e60.v

DOI: https://doi.org/10.21775/9781912530243.10

Internal Amplification Controls in Real-time Polymerase Chain Reaction-based Methods for Pathogen Detection

10

Nigel Cook, Gabriel A. de Ridder, Martin D'Agostino and Maureen B. Taylor

Abstract

Assays based on nucleic acid amplification are highly efficient, but they can be affected by the presence of matrix-derived substances which can interfere or prevent the reaction from performing correctly. Careful sample treatment must be applied/used to remove these inhibitory substances. However no sample treatment can be relied on completely, thus an amplification control should be employed to be able to verify that the assay has performed correctly. An internal amplification control (IAC) is a non-target NA sequence present in the very same reaction as the sample or target nucleic acid extract. If it is successfully amplified to produce a signal, any non-production of a target signal in the reaction is considered to signify that the sample did not contain the target pathogen or organism. If, however, the reaction produces neither a signal from the target nor the IAC, it signifies that the reaction has failed. An IAC concentration optimisation protocol is described.

Introduction

The technological progression from the end-point detection of polymerase chain reaction (PCR) products to their detection in real-time has been widely taken up by the research community, and now nearly all studies which involve nucleic acid amplification-based pathogen detection utilize real-time PCR or quantitative PCR (qPCR). The inherent advantages of qPCR-based methods for pathogen detection over conventional PCR and/or culturing, e.g. shorter turnaround, improved detection limits, specificity and potential for automation. (Rutjes *et al.*, 2005; Bosch *et al.*, 2011) have fostered their widespread implementation in analytical laboratories, especially those dedicated to analysis of food and water (de Paula *et al.*, 2007; Le Guyader *et al.*, 2009; Nenonen *et al.*, 2009; Baque *et al.*, 2011). Currently PCR and RT-PCR is considered to be the 'gold standard' for virus detection (Rolfe *et al.*, 2007; Bosch *et al.*, 2008) and PCR was predicted to be established as a routine reference by 2010 (Hoorfar and Cook, 2003). This however has not happened for a variety of reasons (D'Agostino and Rodriguez-Lazaro 2009), including the fact that there are currently no standard procedures for the detection of viruses in environmental samples (Mattison and Bidawid, 2009; Verheyen *et al.*, 2009). Among the main issues that must be addressed for the effective adoption of PCR-based techniques in analytical laboratories is the concern that they can not always be used effectively for analysis of complex matrices such as foods because of substances inimical to the amplification reaction which may exist in these matrices (Abu Al Soud and Radstrom, 2000; Rossen *et al.*, 1992; Mattison and Bidawid, 2009; Plante *et al.*, 2011). This may cause a negative signal to be obtained when the sample did in fact contain the target microorganism. A false-negative result of analysis of a contaminated sample could result in the exposure of the public to a potential pathogen, with serious consequences.

qPCRs require that nucleic acids be extracted from a sample then delivered to the assay. As qPCR assays use small (microliter) volumes of nucleic acid as template, the microorganism and/or extracted nucleic acids must be concentrated during sample preparation to ensure that target nucleic acids are represented within the template. This process however can also concentrate inhibitory substances with the target nucleic acids. Among the substances which can cause inhibition of a qPCR are bile salts (Widjojoatmodjo *et al.* 1992), haem from blood and humic acids (Hata *et al.*, 2011), polysaccharides and polyphenols (Wei *et al.*, 2008; Plante *et al.*, 2011) which may be co-extracted with the nucleic acids. No process can be relied on to completely remove all of the inhibitory substances (Parshionikar *et al.*, 2004) and thus effective quality control procedures need to be implemented to ensure the validity of the qPCR result (Wong *et al.*, 2007; Pintó and Bosch, 2008; Bosch *et al.*, 2011). An important control for verification of the correct operation of the qPCR assay is an amplification control (AC) (Rolfe *et al.*, 2007; Hoffmann *et al.*, 2009; Lees, 2010; Bosch *et al.*, 2011). From the literature it is evident that different AC systems are being applied. One of the approaches is to use a house-keeping gene which is present in the test sample (Wei *et al.*, 2008) and this is referred to as an endogenous internal control (IC) (Hoffmann *et al.*, 2009). The other approach, which is more suitable for the analysis of water and food samples, is to use an exogenous AC which comprises a nucleic acid sequence that does not occur naturally in the sample under investigation and is unrelated to the target nucleic acid sequence (Hoorfar *et al.*, 2004; Hoffmann *et al.*, 2009; Rip and Gouws, 2009) or is a modification of the target sequence (mimic ICs)(Escobar-Herrera *et al.*, 2006; Hoffmann *et al.*, 2009). The exogenous AC can be constructed in various ways (Rodríguez-Lázaro *et al.*, 2004; Oikonomou *et al.*, 2008; Rip and Gouws, 2009; Deer *et al.*, 2010; Diez-Valcarce *et al.*, 2011; Gregory *et al.*, 2011) and applied in different formats, namely as a competitive or non-competitive internal amplification control (IAC) (Hoorfar *et al.*, 2004; Parshionikar *et al.*, 2004; Deer *et al.*, 2010; Hata *et al.*, 2011) or as a external amplification control (EAC) (Costafreda *et al.*, 2006; Lees, 2010). In the literature the IAC is also referred to as an IC (Rolfe *et al.*, 2007; Hoffmann *et al.*, 2009) or internal positive control (IPC) (Lauri and Mariani, 2009).

IAC rationale

An IAC is a non-target nucleic acid sequence present in the same reaction as the sample nucleic acid extract, and which can be amplified simultaneously with the target. An IAC signal should always be produced when there are no target sequences present in the qPCR. In a qPCR without an IAC, the absence of a target signal can be interpreted as signifying that the sample did not contain the target. However, it could have been that the qPCR had failed through the presence of inhibitory substances, or for other reasons such as thermocycler failure or incorrect mastermix preparation. In this case a false-negative interpretation of the result is possible. The use of an IAC will prevent false-negative interpretations.

To be most effective, the IAC should possess the same primer complementary sequences as the target, in other words, it should be amplifiable by the same primers as are used to amplify the target. However, it should be readily distinguishable from the target amplicon. In conventional PCR this is achieved by making the IAC amplicon a different size to the target, but in qPCR an IAC amplicon-specific probe is used containing a different fluorescent dye from that used to detect the target (Rodríguez-Lázaro *et al.*, 2005a,b). The use of the same primers for amplification of IAC and target will result in competition between the IAC and the target for the primers. This competition means that if there is a greatly higher concentration of target sequences within the qPCR then the IAC may not be amplified, or be amplified to a lesser extent than the target. In such a case, only the target signal will appear. It is not necessary in that case for an IAC signal to be obtained, for of course the presence of a target signal means that the qPCR has operated correctly.

Table 4.1 shows the pattern of results which can be obtained when using an IAC in an qPCR, and how to interpret them.

Table 4.1 Interpretation of pattern of signals which can be observed when an IAC is incorporated into a qPCR-based method for pathogen detection

Observed result	Meaning	Action
Target signal and IAC signal present	qPCR has operated correctly. Sample is target-positive	Record sample as contaminated
Only target signal present	qPCR has operated correctly. Sample is target positive with high amount of target	Record sample as contaminated
IAC signal present and target signal absent	qPCR has operated correctly. Sample is target-negative	Record sample as uncontaminated
IAC and target signal absent	qPCR has failed	Repeat using new mastermix and/or test a dilution of sample nucleic acid extract

IAC construction for qPCR

The basic steps in IAC construction are described below.

Construction of internal amplification control

1. Design primers containing target, non-target, and RNA polymerase sequences.
2. Get non-target DNA.
3. Amplify non-template DNA with hybrid primers.
4. Purify amplicon.
5. Transcribe amplicon (if RNA IAC required).
6. Purify IAC.
7. Quantify IAC.

General procedures for construction of IACs are given in Hoorfar *et al.* (2004). As mentioned, the signal from amplification of an IAC for qPCR needs to be distinguishable from the signal from the target, and this is most often achieved by using different probes. Therefore the IAC oligonucleotide must contain specific sequences to which its probe can bind, and these sequences must be totally different to those targeted for amplification within the pathogen's genome.

An IAC construction procedure that has been widely used involves utilizing probe–sequence combinations from qPCR assays for microorganisms other than the target. For example, when constructing an IAC for a qPCR for *Mycobacterium avium* subsp. *paratuberculosis*, Rodriguez-Lazaro *et al.* (2005a) incorporated sequences from *Listeria monocytogenes* which hybridized with a probe which incorporated the VIC dye. The amplicon could thus be distinguished from the target amplicon, which hybridized with a probe incorporating the FAM dye.

To construct the IAC oligonucleotide containing these non-target sequences along with sequences complementary to the target primers, it is necessary to first construct chimeric primers which link the target primer sequences to primers for the non-target sequences. These chimeric primers are then used to amplify the non-target sequences, and the resulting amplicon contains the full IAC sequence. The amplicon can then be purified and used in the intended qPCR, or cloned into a plasmid which is subsequently used as the IAC.

It is advisable to construct the IAC oligonucleotide such that it is ~100–200 bp larger than the target sequence. In this way, competition between the IAC and target is more favourable to the target.

Fig. 4.1 shows a schematic diagram of this procedure.

Optimization of IAC concentration

It is essential that a competitive IAC be incorporated in an qPCR in an amount which will not out-compete the amplification of low copies of target. To achieve this, it is necessary to optimize the quantity of IAC which is used in the qPCR. The protocol below will mediate optimization of the IAC concentration.

1. Dilute the quantified IAC solution to, for example, 10^{-10} using nuclease-free water with 0.1 mg /ml bovine serum albumin (BSA) added.

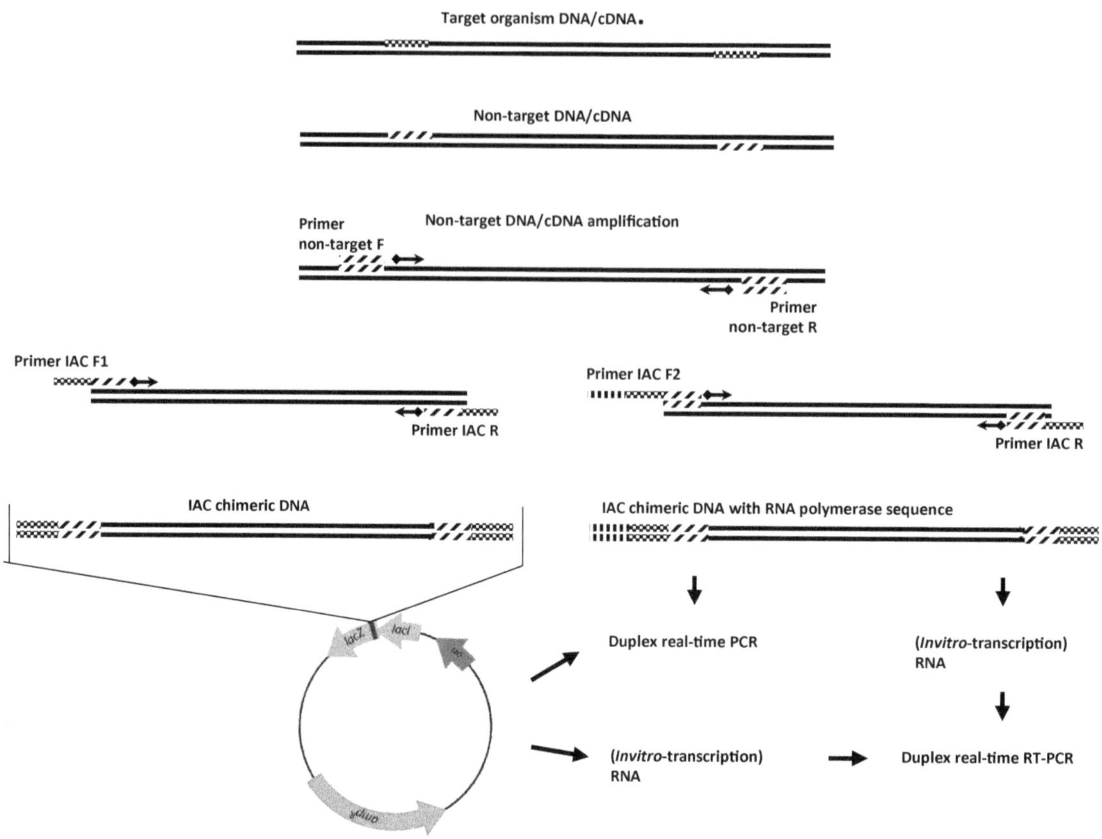

Figure 4.1 Representation of two methods to construct an IAC. (A) Primer IAC F1 comprises the target non-target ( ) forward primer sequences. (B) Primer IAC F2 comprises the target/non-target forward primer sequences as well as an RNA polymerase promoter sequence ( ). Primer IAC R comprises the target/non-target reverse primer sequences.

2 Prepare a mastermix for 11× reactions. Do not add IAC to this mastermix. Replace the volume of target as quoted in the relevant protocol, with the equivalent volume of water.
3 Add n^* μl of each IAC dilution from step 1, to 10 reactions. To the 11th reaction, instead of the IAC, add the same amount of water (negative control). *The volume of IAC which is specified in the relevant protocol.
4 Perform PCR. If it is a nested PCR, perform both rounds.
5 Choose the highest dilution of IAC which produces a positive signal.
6 Prepare a mastermix for four reactions. Again, do not add IAC to this mastermix. Replace the volume of target as quoted in the relevant protocol, with the equivalent volume of water.
7 Add n μl of the dilution chosen in step 5 to the first reaction.
8 Add n μl of the next highest dilution (e.g. if the dilution chosen in step 5 is 10^{-6}, then use 10 μl this step) to the second reaction.
9 Add n μl of the next lowest dilution (e.g. if the dilution chosen in step 5 is 10^{-6}, then use 10^{-5} in this step) to the third reaction.
10 In the fourth reaction use water instead IAC dilution sample.
11 Perform PCR. If it is a nested PCR, perform both rounds.
12 Repeat steps 6–11 twice more.
13 Choose the highest dilution of IAC which consistently gives three positive signals from the three separate mastermixes. This is the working dilution of IAC which is to be used in all subsequent tests which include this particular batch of IAC.

14 Determine quantity of working IAC by extrapolating dilution from the quantity in the neat stock.

It will be found that the concentration of IAC set using this protocol will not out-compete low copies of the target.

Examples of IAC use in qPCR-based methods for pathogen detection

IACs are especially important when a culture-based detection of a bacterium can be very slow, as with mycobacteria, and therefore PCR must be used on nucleic acids extracted directly from a sample, which may contain inhibitors. Rodriguez-Lazaro *et al.* (2005a) developed a real-time PCR assay for quantitative detection of *Mycobacterium avium* subsp. *paratuberculosis* (MAP) which amplified sequences from the IS*900* insertion element which is specific for this bacterium, and included an IAC. The IAC consisted of a 64-bp fragment of a listerolysin (*hly*) gene from *L. monocytogenes* (Mengaud *et al.*, 1988) flanked by primers for IS*900* amplification. It was constructed by amplifying *L. monocytogenes* sequences using hybrid primers where each IS900 sequence flanked 5′ a *hly* sequence. A second PCR reaction performed with the *IS900*-specific primers confirmed that the chimeric DNA produced in the first round PCR contains the complete sequence of the target primers. One hundred molecules of IAC were used per reaction in subsequent tests. The final assay consistently could detect 12 MAP cells with 99% probability, demonstrating its sensitivity and the lack of interference with target amplification by the IAC. Harnessing the assay to a sample pretreatment for milk based on protein and fat removal allowed the resulting method to be able to detect down to 100 MAP cells in 20 ml milk, with no failure of any reaction.

The most significant food-transmitted or environmentally transmitted viruses, such as norovirus, hepatitis A virus or hepatitis E virus, can not be grown, or can be grown only with difficulty, *in vitro* in cell culture, and therefore direct extraction of nucleic acids from a sample followed by PCR are essential for their detection. This being the case, an amplification control is an absolute necessity to ensure accurate interpretation of results. Diez-Valcarce *et al.* (2011) constructed IACs for existing qPCR or reverse transcription qPCR (RTqPCR) assays for a range of viruses which could be found in the environment or foods: bovine polyomavirus, hepatitis A virus, hepatitis E virus, human adenovirus, human norovirus genogroup I, human norovirus genogroup II, murine norovirus and porcine adenovirus. The IACs for the assays for the DNA viruses were plasmid-based, and those for the RNA virus assays were themselves RNA molecules which were transcribed from chimeric sequences cloned into a plasmid transcription vector, following the principles for design of RNA IACs devised by Rodriguez-Lazaro *et al.* (2004). Addition of optimized (following the procedure outlined above) amounts of IAC into the assays did not affect the limits of detection for each specific target virus. To demonstrate how the IACs would indicate reaction failure, a situation in which purification of nucleic acid from a food sample was poorly performed was simulated, by adding 50 μl of non-extracted strawberry puree to 50 μl of nucleic acid extract prior to nucleic acid amplification; no IAC signal was obtained, whilst robust signals were obtained from nucleic acid extract correctly obtained from strawberry samples (by the method of Dubois *et al.*, 2006).

The study of Radin and D'Souza (2011), in which IACs were developed for two broadly reactive qRTPCR assays for noroviruses, provided further demonstration that properly optimized IAC-containing reactions have the same limit of detection as the corresponding assays without IAC. This shows that concerns over competition between IAC and target can be overcome by careful optimization.

The other approach to the use of ACs is the EAC. This involves performing two separate reactions for each sample – one (the test) reaction contains only the sample nucleic acid, but the other (the control reaction) contains the sample nucleic acid plus the amplification control (Costafreda *et al.*, 2006). If it is successfully amplified to produce a signal, any non-production of a target signal in the test reaction is considered to signify that the sample did not contain the target.

If, however, no signal is produced in both the test and control reactions, it signifies that the nucleic acid extract contains inhibitory substances and the reaction has failed. The potential competition issue (see above) has led some workers to adopt the EAC approach. However, using EACs one can never be completely certain that the test reaction has not individually failed, for example through pipetting error or non-homogeneous contamination by inhibitory substances. For example, if an EAC signal is produced in the control reaction, but no target signal is produced in the test reaction, can one be completely certain that the test reaction has not failed? However, using an IAC eliminates this ambiguity, since it is present in the mastermix and a signal will always appear when the reaction has not failed or high levels of competing target are not present (if they are, a target signal will be produced anyway). The concern of the proponents of the EAC approach regards the possibility that a low level of target may be outcompeted by the IAC, leading to a false negative result. However, a thoroughly optimized assay should not present these problems (D'Agostino *et al.*, 2004). A recent study has moreover demonstrated, by a direct comparison between an IAC- and EAC-containing qPCR-based method for detection of NoV GII in foods, that an optimized IAC-containing PCR assay will mediate the same level of detection as an EAC-containing qPCR assay and that it is more cost-effective than an EAC-containing method (de Ridder *et al.*, unpublished data).

A recently proposed (Rossmanith *et al.*, 2011) approach is the use of internal sample process controls (ISPCs). ISPCs are added at the beginning of the sample treatment, like normal sample process controls (SPCs; D'Agostino *et al.*, 2011), then passaged through all subsequent analytical steps. An ISPC can be a recombinant organism containing a single-copy artificial DNA sequence for coamplification. This approach would appear convenient, in that it combines two controls (the SPC and the IAC) in one. However, in routine analysis it would not allow for troubleshooting, e.g. if no ISPC signal is obtained, it will not be clear whether it was the sample process or the amplification which has failed.

Concluding remarks

If qPCR-based monitoring of food supply chains for pathogens is to be effectively performed as part of a food safety programme or an epidemiological investigation, then it is vitally necessary that the reliability of the analytical results can be verified (Rodriguez Lazaro *et al.* 2007). Therefore, during routine monitoring of foodstuffs with a molecular screening method an appropriate suite of controls must be used to verify that the method has performed correctly (D'Agostino *et al.*, 2011), and the AC is an essential component of this suite.

Although an IAC in a qPCR offers an unambiguous control to conclusively show that the assay has performed correctly it must however be borne in mind that the IAC or EAC may be detected in the qPCR even though the nucleic extraction procedure has failed. It is therefore important to implement effective quality control procedures to control the entire analytical process from elution, concentration, nucleic acid extraction to detection (Mattison and Bidawid, 2009; Bosch *et al.*, 2011; D'Agostino *et al.*, 2011) to ensure the validity of results.

References

Anonymous (2005). ISO 22174:2005, 'Microbiology of food and animal feeding stuffs – polymerase chain reaction (PCR) for the detection of food-borne pathogens – General requirements and definitions'. Geneva, International Standards Organization.

Abu AlSoud, W.A., and Rådström, P. (2000). Effect of amplification facilitators on diagnostic PCR in the presence of blood, feces and meat. J. Clin. Microbiol. *38*, 4463–4470.

Baque, R.H., Gilliam, A.O., Robles, L.D., Jakubowski, W., and Slifko, T.R. (2011). A real-time RT-PCR method to detect viable *Giardia lamblia* cysts in environmental waters. Water Res. *45*, 3175–3184.

Bosch A., Guix, S., Sano, D., and Pintó, R.M. (2008). New tools for the study and direct surveillance of viral pathogens in water. Curr. Opin. Biotechnol. *19*, 295–301.

Bosch, A., Sánchez, G., Abbaszadegan, M., Carducci, A., Guix, S., Le Guyader, F.S., Netshikweta, R., Pintó, R.M., van der Poel, W.H.M., Rutjes, S., *et al.* (2011). Analytical methods for virus detection in water and food. Food Anal. Meth. *4*, 4–12.

D'Agostino, M., and Rodriguez-Lazaro, D. (2009). Harmonization and validation of methods in food safety – FOOD-PCR: a case study. In: Global Issues in Food Science and Technology, G.V. Barbosa-Cánovas, A. Mortimer, D. Lineback, W. Spiess, K. Buckle, P. Colonna (eds.). Oxford, Academic Press.

Deer, D.M., Lampel, K.A., and González-Escalona, N. (2010). A versatile internal control for use as DNA in real-time PCR and as RNA in real-time reverse transcription PCR assays. Lett. in Appl. Microbiol. *50*, 366–372.

De Paula, V.S., Diniz-Mendes, L., Villar, L.M., Luz, S.L.B., Silva, L.A., Jesus, M.S., Da Silva, N.M.V.S., and Gaspar, A.M.C. (2007). Hepatitis A virus in environmental water samples from the Amazon Basin. Water Res. *41*, 1169–1176.

Diez-Valcarce, M., Kovač, K., Cook, N., Hernández, M., and Rodríguez-Lázaro, D. (2011). Construction and analytical application of internal amplification controls (IAC) for detection of food supply chain-relevant viruses by real-time PCR-based assays. Food Anal. Meth. *4*, 437–445.

Dubois, E., Hennechart, C., Deboosère, N., Merle, G., Legeay, O., Burger, C., Le Calvé, M., Lombard, B., Ferré, V., and Traoré, O. (2006). Intra-laboratory validation of a concentration method adapted for the enumeration of infectious F-specific RNA coliphage, enterovirus, and hepatitis A virus from inoculated leaves of salad vegetables. Int. J. Food Microbiol. *108*, 164–171.

Escobar-Herrera, J., Cancio, C., Guzmán, G.I., Villegas-Sepulveda, N., Estrada-García, T., García-Lozana, H., Gómez-Santiago, F., and Gutiérrez-Escolano, A.L. (2006). Construction of an internal RT-PCR standard control for the detection of human caliciviruses in stool. J. Virol. Meth. *137*, 334–338.

Frühwirth, K., Fuchs, S., Mester, P., Wagner, M., and Rossmanith, P. (2012). Cloning and Characterization of a Δ-*prfA Listeria monocytogenes* strain containing an artificial single copy genomic internal amplification control (IAC) for use as internal sample process control (ISPC). Food Anal. Methods *5*, 8–18.

Gregory, J.B., Webster, L.F., Griffith, J.F., and Stewart, J.R. (2011). Improved detection and quantitation of noroviruses from water. J. Virol. Meth. *172*, 38–45.

Hata, A., Katayama, H., Kitajima, M., Visvanathan, C., Nol, C., and Furumai, H. (2011). Validation of internal controls for extraction and amplification of nucleic acids from enteric viruses in water samples. Appl. Env. Microbiol. *77*, 4336–4343.

Hoffmann, B., Beer, M., Reid, S.M., Mertens, P., Oura, C.A.L., van Rijn, P.A., Slomka, M.J., Banks, J., Brown, I.H., Alexander, D.J., and King, D.P. (2009). A review of RT-PCR technologies used in veterinary virology and disease control: Sensitive and specific diagnosis of five livestock diseases notifiable to the World Organization for Animal Health. Vet. Microbiol. *139*, 1–23.

Hoorfar, J., and Cook, N. (2003). Critical aspects of standardization of PCR. In: Methods in Molecular Biology: PCR Detection of Microbial Pathogens (K. Sachse and J. Frey, eds.), Humana Press, Totowa. pp. 51–64.

Hoorfar, J., Malorny, B., Abdulmawjood, A., Cook, N., Wagner, M., and Fach, P. (2004). Practical considerations in design of internal amplification control for diagnostic PCR assays. J. Clin. Microbiol. *42*, 1863–1868.

Lauri, A., and Mariani, P.O. (2009). Potentials and limitations of molecular diagnostic methods in food safety. Genes Nutr. *4*, 1–12.

Lees, D., CEN WG6 TAG4. (2010). International standardization of a method for detection of human pathogenic viruses in molluscan shellfish. Food Env. Virol. *2*, 146–155.

Le Guyader, F.S., Parnaudeau, S., Schaeffer, J., Bosch, A., Loisy, F., Pommepuy, M., and Atmar, R.L. (2009). Detection and quantification of noroviruses in shellfish Appl. Env. Microbiol. *75*, 618–624.

Mattison, K., and Bidawid, S. (2009). Analytical methods for food and environmental viruses. Food Env. Virol. *1*, 107–122.

Mengaud, J., Vicente, M.F., Chenevert, J., Pereira, J.M., Geoffroy, C., Gicquel-Sanzey, B., Baquero, F., Perez-Diaz, J.C., and Cossart, P. (1988). Expression in *Escherichia coli* and sequence analysis of the listeriolysin O determinant of *Listeria monocytogenes*. Infect. Immun. *56*, 766–772.

Nenonem, N.P., Hannoun, C., Olsson, M.B., and Bergström, T. (2009). Molecular analysis of an oyster-related norovirus outbreak. J. Clin. Virol. *45*, 105–108.

Parshionikar, S.U., Cashdollar, J., and Fout, G.S. (2004). Development of homologous viral internal controls for use in RT-PCR assays of water-borne enteric viruses. J. Virol. Meth. *121*, 39–48.

Pintó, R.M., and Bosch, A. (2008). Rethinking virus detection in food. In: Koopmans, M.P.G., Cliver, D.O., and Bosch, A. (eds.) Food-borne Viruses: Progress and Challenges. Washington DC: ASM Press, pp. 171–188.

Plante, D., Bélanger, G., Leblanc, D., Ward, P., Houde, A., Trottier, Y.-L. (2011). The use of bovine serum albumin to improve the RT-qPCR detection of food-borne viruses rinsed from vegetable surfaces. Lett. Appl. Microbiol. *52*, 239–244.

Radin, D., and D'Souza, D.H. (2011). Evaluation of two primer sets using newly developed internal amplification controls for rapid human norovirus detection by SYBR green I based real-time RT-PCR. Food Env. Virol. *3*, 61–69.

Rutjes, S.A., Italiaander, R., van der Berg, H.H.J.L., Lodder, W.J., and de Roda Husman, A.M. (2005). Isolation and detection of enterovirus RNA from large-volume water samples by using the Nuclisens miniMAG system and real-time nucleic acid sequence-based amplification. Appl. Env. Microbiol. *71*, 3734–3740.

Rodríguez-Lázaro, D., D'Agostino, D., Pla, M., and Cook, N. (2004). A construction strategy for an internal amplification control (IAC) for real time NASBA-based diagnostic assays: development and clinical application. J. Clin. Microbiol. *42*, 5832–5836.

Rodríguez-Lázaro, D., D'Agostino, M., Herrewegh, A., Pla, M., Cook, N., and Ikonomopoulos, J. (2005a). Real-time PCR-based methods for detection of *Mycobacterium avium* subsp. *paratuberculosis* in water and milk. Int. J. Food Microbiol. *101*, 93–94.

Rolfe, K.J., Parmar, S., Mururi, D., Wreghitt, T.G., Jalal, H., Zhang, H., and Curran, M.D. (2007). An internally controlled, one-step, real-time RT-PCR assay for norovirus detection and genogrouping. J. Clin. Virol. *39*, 319–321.

Rossen, L., Nøskov, P., Holmstrøm, K., and Rasmussen, O.F. (1992). Inhibition of PCR by components of food samples, microbial diagnostic assays and DNA extraction solution. Int. J. Food Microbiol. *17*, 37–45.

Verheyen, J., Timmen-Wego, M., Laudien, R., Boussaad, I., Sen, S., Koc, A., Uesbeck, A., Mazou, F., and Pfister, H. (2009). Detection of adenoviruses and rotaviruses in drinking water sources used in rural areas of Benin, West Africa. Appl. Env. Microbiol. *75*, 2798–2801.

Wei, T., Lu, G., and Clover, G. (2008). Novel approaches to mitigate primer interaction and eliminate inhibitors in multiplex PCR, demonstrated using an assay for the detection of three strawberry viruses. J. Virol. Meth. *151*, 132–139.

Widjojoatmodjo, M.N., Fluit, A.C., Torensma, R., Verdonk, G.P.H.T., and Verhoef, J. (1992). The magnetic immuno polymerase chain reaction assay for direct detection of salmonellae in fecal samples. J. Clin. Microbiol. *30*, 3195– 3199.

Wong, M., Xagoraraki, I., and Rose, J.B. (2007). Recent advances and future needs in environmental virology. In: Bosch A, editor. Human Viruses in Water. Amsterdam: Elsevier BV, pp. 259–283.

DOI: https://doi.org/10.21775/9781912530243.11

Standardization of Real-time PCR Methods in Food Microbiology

11

Kornelia Berghof-Jäger

Abstract

International law requires that only food suitable for consumption may reach the market. To meet this demand, thorough microbiological testing must be performed on raw materials, the manufacturing process and finished products. Real-time PCR methods are particularly well-suited for this testing as they are fast, precise and very specific. Multiple methods, including real-time PCR, exist for testing the same analyte. These are favoured according to regional preferences and regulatory requirements. However, global trade could be simplified if there was an international consensus on a set of analytical standards.

The International Organization for Standardization (ISO) and the European Organization for Standardization (CEN) are platforms for generating standards through open, balanced and consensus-driven processes. To avoid duplication of work and structures, an agreement on technical co-operation between ISO and CEN (Vienna Agreement) was approved in 1991. This allows for focused expertise to be used efficiently to benefit international standardization. Currently, a few general standards exist which describe the basic requirements of PCR methods. General standardization documents focusing on performance characteristics as well as basic requirements and definitions of real-time PCR are in development. Standards for specific detection of the food-borne pathogens *Clostridium botulinum*, *Yersinia*, STEC, *Vibrio* and viruses are also in progress. In parallel, standardization of real-time PCR-based methods for the detection of genetically modified organisms (GMO) and allergens in food are ongoing.

Introduction

Standards help reduce worldwide trade barriers and act as an important instrument for transforming innovative ideas into marketable products. In food testing standardization enables us to select and spread appropriate testing methods which are reproducible and reliable. This is mainly achieved by the involvement of all parties with the aim of unification and validation. The consideration and inclusion of new procedures into standards makes them a powerful tool to provide even greater food safety, while at the same time monitoring regulatory compliance. Hence, the process of standardization is tightly linked at the national, European and global levels. Moreover, established national standards provide the basis for development of standards at the European and global levels.

Standardization in CEN – European Organization for Standardization

Located in Brussels, CEN's national members (Fig. 5.1) are the national standards organizations of the 27 European Union countries, Croatia as well as three countries of the European Free Trade Association (Iceland, Norway and Switzerland) (CEN/CENELEC Internal Regulations – Part 2, 2008, updated January 2010).

It is the responsibility of the CEN National members to implement European standards as national standards, which is unique in the world. The national standards bodies distribute and sell the implemented European standards and must withdraw conflicting national standards.

The technical committee CEN/TC 275 'Horizontal methods – Food analysis' is the responsible

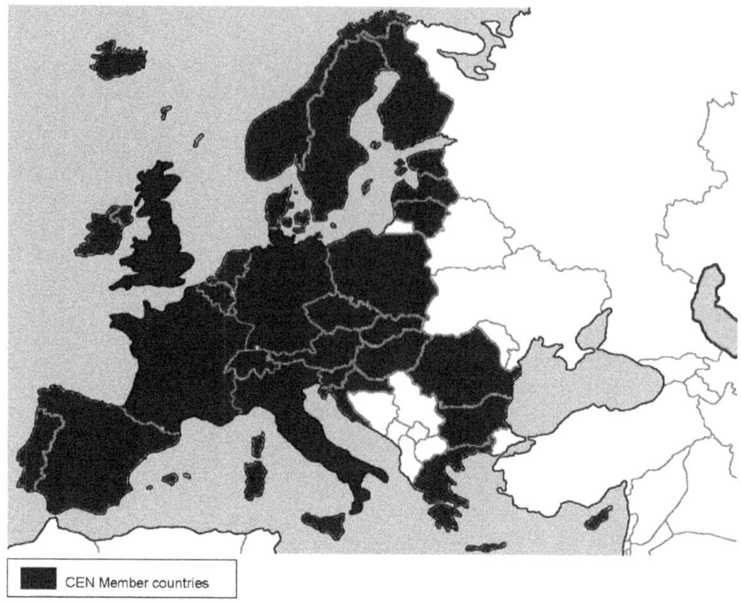

Figure 5.1 Map of CEN member states.

body for the food segment. It was founded in 1990 and began its work in 1991. Standardization issues are directly related to the European Directives in the food branch. Thus, the standardization work of CEN/TC 275 in the field of food analysis is of major importance. The key issues for consumer health protection are represented in the working areas of CEN/TC 275. The main objective of this technical committee is to elaborate the standardized horizontal test methods for the analysis of residues and contaminants in foods. Table 5.1 provides an overview of the different topics processed by the 12 CEN working groups (WG) (CEN TC 275 Food Analysis - Horizontal Methods 2011.

The creation of a new European standard (EN) follows a defined and structured process, which usually takes 36 months from application to completion. A standard can be proposed by everybody who has a need. The standardization process includes the following steps (Fig. 5.2) (CEN-infodesk @ cencenelec.eu-www.cen.eu):

- *Proposal/starting the work.* In principle, any interested party is able to introduce a proposal for new work in CEN. Most standardization work is proposed through the national standards bodies and, where European legislation is concerned, the European Commission (EC) and the European Free Trade Association (EFTA).
- *Acceptance.* The appropriate CEN Technical Committee makes a decision on the adoption of the proposal. An accepted standardization project is allocated to one of its working groups for the drafting of a standard. If the proposal is for a new field of standardization activity, a decision is first made by the CEN Technical Board. The participants in the technical board include all national standards bodies and all standardization partners such as consumers, trade unions, European industry and the counsellors, namely the European Commission (EC) and European Free Trade Association Secretariat (EFTA). One of the principles of standardization in CEN is that, once a project to develop a standard has been accepted, the national standards bodies put all national activity within the scope of the project on hold. This means that they do not initiate new projects, nor revise existing standards at a national level. This obligation is called 'standstill' and allows efforts to be focused on the development of European Standards.
- *Drafting.* The EN is then developed by experts appointed in a working group of the Technical Committee.

Table 5.1 Organization structure of CEN/Technical Committee 275, 'Horizontal methods – Food analysis'

Working group	Title	Secretariat
CEN/TC 275 WG 1	Sulphites	Sweden
CEN/TC 275 WG 2	Sweeteners	Germany
CEN/TC 275 WG 3	Pesticides in fatty food	Germany
CEN/TC 275 WG 4	Pesticides in non fatty food	Germany
CEN/TC 275 WG 5	Biotoxins	Netherlands
CEN/TC 275 WG 6	Microbial contaminants	France
CEN/TC 275 WG 7	Nitrates/nitrites	Germany
CEN/TC 275 WG 8	Irradiated food	Germany
CEN/TC 275 WG 9	Vitamins	Netherlands
CEN/TC 275 WG 10	Heavy metals – trace elements	Sweden
CEN/TC 275 WG 11	Genetically modified organisms and derived products	Germany
CEN/TC 275 WG 12	Food allergens	Germany
CEN/TC 275 WG 13	Process contamination	Netherlands

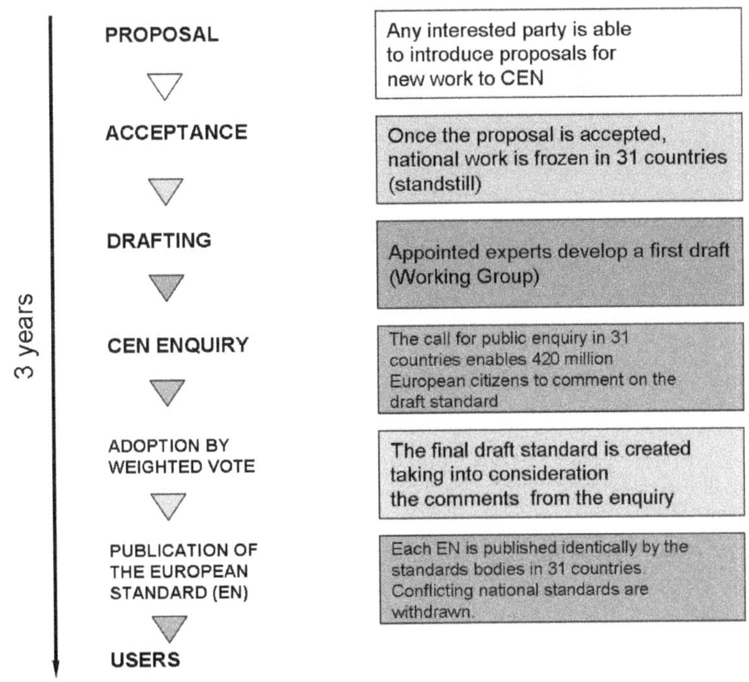

Figure 5.2 The standards-making process.

- *CEN enquiry/public comment at national level.* Once the draft of a European Standard is prepared, it is released for public comment, a process known in CEN as the 'CEN Enquiry'. During this period, anyone (e.g. manufacturers, public authorities, consumers, etc.) may comment on the draft. These views are collated by the 30 national standards bodies and analysed by the CEN Technical Committee.
- *Adoption by weighted vote.* Taking into account the comments generated during the CEN Enquiry, a final version is drafted, which is then submitted to the 31 CEN members for a weighted formal vote, as shown in Fig. 5.3

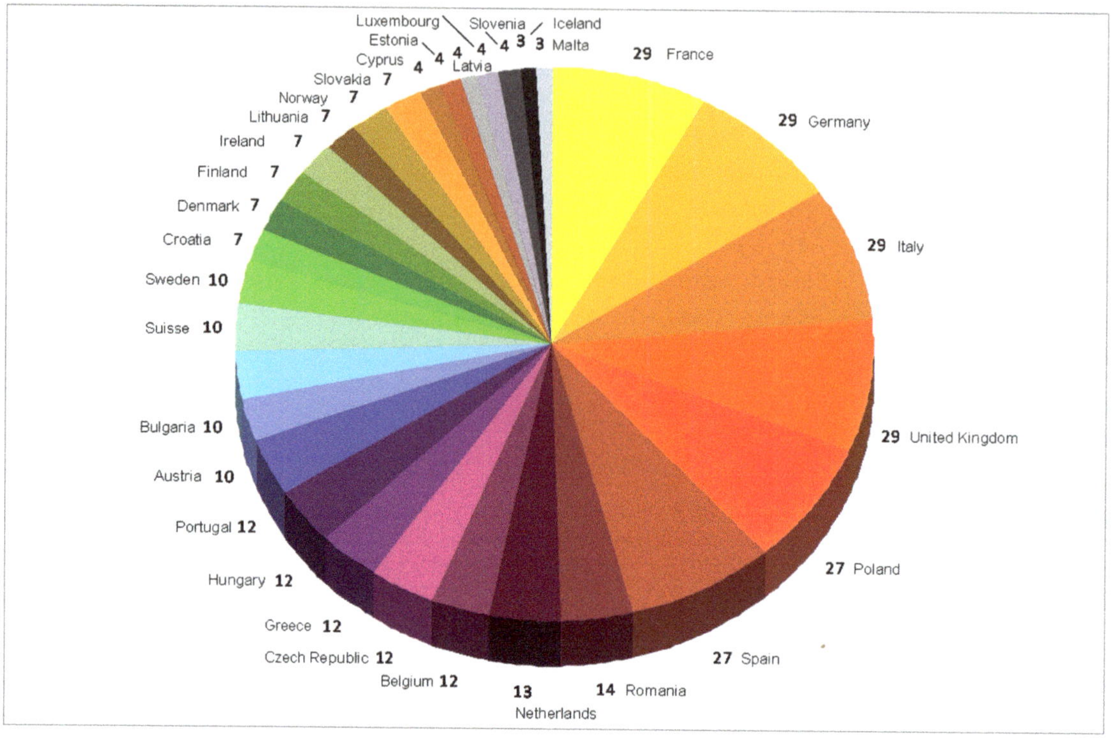

Figure 5.3 Distribution of votes among the CEN members.

(CEN/CENELEC Internal Regulations – Part 2, 2008, updated January 2010).

- *Publication of the European Standard (EN).* After its adoption, each of the 31 national standards bodies publishes the new EN as an identical national standard and withdraws any conflicting national standards. Hence, one EN becomes a national standard in the 31 member countries of CEN. This is unique in the world and results in benefits such as easier access for manufacturers to the market of all European countries.
- *Review of the EN.* To ensure that a European Standard is still current, it is reviewed at least once every 5 years. As a result, the EN may be confirmed, withdrawn, amended or revised.

In addition to the European norms, CEN produces CEN workshop agreements and CEN technical specifications. The development of these documents undergoes streamlined processes with quicker adaptation procedures.

CWAs are developed in CEN workshops open to anyone with the interest in the development of such documents. There is no geographical limit on the participation and hence participants may come from outside Europe. The development of a CWA takes, on average, between 10 and 12 months. CWAs do not have the status of a European Standard.

Technical Specifications (CEN/TS) can be used as a European Pre-Standard for innovative features of upstream technology, or when various alternatives need to coexist in anticipation of future harmonization. These documents do not have the status of a European Standard and, as with the CWAs, are not adopted as national standards.

Standardization in ISO – International Organization for Standardization

ISO is a worldwide federation of national standards bodies (ISO member bodies) (Fig. 5.4). The work of preparing International Standards is normally carried out through ISO technical committees. Each member body interested in a

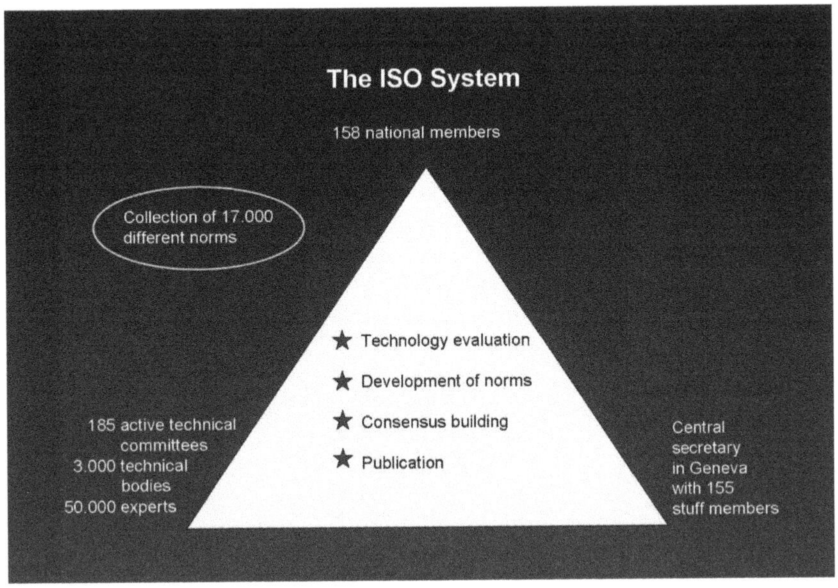

Figure 5.4 The ISO system.

subject for which a technical committee has been established has the right to be represented on that committee. International organizations, governmental and non-governmental, in liaison with ISO, also take part in the work.

ISO was founded in 1947. The number of member countries has increased massively since then, and 2006 were already 158 countries with their national standardization organizations in the work involved with ISO.

There are 185 active technical committees in ISO, and within the numerous working groups 17,000 standards have been elaborated. The central secretariat in Geneva has 155 employees to coordinate and promote these standardization efforts. Each year about 1000 new standards emerge, which require major administrative and coordinating work.

The ISO structure consists of the General Assembly, the Council and the Central Secretariat in Geneva. The main work to develop standards is performed under the coordination of the Technical Committees (TC) and subcommittees (SC) in the individual working groups (WGs) (Fig. 5.5).

In the elaboration process of an International Standard, a defined multistep process must be observed, which is similar to the CEN (Table 5.2).

Generally, for worldwide acceptance, ISO standards have been adopted as trade standards. CEN standards are less effective than global standards in their acceptance by trading partners.

Table 5.2 The six stages of ISO standard development

Stage	Status of document
1. Proposal	NP: New Work Item
2. Preparatory	WD: Working Draft
3. Committee	CD: Committee Draft
4. Enquiry	DIS: Draft International Standard
5. Approval	FDIS: Final Draft International Standard
6. Publication	IS: International Standard

Vienna agreement – the agreement on technical co-operation between ISO and CEN

The Vienna Agreement is an agreement on technical co-operation between ISO and the European Committee for Standardization (CEN). It was formally approved on 27 June 1991 in Vienna by the CEN Administrative Board following its approval by the ISO Executive Board at its meeting on

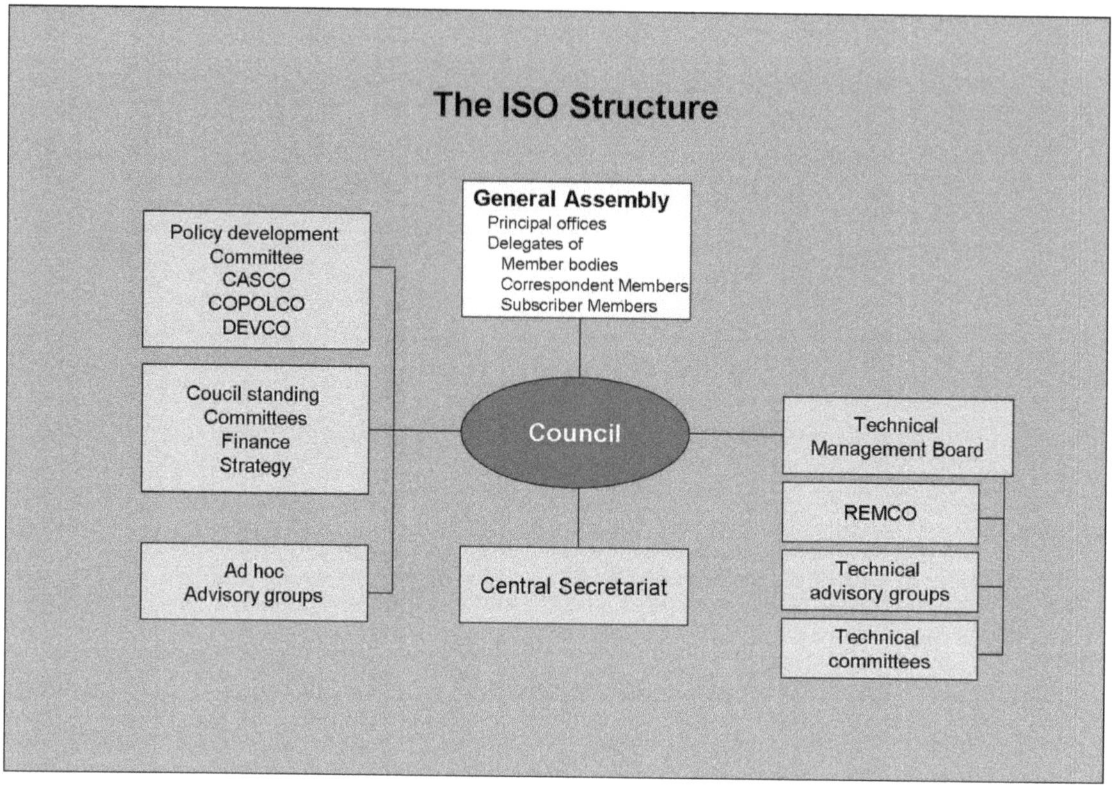

Figure 5.5 Structure of ISO.

16 and 17 May 1991 in Geneva. It replaced the Agreement on exchange of technical information between ISO and CEN (Lisbon Agreement) concluded in 1989. The 'codified' Vienna Agreement was approved by ISO Council and the CEN Administrative Board in 2001.

Essentially, the agreement recognizes the primacy of International Standards, but the agreement also recognizes that particular needs (e.g. of the Single European Market) may require the development of standards for which a need has not been recognized on a global level.

As a result, the agreement sets out two essential modes for collaborative development of standards: the mode under ISO lead and the mode under CEN lead, in which documents developed within one body, are notified for the simultaneous approval by the other. The benefits expected from the use of this agreement include:

- increasing transparency of ongoing work in CEN to ISO members, and their possibility to influence the content of CEN standards;
- avoiding duplication of work and structures, thus allowing focused expertise to be used efficiently to benefit international standardization;
- increasing the speed of elaboration, availability and maintenance of standards through a single consensus (ISO – Standards development procedures – Cooperation with CEN, VA codified, Version 3.3, 2001).

Standardization of molecular methods

Standardization of molecular methods for the detection of food pathogens

Until the end of the 1980s, regular testing of food for pathogenic microorganisms was largely accomplished using cultural methods. Since 1990, alternative methods were developed and were used with increasing frequency. The need for rapid and specific methods led to the foundation of a new working group. This was initiated in the year 2000 by the German Standardization Organization (DIN) in the CEN/TC 275 in WG 6. The aim was to establish standards for molecular testing methods for the detection of pathogens in food. WG 6 mostly agreed to the proposal. The decision to stay within CEN/TC 275 WG 6 was made because the new activities were covered by the scope of WG 6, which is the standardization of horizontal microbiological analysis methods for all food and animal feeding stuffs as well as for any other sample that may be a source of microbiological contamination of food products.

The Chair for the newly formed task force 3 (TAG 3) was taken over by Germany and the Secretariat by France. As of 2011, the working group consists of 14 European countries.

The TAG 3 first focused on the elaboration of general documents. After this initial phase, specific methods were then included in the process of standardization. The elaborated International Standards are applicable to the detection of food-borne pathogens in foods using molecular detection methods, e.g. polymerase chain reaction (PCR). The standards may also be used for other purposes in the food sector (e.g. the detection of food-borne pathogens in environmental samples), and for animal feed.

During initial development of specific standards TAG 3 implements a unique structure for all specific PCR-based standards. It consists of a normative main part and several informative annexes. In the main part, general advice is given regarding enrichment and PCR analytics. The concrete procedures, including specific DNA sequences, are located in the informative annexes. This structure allows the users of the standard to also use validated methods which are based on DNA regions different to those mentioned in the annexes. In view of the rapid development and constant knowledge progression in molecular diagnostics, as well as the ubiquitous constrained caused by numerous patents, this represents a pragmatic solution. In addition, such a structure allows for fast revision of standards. When new validated methods become available, these methods can be easily included as an additional annex. This structure is also used for standards describing PCR methods which were elaborated outside of TAG 3.

Owing to the increasing demand for validated molecular detection methods, in 2003, TC275 WG 6 decided to create task force 4 (TAG 4) to develop and standardize a PCR-based quantitative method for the detection of food-related viruses. One year later in 2004, an ad hoc group was established to investigate developing a standardized method for STEC other than O157 (Fig. 5.6). As of 2011, the elaborated documents are in different stages of the standardization process (Table 5.3).

In 2005, CEN TC275 WG 6 agreed upon a procedure within CEN on how to integrate innovative reference methods, e.g. PCR-based methods in existing standards. It was decided that each time a standard reference method was established or revised, the possible use of new technologies, including PCR, were to be reviewed alongside the existing culture method to determine the most suitable reference method. In addition to the existing reference standard for a given microorganism, the need to develop a standard method for different objectives (such as pathogenicity, etc.) was to be considered in the same way. This WG 6 agreement is highly relevant as it is the basis for the inclusion of new specific and sensitive technologies into existing standards. This agreement was also adopted by the ISO SC9 Committee.

Standardization of molecular methods for the detection of genetically modified organisms (GMO)

Since May 1997, Regulation (EC) No 258/97 of the European Parliament and of the Council concerning novel foods and novel food ingredients applies to the European Community. According to this regulation, genetically modified foods must be officially authorized and labelled. The implementation of this regulation requires the availability of validated detection methods because only then can it be checked if food on the market is duly authorized and labelled.

In 1998, due to the 'Novel Food Regulation' and its labelling requirements, a decision was made in CEN to implement a new working group

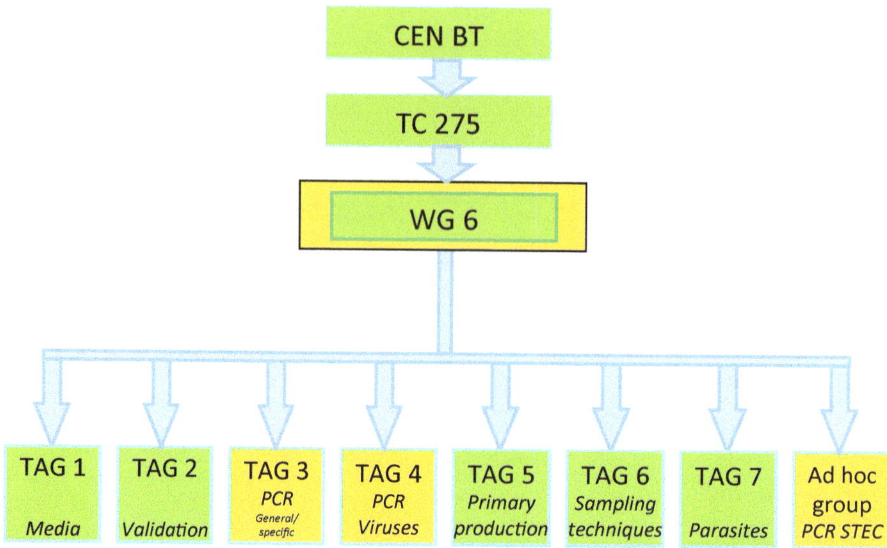

Figure 5.6 Structure of CEN TC275/WG 6 including the working groups for molecular methods.

for the elaboration of standards for the detection of genetically modified organisms within CEN/TC 275. This working group, CEN/TC 275/TC 275/WG 11, 'Genetically modified foods', was founded in February 1999. Since foundation, the Secretariat of the Working Group 11, in which 15 European countries are represented, is led by DIN. WG 11 examines the methods suggested by the European Member States and selects appropriate qualitative and quantitative methods for introduction into the standardization process.

Since 2002, countries outside the CEN have shown interest in these standardization activities. To allow monitoring of the European standardization work, a mirror group was created in ISO within the technical committee responsible for foods, the ISO/TC 34/WG 7. Since then, all projects of WG 11 have been elaborated under the 'Vienna Agreement' led by CEN to allow simultaneous publication as a European and International Standard (ISO).

In 2008, a new committee, SC 16, 'Horizontal methods for molecular biomarker analysis', was established in which ISO/TC 34/WG 7 was embedded. This was mainly due to the request of the American National Standards Institute (ANSI). The development of new European and International Standards for GMO detection mainly occurs in this committee. In Table 5.4, the previously published standards for GMO detection are summarized.

WG 11 agreed on a structure of standards similar to that used for testing of food-relevant microorganisms. Such a structure allows for quick revision of standards as new validated methods become available.

The interrelationship of the EN/ISO standards for analysis of GMO from sampling to result is included in Fig. 5.7 (ISO, 2006c).

Standardization of molecular methods for the detection of food allergens

In the European Community, it is mandatory that potentially allergenic substances are labelled. The legal basis is 'Directive 2000/13/EC of the European Parliament and of the Council of 20 March 2000 on the approximation of the laws of the Member States relating to the labelling, presentation and advertising of foodstuffs'. Hence, the necessity arose to elaborate standard methods on a European level, which can be used according to the Directive for the control of allergens in food. To this end, WG 12 was founded within

Table 5.3 Status of standardization of PCR-based methods for detection of food-borne pathogens

	Status	Working Group
General standards		
ISO 22174:2005 Microbiology of food and animal feeding stuffs – Polymerase chain reaction (PCR) for the detection of food-borne pathogens – General requirements and definitions (ISO, 2005a)	Published 2005	TAG 3
ISO/TS 20836:2005 Performance testing for thermal cyclers- Microbiology of food and animal feeding stuffs – Polymerase chain reaction (PCR) for the detection of food-borne pathogens (ISO, 2005b)	Published 2005	TAG 3
ISO 20837:2006 Microbiology of food and animal feeding stuffs – Polymerase chain reaction (PCR) for the detection of food-borne pathogens – Requirements of sample preparation for qualitative detection (ISO, 2006a)	Published 2006	TAG 3
ISO 20838:2006 Microbiology of food and animal feeding stuffs – Polymerase chain reaction (PCR) for the detection of food-borne pathogens Requirements for amplification and detection for qualitative methods (ISO, 2006b)	Published 2006	TAG 3
ISO 22118 Polymerase chain reaction (PCR) for the detection and quantification of food-borne pathogens – Performance characteristics (ISO, 2011a)	Published 2011	TAG 3
ISO 22119 Real-time polymerase chain reaction (PCR) for the detection of food-borne pathogens – General requirements and definitions (ISO, 2011b)	Published 2011	TAG 3
Specific standards		
Detection of botulinum type A, B, E and F neurotoxin-producing clostridia	Draft	TAG 3
Detection of pathogenic *Yersinia enterocolitica* and *Yersinia pseudotuberculosis*	Draft	TAG 3
Detection of *Vibrio parahaemolyticus* in seafoods: Part 1: Quantitative determination of total and thermostable direct haemolysin (TDH) and thermo-stable direct-related haemolysin (TRH) positive *Vibrio parahaemolyticus* using nucleic acid hybridization	Draft	TAG 3
Development of real-time PCR for the detection and numeration of pathogenic *Vibrio* spp. in shellfish	Draft	TAG 3
Horizontal method for detection of hepatitis A virus and norovirus in food using real-time RT-PCR- Part 1: Method for quantitative determination	Draft	TAG 4
Horizontal method for the detection of shigatoxin-producing *E. coli* (STEC) belonging to O157, O111, O26, O103 and O145 serogroups – qualitative method	Final vote	Ad hoc group

Table 5.4 Status of standardization of PCR based and immunological methods for detection of GMOs in food

Standard	**Status**
CEN/TS 21568:2006: Foodstuffs – Methods of analysis for the detection of genetically modified organisms and derived products – Sampling strategies (European Committee for Standardization, 2006).	Published 2006
ISO 21569: 2005: Foodstuffs – Methods of analysis for the detection of genetically modified organisms and derived products – Qualitative nucleic acid based methods (ISO, 2005c)	Published 2005
ISO 21570:2005 Foodstuffs – Methods of analysis for the detection of genetically modified organisms and derived products – Quantitative nucleic acid based methods (ISO, 2005d)	Published 2005
ISO 21571:2005 Foodstuffs – Methods of analysis for the detection of genetically modified organisms and derived products – Nucleic acid extraction (ISO, 2005e)	Published 2005
ISO 21572:2004 Foodstuffs – Methods for the detection of genetically modified organisms and derived products – Protein based method (ISO, 2004)	Published 2004
ISO 24276:2006 Foodstuffs – Methods of analysis for the detection of genetically modified organisms and derived products – General requirements and definitions [(ISO, 2006c)	Published 2006
ISO/TS 21098:2005 Foodstuffs – Nucleic acid based methods of analysis of genetically modified organisms and derived products – Information to be supplied and procedure for the addition of methods to ISO 21569, ISO 21570 or ISO 21571 (ISO, 2005f)	Published 2005

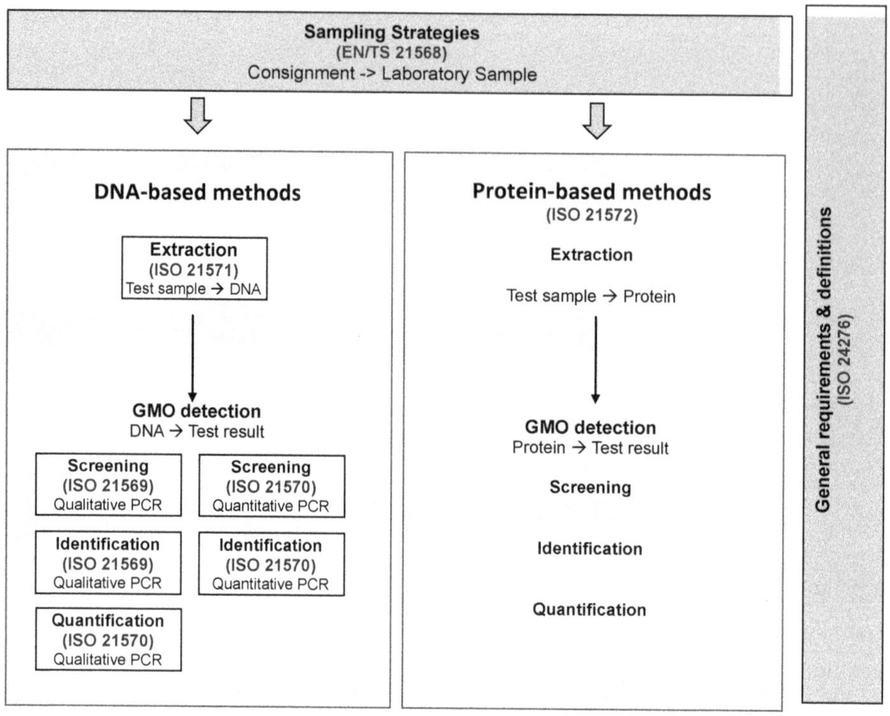

Figure 5.7 Interrelationship of the EN/ISO GMO standards.

Table 5.5 Status of standardization of PCR-based methods for detection of food-borne allergens within WG 12

Standard	Status
EN 15634-1:2009 Foodstuffs – Detection of food allergens by molecular biological methods- Part 1: General considerations (European Committee on Standardization, 2009)	Published 2009
EN 15842:2010 Foodstuffs – Detection of food allergens – General considerations and validation of methods (European Committee on Standardization, 2010)	Published 2010
Foodstuffs – Detection of food allergens by molecular biological methods. Part 2: Celery (*Apium graveolens*) – qualitative determination of a specific DNA sequence in cooked sausages by real-time PCR	CEN enquiry
Foodstuffs – Detection of food allergens – Template for supplying information about immunological methods and molecular biological methods	Draft

the CEN/TC275. The Secretariat and the Chair were selected from DIN. In 2003, the working group began with the aim of evaluating method proposals that had been collected in advance. As a result of these investigations, a list was established which defines the most relevant food allergens:

- peanut
- hazelnut
- milk proteins
- eggs
- gluten
- soy beans.

WG 12 has since created general documents as well as specific immunological and PCR-based methods. The group still continues to be concerned with obligatory technical requirements for sampling and reference material for method validation. Table 5.5 gives an overview of the molecular biological methods that are published or still in development, as well as all general documents for the detection of food allergens.

In addition to the molecular methods, WG 12 is currently creating two immunological methods for the detection of hazelnut.

References

European Committee for Standardization (2006). EN/TS 21568:2006: Foodstuffs – Methods of analysis for the detection of genetically modified organisms and derived products – Sampling strategies. Brussels, European Committee for Standardization.

European Committee for Standardization (2009). EN 15634-1:2009 Foodstuffs- Detection of food allergens by molecular biological methods- Part 1: General considerations. Brussels, European Committee for Standardization.

European Committee for Standardization (2010). EN 15842:2010 Foodstuffs – Detection of food allergens – General considerations and validation of methods. Brussels, European Committee for Standardization.

ISO (2004). EN ISO 21572:2004: Foodstuffs – Methods for the detection of genetically modified organisms and derived products – Protein based method. Geneva, International Organization for Standardization.

ISO (2005a). ISO 22174:2005: Microbiology of food and animal feeding stuffs – Polymerase chain reaction (PCR) for the detection of food-borne pathogens – General requirements and definitions. Geneva, International Organization for Standardization.

ISO (2005b). ISO/TS 20836:2005: Performance testing for thermal cyclers – Microbiology of food and animal feeding stuffs – Polymerase chain reaction (PCR) for the detection of food-borne pathogens. Geneva, International Organization for Standardization.

ISO (2005c). ISO 21569:2005: Foodstuffs – Methods of analysis for the detection of genetically modified organisms and derived products – Qualitative nucleic acid based methods. Geneva, International Organization for Standardization.

ISO (2005d). ISO 21570:2005: Foodstuffs – Methods of analysis for the detection of genetically modified organisms and derived products – Quantitative nucleic acid based methods. Geneva, International Organization for Standardization.

ISO (2005e). ISO 21571:2005, Foodstuffs – Methods of analysis for the detection of genetically modified organisms and derived products – Nucleic acid extraction. Geneva, International Organization for Standardization.

ISO (2005f). ISO/TS 21098:2005, Foodstuffs – Nucleic acid based methods of analysis of genetically modified organisms and derived products – Information to be supplied and procedure for the addition of methods to ISO 21569, ISO 21570 or ISO 21571. Geneva, International Organization for Standardization.

ISO (2006a). ISO 20837:2006: Microbiology of food and animal feeding stuffs – Polymerase chain reaction (PCR) for the detection of food-borne pathogens – Requirements of sample preparation for qualitative detection. Geneva, International Organization for Standardization.

ISO (2006b). ISO 20838:2006: Microbiology of food and animal feeding stuffs – Polymerase chain reaction (PCR) for the detection of food-borne pathogens Requirements for amplification and detection for qualitative methods. Geneva, International Organization for Standardization.

ISO (2006c). ISO 24276:2006, Foodstuffs – Methods of analysis for the detection of genetically modified organisms and derived products – General requirements and definitions. Geneva, International Organization for Standardization.

DOI: https://doi.org/10.21775/9781912530243.12

MIQE: Guidelines for the Design and Publication of a Reliable Real-time PCR Assay

12

Jim Huggett, Tania Nolan and Stephen A. Bustin

Abstract

The capacity to amplify and detect trace amounts of nucleic acids has made the polymerase chain reaction (PCR) the most formidable molecular technology in use today. Its versatility and scope was further broadened first with the development of reverse transcription (RT)-PCR, which opened up the entire RNA field to thorough exploration and then, most conspicuously, with its evolution into real-time quantitative PCR (qPCR). Speed, simplicity, specificity, wide linear dynamic range, multiplexing and high-throughput potential, reduced contamination risk, simplified detection and data analysis procedures as well as availability of increasingly affordable instrumentation and reduced reagent cost have made qPCR the molecular method of choice when quantifying nucleic acids. Detection of pathogens, SNP analyses and quantification of RNA, even real-time analysis of gene expression *in vivo* have become routine applications and constant enhancements of chemistries, enzymes, mastermixes and instruments continue to extend the scope of qPCR technology by promising added benefits such as extremely short assay times measured in minutes, low reagent usage and exceptionally rapid heating/cooling rates. The whole process is driven by the insatiable demand for ever-more specific, sensitive, convenient and cost-effective protocols.

However, it has also become clear that variable pre-assay conditions, poor assay design and incorrect data analysis have resulted in the regular publication of data that are inconsistent, inaccurate and often simply wrong. The problem is exacerbated by a lack of transparency of reporting, with the details of technical information wholly inadequate for the purpose of assessing the validity of reported qPCR data. This has serious consequences for basic research, reducing the potential for translating findings into valuable applications and potentially devastating implications for clinical practice. In response, guidelines proposing a minimum standard for the provision of information for qPCR experiments ('MIQE') have been launched. These aim to establish a standard for accurate and reliable qPCR experimental design as well as recommendations to ensure comprehensive reporting of technical detail, indispensable conditions for the maturing of qPCR into a robust, accurate and reliable nucleic acid quantification technology.

Introduction

Quantitative real-time PCR (qPCR) has become a ubiquitous, mainstream technology widely referred to as the gold standard for nucleic acid detection and quantification in basic research as well as for clinical applications (Bustin, 2010a). Awkwardly, the remarkable transparency and practical simplicity of qPCR has been replaced by numerous conceptual, reagent, protocol and analysis alternatives that continue to yield quantitative results, but require a detailed understanding of the methodology as well as the methods involved to allow an assessment of the data validity. In its simplest form, qPCR is extensively used as a low throughput method for the detection or quantification of limited numbers of individual SNPs or RNAs in relatively few samples or for corroborating fluorescent microarray analyses, as it complements their relatively limited sensitivity, and dynamic range. However, its power and versatility have propelled a growing enthusiasm towards more complex applications. Novel chemistries, both fluorescence- and non-fluorescence-based, fast cyclers with rapid ramping and cooling capability and able to process thousands of assays at the same time, and fast reagents allowing minimal cycling times have combined with advances in automation technology to convert qPCR into a very high-throughput technology generating results in very short time. One of the most remarkable technical developments is the integration of qPCR with nanotechnology. The emergence of high-throughput nanolitre qPCR (Morrison, 1995) and microfluidic digital PCR (Zhang and Xing, 2007) is coupled to advances such as the integration of DNA purification, concentration and qPCR into a real time micro-PCR chip (Min *et al.*, 2011) and the development of innovative, portable PCR instruments (Qiu *et al.*, 2011). Amongst many novel applications, qPCR is used for the rapid and reliable detection of biological threat agents (Christensen *et al.*, 2006), custom-made commercial detection systems are available for the detection of clinical pathogens (Rossney *et al.*, 2008; Novak-Weekley *et al.*, 2010; Zidaric *et al.*, 2011), genetically modified organisms (Dorries *et al.*, 2010) and a combination of nanotechnology and electrochemical qPCR (Yeung *et al.*, 2008) has resulted in the development of hand-held devices that allow local and immediate assays for point-of-care testing applications (Won *et al.*, 2011). This has channelled the technology from a research setting, where, in general, qPCR assays are conceived, designed, performed and controlled by experienced researchers into a wider arena, with high volume 'black box' assays performed as rapidly as possible by less proficient staff.

Current qPCR assay design and reporting are flawed

Unsurprisingly, the ubiquity of the assay itself, combined with abundant choices of methodology, methods and applications, has given rise to numerous protocol alternatives with the potential to produce conflicting data (Bustin and Nolan, 2004; Bustin, 2008a,b; Murphy and Bustin, 2009). Any latent tendency towards inconsistency is exacerbated by the significant probability of variability associated with the numerous steps that make up a qPCR assay (Bustin, 2010a) (Fig. 11.1). Pre-analysis variability derives from issues such as poorly defined sample selection or handling, patchy nucleic acid quality, inconsistent use of controls, poor assay design as well as non-existent optimization and validation. Inconsistencies with post-qPCR data processing are caused by a lack of quality control of the post-qPCR data, especially unsuitable methods of normalization, misguided data analysis procedures and challenges associated with applying the correct statistical methods (Hellemans *et al.*, 2007).

For RT-qPCR, in particular, these problems are critical and have been well documented over many years (Bustin, 2000, 2002, 2004; Bustin and Nolan, 2004; Bustin *et al.*, 2005; Nolan *et al.*, 2006a). Poorly designed, validated and executed assays may, and indeed do,

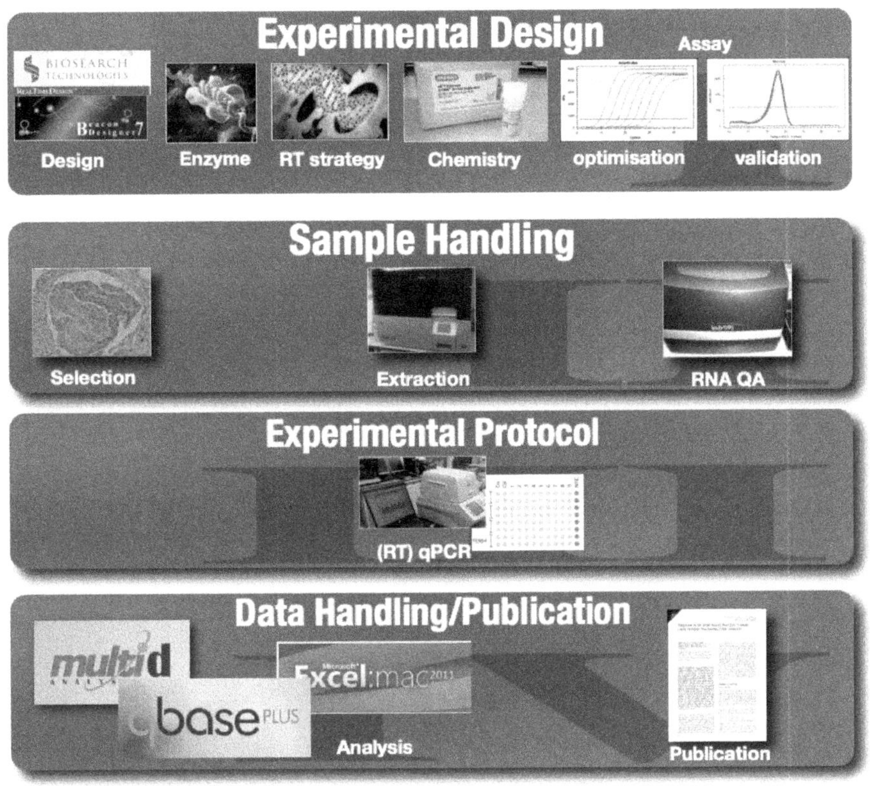

Figure 11.1 RT-qPCR assay. A complete RT-qPCR assay consists of four key steps: (1) Experimental design which includes in silico and empirical validation of each assay. (2) Consistent sample handling and quality assessment. (3) The RT-qPCR steps themselves. (4) Data analysis and publication.

generate a wealth of measurements, but the results can be highly variable and thus inaccurate, ultimately reducing the precision of the measurement or, worse, introducing biases leading to statistically positive but incorrect results. Hence the persistent appeal of developing a set of standard and objective quality control measures (Bustin and Mueller, 2005, 2006; Bustin, 2006, 2008a,b,d, 2010; Murphy *et al.*, 2007; Murphy and Bustin, 2009). The most reliable means of judging the plausibility of a body of work is to study the 'Materials and Methods' section for the relevant information, particularly that relating to the latter aspect of the experimental protocol. A recent survey confirms that this section is avidly read by researchers (Bustin, 2010b), but a second survey shows that the detail of the information provided is often wholly inadequate (Huggett and Bustin, 2011). Unfortunately, not only are many assays designed poorly but also many publications utilizing qPCR technology provide insufficient information to allow a reader to assess the assay and evaluate the validity of conclusions derived from the qPCR data (Bustin, 2010b).

Common omissions that could easily be remedied include simple things like the accession numbers for mRNA or genomic sequences or specifics of the reverse transcription step (Huggett and Bustin, 2011). The latter is especially important, since cDNA priming method and choice of RT have long been known to have a significant impact on results (Stahlberg *et al.*, 2004a,b). Even if the claim is that the experimental protocol recommended by the manufacturer was followed, experience shows that individual researchers frequently introduce subtle, yet consequential variations. Several other pieces of key information are often omitted:

1 Detailed sample information: It remains a remarkable feature of many publications utilizing qPCR technology that very basic information with respect to the samples under investigation is not published. This is of particular importance when considering gene expression analyses from tissue biopsies, where sample selection, acquisition, handling and storage can significantly affect quantification results. Since it should be taken for granted that the researchers have gathered as much information as possible about their samples, the release of that information should add no additional burden to the exacting task of gathering data for publication. Furthermore, it is also necessary to provide details of sample processing procedures, since samples pass through numerous preparative steps prior to the qPCR assay, every one of which can introduce additional variability (Hammerle-Fickinger *et al.*, 2009; Tichopad *et al.*, 2009).
2 Sample quality: it is accepted that RNA integrity and quality assessment is essential for reliable quantification of RNA using RT-qPCR assays (Perez-Novo *et al.*, 2005; Fleige and Pfaffl, 2006; Fleige *et al.*, 2006) and a recent, comprehensive study demonstrates that compromised RNA quality has a measurable impact on the significance of differential expression of prognostic marker genes between cancer patient risk groups, and on risk classification performance using a multigene signature (Vermeulen *et al.*, 2011). The availability of microfluidics-based devices for nucleic acid quality assessment allows automated, rapid and standardized quality assessment of very small amounts of total RNA with quality metrics such as the RIN (Agilent), RQI (BioRad) or SDV (lab901) to represent the level of degradation in a sample. However, it must be remembered that assessing the integrity of rRNA does not necessarily equate with assessing mRNA integrity, and specifically did not in the study by Vermeulen *et al.* (2011), hence the suggestion of introducing a 3′:5′ mRNA-specific integrity assessment (Nolan *et al.*, 2006a) as a more suitable alternative. Importantly, inhibition of reverse-transcription or PCR should be checked by dilution of the sample or use of a universal inhibition assay such as SPUD (Nolan *et al.*, 2006b). Yet very few publications mention the term RNA quality (Huggett and Bustin, 2011).
3 Normalization: normalization strategies are aimed at selecting stable and minimally regulated reference genes and identifying the optimal number of reference genes for the calculation of a normalization factor. This reduces technical variability and increases confidence in data by controlling for experimental error introduced during the multistage process required to extract and process the RNA. Different strategies are not mutually exclusive and the general recommendation is to match sample size, RNA quantity and protocol around three validated internal controls for the final normalization (Huggett *et al.*, 2005). Certainly, there is abundant published evidence that the use of a single reference gene to show small changes in target copy numbers generates unreliable data (Tricarico *et al.*, 2002; Dheda *et al.*, 2004). However, most papers normalize target gene copy numbers against single, invalidated reference genes, even though there are several published methods clearly detailing methods for selection of appropriate reference genes sets (Andersen *et al.*, 2004; Pfaffl *et al.*, 2004; Vandesompele *et al.*, 2002). When this is combined with the reporting of small changes (e.g. differences of less than threefold) it is impossible to conclude whether findings are

differences in expression of the gene of interest, the references gene or a combination of both; consequently statistically significant measurement bias can be introduced leading to incorrect findings (Dheda *et al.*, 2005).

4 PCR efficiency: relative expression levels of mRNA of genes of interest (GOI) are frequently reported relative to those of one or, ideally, several reference genes using the comparative *Cq* method (Livak and Schmittgen, 2001). This involves normalizing the expression of GOI against the internal reference gene(s) *within* two samples, one of which acts as control (ΔCq), followed by a comparison of the normalized difference in expression *between* the two samples ($\Delta\Delta Cq$). In order to be reliable, this requires constant doubling of amplicon with each cycle for each of the targets, or equal amplification efficiency at the very least. However, actual efficiency can range between 60% and 100%, depending on assay quality, sample and target characteristics, reagents as well as instrument variability. Hence small differences between GOI and reference genes can lead to huge differences in relative expression ratios and generate distorted results. Again it is remarkable that not only is this a well-known fact, but that there are a number of efficiency-corrected relative quantification models that provide an efficient and reliable means of quantifying nucleic acids (Pfaffl, 2001; Peirson *et al.*, 2003). They depend on the calculation of PCR efficiency for every target and incorporate this information into relative expression algorithms. Popular software packages such as Genex (Bergkvist *et al.*, 2010), qBase (Hellemans *et al.*, 2007) and REST (Relative Expression Software Tool) (Pfaffl *et al.*, 2002) all make use of such models.

Since all of this information is essential to assay performance, omitting it in a publication makes it impossible to judge the technical quality of published data.

One of the most egregious examples of the enormous implications for the health and lives of individuals that result from inappropriate use of this technology is provided by the use of RT-qPCR data that purported to demonstrate the presence of measles virus (MV) RNA in children with developmental disorders (Uhlmann *et al.*, 2002). It provided sustenance to the controversy surrounding the triple measles mumps and rubella (MMR) virus vaccine, as the data were interpreted as providing evidence for a link between MMR, gut pathology and autism. However, a detailed analysis of the raw data underlying that report carried out by one of the authors (SAB) acting as an expert witness to the UK High Court and the US Vaccine Court, revealed that these data were obtained amongst a catalogue of mistakes, inaccuracies and inappropriate analysis methods as well as contamination and poor assay performance (Bustin, 2008c). A reanalysis of the data concluded that the assay had been detecting DNA and since MV is an RNA-only virus, the RT-qPCR data had been erroneously interpreted, a conclusion confirmed elsewhere (Afzal *et al.*, 2006; D'Souza *et al.*, 2006; Hornig *et al.*, 2008).

MIQE

A growing consensus has been developing around the need to improve published information with relevant experimental detail that covers every aspect important to the qPCR assay itself, as well as issues relating to pre- and post-assay parameters. Specifically, it became clear that there is a requirement for a set of recommendations that can be used by journal reviewers, who need to be able to evaluate the reliability of the experimental protocols and ensure the inclusion of all essential information in the final publication. This need was addressed by the 'Minimum Information for Publication of Quantitative Real-Time PCR Experiments' (MIQE) guidelines (Bustin *et al.*, 2009), with a recent amendment clarifying the disclosure of primer sequences (Bustin *et al.*, 2011). The last year has seen a rapid expansion in the number of researchers aware of the existence of these guidelines (Fig. 11.2) as well as an increasing number of citations of the original publication in the per-reviewed literature (Fig. 11.3). There even is an iPAD/iPOD/iPhone as well as an Android app available from the iTunes store.

The ultimate goal of the MIQE guidelines is to make all technical aspects of the qPCR assay sufficiently transparent, so that the reader of a publication can take technical excellence for granted and focus on the biological relevance of the data. MIQE is made up from nine sections that list 85 parameters that constitute the minimum information required to allow potential reproduction as well as unambiguous quality assessment of a qPCR-based experiment. The nine key areas are 'Experimental design, sample, nucleic acids, reverse transcription, target, primers and probes, assay details, PCR cycling and data analysis'. At first sight, these look overwhelming, arduous and over-exacting. In practice, it is clear that most, if not all of these parameters describe information that would be obtained as a matter of course during the experimental design, optimization and validation stages. Importantly, there is a clear hierarchy with some parameters, labelled 'E' (essential) in the published guidelines, indispensable for an adequate description of the qPCR assay, whereas other components, labelled 'D' (desirable) more peripheral, yet constituting an effective foundation for the realization of best practice protocols. There is increasing recognition that the MIQE guidelines provide the basis for much-needed standardization as well as encourage the publication of essential information that should be accessible to reviewer and reader. Of course, these parameters are not set in stone and are open for discussion; indeed, a core set of sections is being implemented by the BMC group of open-access journals (Bustin *et al.*, 2010).

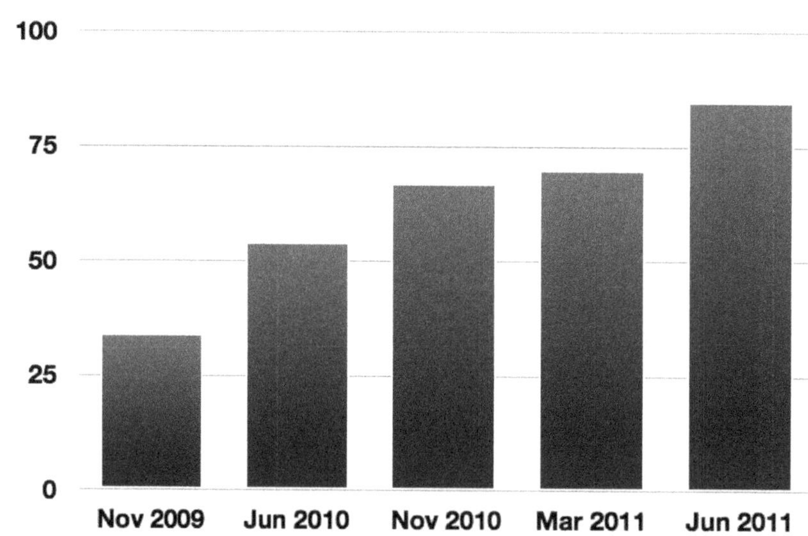

Figure 11.2 MIQE awareness. Attendees of qPCR meetings (2009: qPCR USA, San Francisco; 2010: Gothenburg and qPCR USA; 2011: Freising and Prague) were asked whether they had heard of the MIQE guidelines. The graph shows the percentage of 'yes' responses.

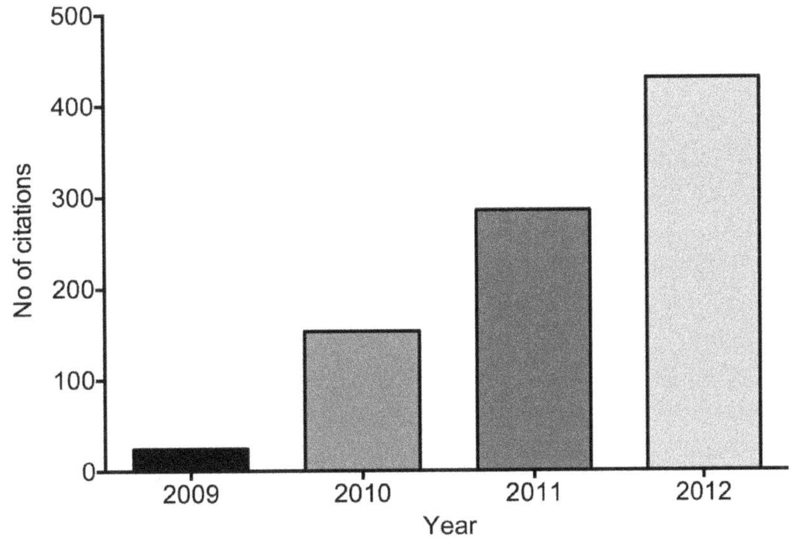

Figure 11.3 Citations of the MIQE paper in the peer-reviewed literature. The number of citations were obtained from Web of Knowledge (www.webofknowledge.com) and stood at 453 at the end of November 2011.

The most controversial aspects of the original MIQE guidelines related to the proposal that publications must divulge the sequences of any primers used and especially should also report the sequences of any probes. The rationale behind releasing the primer sequences was straightforward: an experiment cannot be reproduced if one of the principal reagents is unavailable. Lack of access to a probe sequence, on the other hand, does not preclude analysis of the specificity, efficiency and sensitivity of an assay; however, for completeness' sake it is but a small step to take for most researchers. Many commercial qPCR assays are not supplied with the primer/probe sequences, since most vendors consider this commercially sensitive information; usually there are also no details provided on empirical validation of each individual assay. Publications utilizing such assays could not satisfy the original MIQE requirements, placing limits on a universal acceptance of MIQE. Consequently, an amendment of the original guidelines now requires either primer sequences or a clearly defined amplicon context sequence. This guidance was issued based on the assessment that in the absence of full primer sequence disclosure it is possible to achieve an adequate level of transparency, but only if there is an appropriate level of background information and disclosure of validation results on the qPCR assay. Consequently, if primer sequences are not disclosed, a MIQE-compliant publication should institute the same validation criteria used for assays reporting primer/probe sequences. Specifically, when reporting a precise fold-change for a transcript it remains an essential requirement that the PCR efficiency, analytical sensitivity and specificity of each individual assay be determined. This information should be verified by the investigator for the actual assay that is being reported using the conditions and personnel in their laboratory and not extrapolated from commercial assays validated by the vendors.

The implementation of appropriate statistical methodologies for data handling and processing is an essential complement to any improvements introduced to the practical workflow. Obviously, there are a large number of statistical tools that can be used to address and minimize the variability discussed above and amongst many, specific studies have been published that look at the identification and handling of outliers and precision associated with calibration curves (Burns *et al.*, 2005), the relative merits of obtaining *Cqs* from the threshold method or sigmoidal functions (Rutledge and Cote, 2003; Rutledge, 2004; Rutledge and Stewart, 2008) and limits of detection modelled from sample replication and *Cq* values (Burns and Valdivia, 2008). Furthermore, tools have become available that allow management and analysis of qPCR data (Muller *et al.*, 2002; Jin *et al.*, 2006; Hellemans *et al.*, 2007; Gallup and Ackermann, 2008; Ritz and Spiess, 2008). However, since there are no guidelines or universally accepted standards for data handling and interpretation, the use of multiple statistical tools adds to variability and discordance. Hence MIQE guidelines propose the disclosure of key statistical features associated with a qPCR assay, as well as reporting of what software was used to analyse results.

Appropriate statistical modelling and analysis for the interpretation of qPCR data is of particular importance for clinical applications, where false-positive or negative results can have serious implications. A systematic evaluation of the various qPCR data analysis methods has shown that they differ substantially in their performance (Karlen *et al.*, 2007); hence MIQE guidelines specify the importance of providing detailed information on the methods of data analysis and confidence estimation, especially identifying the statistical methods used to evaluate variances. Since expanding sample size can increase the power of a statistical test, technical repeats are a much favoured and reported sign of qPCR virility. Whilst these do help reduce measurement error, technical repeats simply provide a commentary on the researchers', or their robots' ability to pipette accurately. Technical replicates are needed when optimizing a PCR reaction to ensure that the assay is optimum. However the high precision of an optimal qPCR assay (typically less than 10% coefficient of variation) means that when this is replicated, far from increasing the reliability of results, it can distort the statistics of determining confidence in experimental data. Furthermore if technical replicates are to be included when conducting RT-qPCR it is essential that the more variable reverse transcription reaction is replicated (a process that is automatic with one step RT-qPCR), as a frequent conceptual error is to perform a single reverse transcription and replicate the PCR step on the cDNA sample. This action has surely led to numerous publications of very precise yet totally biased results.

Appropriate biological replication is essential if findings are meant to be valid in the context of a conceptually large population from which the subjects were sampled, rather than only for the particular individuals considered in the experiment (Mehta *et al.*, 2004). Since biological variability is larger than technical variation, increasing biological replication usually translates into more effective gains in power. However, increasing sample size generally leads to added cost and increased time for performing the experiments. In addition, some biological replication cannot be increased, e.g. when comparing large numbers of healthy individuals with a limited number of patients with a particular disease.

It is important to note that the need for appropriate biological replication and other aspects of the MIQE recommendations do not reflect the deliberations of a handful of purists frequenting the whitest of ivory towers. MIQE matters because it is more important that findings truly reflect what is being investigated than it is to publish a Cell or Nature paper. When, as is often the case, high impact publications are the sole aim of a study a poor design can lead to the embarrassing process of a retraction.

Conclusions

qPCR and RT-qPCR are powerful technologies that have enabled many of the advances made in our understanding of basic biological and disease processes; both are also increasingly used for clinical diagnostic purposes. However, the combination of ease of use and lack of rigorous standards of practice has resulted in widespread misinterpretation of data and consequent publication of erroneous conclusions. Any solution to the challenge of how to make PCR-based assays more reliable requires both an appreciation and an understanding of numerous attributes that include biological concepts, statistics, mathematical modelling, technical know-how and a willingness to share this knowledge. This range of fundamental variables must be addressed by guidelines that permit a shift of focus from questions regarding the technological relevance underlying a publication's conclusion to the actual biological or diagnostic issues being addressed. MIQE constitutes a reference framework for communication within the research community, instrument and reagent manufacturers and publishers that promises to deliver guidelines that promote transparency of experiments and confidence in results and conclusions that advance, rather than impede our knowledge.

References

Afzal, M.A., Ozoemena, L.C., O'Hare, A., Kidger, K.A., Bentley, M.L., and Minor, P.D. (2006). Absence of detectable measles virus genome sequence in blood of autistic children who have had their MMR vaccination during the routine childhood immunization schedule of UK. J. Med. Virol. *78*, 623–630.

Andersen, C.L., Jensen, J.L., and Orntoft, T.F. (2004). Normalization of real-time quantitative reverse transcription-PCR data: a model-based variance estimation approach to identify genes suited for normalization, applied to bladder and colon cancer data sets. Cancer Res. *64*, 5245–5250.

Bergkvist, A., Rusnakova, V., Sindelka, R., Garda, J.M., Sjogreen, B., Lindh, D., Forootan, A., and Kubista, M. (2010). Gene expression profiling – clusters of possibilities. Methods *50*, 323–335.

Burns, M.J., and Valdivia, H. (2008). Modelling the limit of detection in real-time quantitative PCR. Eur. Food Res. Technol. *226*, 1513–1524.

Burns, M.J., Nixon, G.J., Foy, C.A., and Harris, N. (2005). Standardisation of data from real-time quantitative PCR methods – evaluation of outliers and comparison of calibration curves. BMC Biotechnol. *5*, 31.

Bustin, S.A. (2000). Absolute quantification of mRNA using real-time reverse transcription polymerase chain reaction assays. J. Mol. Endocrinol. *25*, 169–193.

Bustin, S.A. (2002). Quantification of mRNA using real-time reverse transcription PCR (RT-PCR): trends and problems. J. Mol. Endocrinol. *29*, 23–39.

Bustin, S.A. (2004). Meaningful quantification of mRNA using real-time PCR. In PCR Technology Current Innovations, Griffin, H.G., and Griffin, A., ed. (CRC Press Boca, Raton, FL), pp. 225–233.

Bustin, S.A. (2006). Nucleic acid quantification and disease outcome prediction in colorectal cancer. Personalized Med. 3, 207–216.

Bustin, S. (2008a). Molecular medicine, gene-expression profiling and molecular diagnostics: putting the cart before the horse. Biomarkers Med. *2*, 201–207.

Bustin, S.A. (2008b). Real-time polymerase chain reaction – towards a more reliable, accurate and relevant assay. Eur. Pharm. Rev. *6*, 19–27.

Bustin, S.A. (2008c). Real-time quantitative PCR-opportunities and pitfalls. Eur. Pharm. Rev. *4*, 18–23.

Bustin, S.A. (2010a). Why the need for qPCR publication guidelines? – The case for MIQE. Methods *50*, 217–226.

Bustin, S.A. (2010b). Developments in real-time PCR research and molecular diagnostics. Expert Rev. Mol. Diagn. *10*, 713–715.

Bustin, S.A., and Mueller, R. (2005). Real-time reverse transcription PCR (qRT-PCR) and its potential use in clinical diagnosis. Clin. Sci. (Lond.) *109*, 365–379.

Bustin, S.A., and Mueller, R. (2006). Real-time reverse transcription PCR and the detection of occult disease in colorectal cancer. Mol. Aspects Med. *27*, 192–223.

Bustin, S.A., and Nolan, T. (2004). Pitfalls of quantitative real-time reverse-transcription polymerase chain reaction. J. Biomol. Tech. *15*, 155–166.

Bustin, S.A., Benes, V., Nolan, T., and Pfaffl, M.W. (2005). Quantitative real-time RT-PCR – a perspective. J. Mol. Endocrinol. *34*, 597–601.

Bustin, S.A., Benes, V., Garson, J.A., Hellemans, J., Huggett, J., Kubista, M., Mueller, R., Nolan, T., Pfaffl, M.W., Shipley, G.L., *et al.* (2009). The MIQE guidelines: minimum information for publication of quantitative real-time PCR experiments. Clin. Chem. *55*, 611–622.

Bustin, S.A., Beaulieu, J.F., Huggett, J., Jaggi, R., Kibenge, F.S., Olsvik, P.A., Penning, L.C., and Toegel, S. (2010). MIQE precis: Practical implementation of minimum standard guidelines for fluorescence-based quantitative real-time PCR experiments. BMC Mol. Biol. *11*, 74.

Bustin, S.A., Benes, V., Garson, J.A., Hellemans, J., Huggett, J., Kubista, M., Mueller, R., Nolan, T., Pfaffl, M.W., Shipley, G.L., *et al.* (2011). Primer sequence disclosure: a clarification of the MIQE guidelines. Clin. Chem. *57*, 919–921.

Christensen, D.R., Hartman, L.J., Loveless, B.M., Frye, M.S., Shipley, M.A., Bridge, D.L., Richards, M.J., Kaplan, R.S., Garrison, J., Baldwin, C.D., *et al.* (2006). Detection of biological threat agents by real-time PCR: comparison of assay performance on the R.A.P.I.D., the LightCycler, and the Smart Cycler platforms. Clin. Chem. *52*, 141–145.

D'Souza, Y., Fombonne, E., and Ward, B.J. (2006). No evidence of persisting measles virus in peripheral blood mononuclear cells from children with autism spectrum disorder. Pediatrics *118*, 1664–1675.

Dheda, K., Huggett, J.F., Bustin, S.A., Johnson, M.A., Rook, G., and Zumla, A. (2004). Validation of housekeeping genes for normalizing RNA expression in real-time PCR. Biotechniques *37*, 112–119.

Dheda, K., Huggett, J.F., Chang, J.S., Kim, L.U., Bustin, S.A., Johnson, M.A., Rook, G.A., and Zumla, A. (2005). The implications of using an inappropriate reference gene for real-time reverse transcription PCR data normalization. Anal. Biochem. *344*, 141–143.

Dorries, H.H., Remus, I., Gronewald, A., Gronewald, C., and Berghof-Jager, K. (2010). Development of a qualitative, multiplex real-time PCR kit for screening of genetically modified organisms (GMOs). Anal. Bioanal. Chem. *396*, 2043–2054.

Fleige, S., and Pfaffl, M.W. (2006). RNA integrity and the effect on the real-time qRT-PCR performance. Mol. Aspects Med. *27*, 126–139.

Fleige, S., Walf, V., Huch, S., Prgomet, C., Sehm, J., and Pfaffl, M.W. (2006). Comparison of relative mRNA quantification models and the impact of RNA integrity in quantitative real-time RT-PCR. Biotechnol. Lett. *28*, 1601–1613.

Gallup, J.M., and Ackermann, M.R. (2008). The 'PREXCEL-Q Method' for qPCR. Int. J. Biomed. Sci. *4*, 273–293.

Hammerle-Fickinger, A., Riedmaier, I., Becker, C., Meyer, H.H., Pfaffl, M.W., and Ulbrich, S.E. (2009). Validation of extraction methods for total RNA and miRNA from bovine blood prior to quantitative gene expression analyses. Biotechnol. Lett. *32*, 35–44.

Hellemans, J., Mortier, G., De Paepe, A., Speleman, F., and Vandesompele, J. (2007). qBase relative quantification framework and software for management and automated analysis of real-time quantitative PCR data. Genome Biol. *8*, R19.

Hornig, M., Briese, T., Buie, T., Bauman, M.L., Lauwers, G., Siemetzki, U., Hummel, K., Rota, P.A., Bellini, W.J., O'Leary, J.J., *et al.* (2008). Lack of association between measles virus vaccine and autism with enteropathy: a case-control study. PLoS One *3*, e3140.

Huggett, J., and Bustin, S.A. (2011). Standardisation and reporting for nucleic acid quantification. Accredit. Qual. Assur. *16*, 399–405.

Huggett, J., Dheda, K., Bustin, S., and Zumla, A. (2005). Real-time RT-PCR normalisation; strategies and considerations. Genes Immun. *6*, 279–284.

Jin, N., He, K., and Liu, L. (2006). qPCR-DAMS: a database tool to analyze, manage, and store both relative and absolute quantitative real-time PCR data. Physiol. Genomics *25*, 525–527.

Karlen, Y., McNair, A., Perseguers, S., Mazza, C., and Mermod, N. (2007). Statistical significance of quantitative PCR. BMC Bioinformatics *8*, 131.

Livak, K.J., and Schmittgen, T.D. (2001). Analysis of relative gene expression data using real-time quantitative PCR and the 2(-Delta Delta C(T)) Method. Methods *25*, 402–408.

Mehta, T., Tanik, M., and Allison, D.B. (2004). Towards sound epistemological foundations of statistical methods for high-dimensional biology. Nat. Genet. *36*, 943–947.

Min, J., Kim, J.H., Lee, Y., Namkoong, K., Im, H.C., Kim, H.N., Kim, H.Y., Huh, N., and Kim, Y.R. (2011). Functional integration of DNA purification and concentration into a real time micro-PCR chip. Lab. Chip *11*, 259–265.

Morrison, D.K. (1995). Mechanisms regulating Raf-1 activity in signal transduction pathways. Mol. Reprod. Dev. *42*, 507–514.

Muller, P.Y., Janovjak, H., Miserez, A.R., and Dobbie, Z. (2002). Processing of gene expression data generated by quantitative real-time RT-PCR. Biotechniques *32*, 1372–1379.

Murphy, J., and Bustin, S.A. (2009). Reliability of real-time reverse-transcription PCR in clinical diagnostics: gold standard or substandard? Expert Rev. Mol. Diagn. *9*, 187–197.

Murphy, J., Dorudi, S., and Bustin, S.A. (2007). Molecular staging of colorectal cancer: new paradigm or waste of time? Expert Opin. Med. Diagn. *1*, 31–45.

Nolan, T., Hands, R.E., and Bustin, S.A. (2006a). Quantification of mRNA using real-time RT-PCR. Nat. Protocols *1*, 1559–1582.

Nolan, T., Hands, R.E., Ogunkolade, B.W., and Bustin, S.A. (2006b). SPUD: a qPCR assay for the detection of inhibitors in nucleic acid preparations. Anal. Biochem. *351*, 308–310.

Novak-Weekley, S.M., Marlowe, E.M., Miller, J.M., Cumpio, J., Nomura, J.H., Vance, P.H., and Weissfeld, A. (2010). *Clostridium difficile* testing in the clinical laboratory by use of multiple testing algorithms. J. Clin. Microbiol. *48*, 889–893.

Peirson, S.N., Butler, J.N., and Foster, R.G. (2003). Experimental validation of novel and conventional approaches to quantitative real-time PCR data analysis. Nucleic Acids Res. *31*, e73.

Perez-Novo, C.A., Claeys, C., Speleman, F., Van Cauwenberge, P., Bachert, C., and Vandesompele, J. (2005). Impact of RNA quality on reference gene expression stability. Biotechniques *39*, 52, 54, 56.

Pfaffl, M.W. (2001). A new mathematical model for relative quantification in real-time RT-PCR. Nucleic Acids Res. *29*, E45.

Pfaffl, M.W., Horgan, G.W., and Dempfle, L. (2002). Relative expression software tool (REST) for group-wise comparison and statistical analysis of relative expression results in real-time PCR. Nucleic Acids Res. *30*, e36.

Pfaffl, M.W., Tichopad, A., Prgomet, C., and Neuvians, T.P. (2004). Determination of stable housekeeping genes, differentially regulated target genes and sample integrity: BestKeeper – Excel-based tool using pair-wise correlations. Biotechnol. Lett. *26*, 509–515.

Qiu, X., Chen, D., Liu, C., Mauk, M.G., Kientz, T., and Bau, H.H. (2011). A portable, integrated analyzer for microfluidic-based molecular analysis. Biomed. Microdevices *13*, 809–817.

Ritz, C., and Spiess, A.N. (2008). qpcR: an R package for sigmoidal model selection in quantitative real-time polymerase chain reaction analysis. Bioinformatics *24*, 1549–1551.

Rossney, A.S., Herra, C.M., Brennan, G.I., Morgan, P.M., and O'Connell, B. (2008). Evaluation of the Xpert methicillin-resistant Staphylococcus aureus (MRSA) assay using the GeneXpert real-time PCR platform for rapid detection of MRSA from screening specimens. J. Clin. Microbiol. *46*, 3285–3290.

Rutledge, R.G. (2004). Sigmoidal curve-fitting redefines quantitative real-time PCR with the prospective of developing automated high-throughput applications. Nucleic Acids Res. *32*, e178.

Rutledge, R.G., and Cote, C. (2003). Mathematics of quantitative kinetic PCR and the application of standard curves. Nucleic Acids Res. *31*, e93.

Rutledge, R.G., and Stewart, D. (2008). A kinetic-based sigmoidal model for the polymerase chain reaction and its application to high-capacity absolute quantitative real-time PCR. BMC Biotechnol. *8*, 47.

Stahlberg, A., Hakansson, J., Xian, X., Semb, H., and Kubista, M. (2004a). Properties of the reverse transcription reaction in mRNA quantification. Clin. Chem. *50*, 509–515.

Stahlberg, A., Kubista, M., and Pfaffl, M. (2004b). Comparison of reverse transcriptases in gene expression analysis. Clin. Chem. *50*, 1678–1680.

Tichopad, A., Kitchen, R., Riedmaier, I., Becker, C., Stahlberg, A., and Kubista, M. (2009). Design and optimization of reverse-transcription quantitative PCR experiments. Clin. Chem. *55*, 1816–1823.

Tricarico, C., Pinzani, P., Bianchi, S., Paglierani, M., Distante, V., Pazzagli, M., Bustin, S.A., and Orlando, C. (2002). Quantitative real-time reverse transcription polymerase chain reaction: normalization to rRNA or single housekeeping genes is inappropriate for human tissue biopsies. Anal. Biochem. *309*, 293–300.

Uhlmann, V., Martin, C.M., Sheils, O., Pilkington, L., Silva, I., Killalea, A., Murch, S.B., Walker-Smith, J., Thomson, M., Wakefield, A.J., *et al.* (2002). Potential viral pathogenic mechanism for new variant inflammatory bowel disease. Mol. Pathol. *55*, 84–90.

Vandesompele, J., De Preter, K., Pattyn, F., Poppe, B., Van Roy, N., De Paepe, A., and Speleman, F. (2002). Accurate normalization of real-time quantitative RT-PCR data by geometric averaging of multiple internal control genes. Genome Biol. *3*, 0034.1–0034.11.

Vermeulen, J., De Preter, K., Lefever, S., Nuytens, J., De Vloed, F., Derveaux, S., Hellemans, J., Speleman, F., and Vandesompele, J. (2011). Measurable impact of RNA quality on gene expression results from quantitative PCR. Nucleic Acids Res. *39*, e63.

Won, B.Y., Shin, S., Baek, S., Jung, Y.L., Li, T., Shin, S.C., Cho, D.Y., Lee, S.B., and Park, H.G. (2011). Investigation of the signaling mechanism and verification of the performance of an electrochemical real-time PCR system based on the interaction of methylene blue with DNA. Analyst *136*, 1573–1579.

Yeung, S.S., Lee, T.M., and Hsing, I.M. (2008). Electrochemistry-based real-time PCR on a microchip. Anal. Chem. *80*, 363–368.

Zhang, C., and Xing, D. (2007). Miniaturized PCR chips for nucleic acid amplification and analysis: latest advances and future trends. Nucleic Acids Res. *35*, 4223–4237.

Zidaric, V., Kotnik Kevorkijan, B., Oresic, N., Janezic, S., and Rupnik, M. (2011). Comparison of two commercial molecular tests for the detection of Clostridium difficile in the routine diagnostic laboratory. J. Med. Microbiol. *60*, 1131–1136.

DOI: https://doi.org/10.21775/9781912530243.13

Analysis of mRNA Expression by Real-time PCR

13

Stephen A. Bustin and Tania Nolan

Abstract

The last few years have witnessed the transformation of the real-time, fluorescence-based reverse transcription polymerase chain reaction (RT-qPCR) from an experimental technology into a mainstream scientific tool for the detection and quantification of RNA with an enormous range of uses in basic research, molecular medicine and biotechnology. The continuous improvement of reagents and instruments, combined with the trend towards high throughput and miniaturization, is likely to reinforce that pre-eminence and continue to open up new application areas. Nonetheless, although in principle undoubtedly a straightforward technology, the reliability of RT-qPCR assays depends on a series of sequential steps that include careful experimental design, optimization and validation, which must be implemented pragmatically to obtain meaningful, biologically relevant data.

Introduction

The real-time quantitative reverse transcription polymerase chain reaction (RT-qPCR) (Gibson *et al.*, 1996) offers a unique combination of conceptual simplicity, cost-efficiency and accuracy, making it relevant for every aspect of biological investigation (Kubista *et al.*, 2006). Its homogeneous (closed tube) format, use of fluorescent reporter molecules and lack of requirement for post-PCR handling makes RT-qPCR assays convenient, easily automated and much less prone to contamination (Bustin, 2000; Ginzinger, 2002). This convenience is enhanced by its accuracy, speed and wide dynamic range, making it today's method of choice for any application that aims to detect RNA, assess its abundance or distinguish small sequence changes (Bustin *et al.*, 2005). Consequently RT-qPCR has become the method of choice for acquiring quantitative data in basic research, detecting genetically manipulated organisms (Buh Gasparic *et al.*, 2010) and is increasingly used for routine clinical diagnostic assays (Bustin and Mueller, 2005) and as a standard exploratory tool in pharmaceutical microbiology and drug development (Brazeau, 2004)

Recent developments

RT-qPCR technology continues to evolve in response to the need for ever-more specific, convenient and cost-effective applications, with several recent developments improving assay performance and accuracy as well as simplifying data analysis:

- The introduction of more reliable and labour-saving, yet less expensive reagents, coupled with the launch of affordable, but more robust, qPCR thermal cyclers simplifies reaction assembly and improves flexibility, performance and consistency of qPCR assays. GMP-compliant manufacturing and formulation facilities permit the production of oligonucleotides with rigorous quality control and the documentation required for clinical trials and diagnostic expression analysis.

- The extension of qPCR into digital PCR has unlocked another vast range of applications for the technology. In digital PCR a sample is terminally diluted across a large number of separate PCR reactions and results are recorded at the reaction endpoint. Most will contain no template molecules and give a negative amplification result. A successful amplification is counted as one molecule; hence recording the number of positive PCR reactions allows accurate quantification of the individual template molecules present in the original sample
- Another recent development has been termed qPCR tomography (Sindelka *et al.*, 2008) which combines the use of laser microdissection with RT-qPCR analysis to obtain expression profiles from within single cells.
- Automated workflows (Grisold and Kessler, 2006; Marshall *et al.*, 2007), more reliable multiplexing assays (Westh *et al.*, 2009; Beck *et al.*, 2010), the launch of instruments using 384- and 1536-well formats and of miniaturized qPCR systems (Morrison *et al.*, 2006), including microfluidic digital PCR (Zhang and Xing, 2007), has resulted in the expansion of qPCR technology into high-throughput applications. Miniaturisation offers several potential advantages such as short assay time, low reagent usage and rapid heating/cooling rates, as well as integration of multiple processing modules to further reduce size and power consumption. (Marcus *et al.*, 2006; Neuzil *et al.*, 2006). High-density array-base formats and support for faster assay conditions enable shorter assay times, allowing the parallel screening of hundreds of targets at vastly reduced reagent cost where RT-qPCR can provide an alternative to cDNA microarrays for accurate expression analysis with high sensitivity (Dixon *et al.*, 2009; Wagner and Arany, 2009) or even be used for routine diagnostic services (Gunson *et al.*, 2008). They also allow powerful multiway profiling, where quantification of reporter genes permits the characterization of samples as functions of multiple factors such as time, genetic make-up or drug load (Stahlberg *et al.*, 2008).
- The increased emphasis on high-throughput profiling ties in with the increased understanding for the need for constant and consistent quality qPCR data control. This has resulted in the development of software programs that aim to minimize subjective human input and maximize objective multilevel analysis. The publication of numerous advanced algorithms (Burns *et al.*, 2005; Larionov *et al.*, 2005; Jin *et al.*, 2006; Nordgard *et al.*, 2006; Smith *et al.*, 2007; Batsch *et al.*, 2008; O'Connor and Runquist, 2008; Platts *et al.*, 2008; Rutledge and Stewart, 2008a; Spiess *et al.*, 2008; Verderio *et al.*, 2008a) and their emergent inclusion in the development of more intuitive analysis software helps with assay set-up (Gallup and Ackermann, 2008), project management and data analysis (Hellemans *et al.*, 2007) to allow robust, accurate and biologically relevant quantification.

Quality considerations

Reliable and reproducible RT-qPCR assays require meticulous adherence to standard operating procedures for sample acquisition and handling, as well as the preparation, quality assessment and storage of RNA (Bustin, 2010; Bustin *et al.*, 2009, 2011b; Huggett and Bustin, 2011).

The consistency and reliability of RT-qPCR assays depends on the proper execution of a number of steps, in particular those relating to sample selection, template quality, assay design and data analysis (Nolan *et al.*, 2006a) as well as the correct application of statistical models and methods for data analysis (Fu *et al.*, 2006). If correct procedures are adhered to, results are generally robust, reproducible and accurate. Specifically, the reliability of the RT-qPCR assay is determined by careful consideration of the quality of the RNA template (Fleige and Pfaffl, 2006; Fleige *et al.*, 2006; Nolan *et al.*, 2006b; Vermeulen *et al.*, 2011), the choice of cDNA priming strategy (Bustin and Nolan, 2004) and reverse transcriptase (Stahlberg *et al.*, 2004a,b), the characteristics of the primers (Pattyn *et al.*, 2003, 2006) and the validity of the normalization method (Pfaffl *et al.*, 2002, 2004; Szabo *et al.*, 2002; Vandesompele *et al.*, 2002a; Akilesh *et al.*, 2003; Andersen *et al.*, 2004; Haller *et al.*, 2004; Abruzzo *et al.*, 2005; Kanno *et al.*, 2006). For diagnostic laboratories, external quality assurance programmes have been shown to be useful (Marubini *et al.*, 2004; Raggi *et al.*, 2005; Ramsden *et al.*, 2006; Verderio *et al.*, 2008b; Raymaekers *et al.*, 2009). Not surprisingly, the pervasive penetration of this technology has led to the development of numerous, distinct and frequently divergent experimental protocols that often generate discordant results (Bustin, 2002, 2005; Bustin *et al.*, 2005). Fortunately, the resulting uncertainty has increased the awareness of a need for common guidelines, in particular those relating to quality assessment of every component of the RT-qPCR assay and appropriate data analysis (Nolan *et al.*, 2006a). This clear requirement for improved consistency of gene expression measurements is particularly relevant in relation to human clinical diagnostic assays (Bustin and Mueller, 2005, 2006). Certainly, it is apparent that while qPCR is driving PCR-based innovations, effective guidelines regulating its use are essential for the future of molecular diagnostics within biomedical sciences (Valasek and Repa, 2005; Picard *et al.*, 2006). The MIQE guidelines, discussed in Chapter 11, are a list of recommendations that cover the critical steps that make up a qPCR assay and provide standards that will help make RT-qPCR a more robust and reliable technique.

Sample selection

Optimal sample quality is a prerequisite for generating valid quantitative data. Hence sample selection and collection as well as RNA quality control, discussed below, are critical parameters in test performance and must be optimized (Muller *et al.*, 2004; Benoy *et al.*, 2006). In general, extraction of RNA from tissue culture, blood and serum is relatively straightforward, while there are significant problems associated with the extraction of RNA from solid tissue, faeces, semen, plants and soil samples. A critical consideration concerns the quantification of RNA from complex tissue samples, as these usually contain several different cell types that may express the target RNA(s) at different levels of abundance. This inevitably results in the averaging of the expression of different cell types and the expression profile of a specific cell type may be masked, lost or ascribed to and dismissed as illegitimate transcription. This is particularly relevant when comparing gene expression profiles between normal and cancer tissue since normal cells adjacent to a tumour may be phenotypically normal but genotypically abnormal or exhibit altered gene expression profiles due to their proximity to the tumour. This problem can be addressed by microdissection, in particular laser capture microdissection (Emmert-Buck *et al.*, 1996), and significant differences have been detected in the gene expression profiles of microdissected and bulk tissue samples (Fink *et al.*, 2002; Oji *et al.*, 2003). Incidentally, when dealing with very small amounts of sample, e.g. single cells, or minute amounts of laser capture microdissected fresh/frozen sample, it is best to carry out the RT-qPCR directly from lysed tissue without going through an extraction process (Hartshorn *et al.*, 2005). A further refinement combines RT-qPCR with in situ hybridization on adjacent tissue sections and so allows simultaneous spatiotemporal monitoring and quantification of gene expression changes (Haupt *et al.*, 2006). However, it is essential to realize that quantification from very few cells also poses problems, as different cells of the same type may not express the same set of mRNAs (Peixoto *et al.*, 2004). In addition, it could be expected that identical cells samples would contain identical copy numbers of a target mRNA or more likely a normal distribution however, it has been demonstrated that mRNA distribution is often in a lognormal fashion (Bengtsson *et al.*, 2008). Microdissection can be carried out on either fresh/frozen or archival samples.

Choice of chemistry

The appropriate detection chemistry depends on the application: The use of a double stranded DNA binding dye in the PCR mix, e.g. SYBR Green I dye, is the most cost-effective chemistry and is widely used for initial investigations, large scale screening studies and for primer optimization steps. While it may be possible to use legacy assays, it is advisable to adopt a new design that is specifically designed for RT-qPCR. These assays are very simple to design (Bustin *et al.*, 2011a) and there are several web-based options that are freely available, including from commercial companies such as OligoArchitect™ (sigma.com/designmyprobe). Although the primers alone determine the specificity of product detection, information regarding product size and amplicon population is determined from a post amplification melting curve analysis. The introduction of high-resolution melting (HRM) can be used to identify very small differences between targets and is sensitive even to single base differences (Wittwer *et al.*, 2003). Other non-fluorescent probe-based chemistries include Plexor (www.promega.com), LUX primer (www.invitrogen.com) and Snapback primers (Zhou *et al.*, 2008).

At very low (< 1000 copies) target concentration there is a greater probability of non-specific amplification and problems with primer dimer products becoming more pronounced. In this case the use of a probe to detect amplicons may be preferable. The most widely used probe system is the dual-labelled fluorescent hydrolysis probe, which can be synthesized with added locked nucleic acid (LNA; Sigma Custom Products), minor groove binder (MGB; Exiqon) or ZNA modifications (Sigma Custom Products). LNA is a nucleic acid analogue that contains a 2′-O–4′-C methylene bridge, which restricts the flexibility of the ribofuranose ring and locks the structure into a rigid bicyclic formation, conferring enhanced hybridization performance. MGBs are dsDNA-binding agents that are attached to the 3′ end of hydrolysis probes and also stabilize the hybridization reaction. In practice, both modifications allow for the design of shorter probes, which increases design flexibility and specificity. This is particularly useful when detecting single nucleotide polymorphisms or when targeting AT-rich sequences and when targets are highly structured or repetitive. Zipped Nucleic Acids (Paris *et al.*, 2010) consist of a stretch of DNA with the addition of spermine units to the 5′ or 3′ end. The development of intercalating nucleic acids

(Christensen and Pedersen, 2002; Christensen *et al.*, 2004; Filichev *et al.*, 2004) provides additional alternatives for the improvement of assay design.

Amplicons and primers

There is no consensus opinion concerning the best amplicon location to target on a mRNA molecule. Where possible, assays should be designed to span intron/exon boundaries, selecting the longest possible intron, as this avoids signal generation from amplified genomic DNA (gDNA). This is essential when targeting specific splice variants of a mRNA. It is most economical to place the downstream primer across the splice junction, as this allows the use of a single (expensive) probe for the detection of possible splice variants in separate assays and, at the same time, minimizes multiple target amplification in any one assay. However, we have noticed that, very often, it is not possible to design an optimal assay around splice junctions and if the choice is between a poor (i.e. inefficient and/or insensitive) assay designed across a splice junction and an efficient one contained within a single exon, it is better to use the latter. In this case it is advisable to treat the sample with DNase I and essential to add a 'no RT' control to determine the contribution of any gDNA contamination to the mRNA quantification. For some specialist applications, e.g. the quantification of mRNA in a single fertilized egg, or cases where the target is a highly expressed gene from a single copy gDNA sequence, the relative difference in the abundance of gDNA and mRNA is such that any gDNA will not interfere significantly with accurate quantification of mRNA. Since there are so many variables to consider when designing an assay it is critical to ensure that as much information is provided as possible.

The amplicon length should be minimized because this tends to increase amplification efficiency. In addition, small amplicons are more tolerant of reaction conditions because they are more likely to be denatured during the 92–95°C step of the PCR, allowing the probes and primers to compete more effectively for binding to their complementary targets. They also require shorter polymerization times for amplicon replication, making amplification of gDNA contaminants containing the intronic sequences less likely. From a practical perspective it is also easier and cheaper to synthesize a shorter oligonucleotide, when a synthetic sense-strand oligonucleotide is used to generate amplicon-specific standard curves and positive controls. When it is impossible to find an acceptable short amplicon the design parameters relating to amplicon size should be changed little by little, until the smallest acceptable amplicon is found. Whenever possible, the amplicon should be around 100 bp, no more than 150 bp in length when using a probe and up to 250 bp when using SYBR Green I binding dye detection. Amplicons should have a G/C content of 40–60% to ensure complete denaturation during thermal cycling, leading to an efficient reaction. Furthermore, highly GC-rich sequences are susceptible to non-specific interactions that result in non-specific signal in assays utilizing SYBR Green I dye.

Primer selection, especially of the reverse primer used in the RT step, is critical since it affects the sensitivity of the RT-PCR assay (Raengsakulrach *et al.*, 2002). It has also been shown that tissue-specific factors such as polysaccharides or proteins can influence amplification kinetics in a sequence-specific manner and that this affect can be mitigated, in part, by appropriate primer selection (Tichopad *et al.*, 2004). The structure of the RNA target at the primer-binding site must be taken into account, as this affects the accessibility of the target to the primers. Even so-called single-stranded RNA is typically extensively folded and selection of a primer binding site at a position that is normally double stranded can result in a very inefficient assay (Fig. 3.1). For RNA viruses there is the additional problem of different viral serotypes resulting in sequence variability and it may be necessary to use a nested RT-qPCR assay with universal primers that bind to target sequences which are shared by all the serotypes followed by a serotype-specific primer pair (Aradaib, 2009; Zaayman *et al.*, 2009).

The choice of 3′-terminal position is crucial for the control of mispriming during the PCR. Primers with one or more G or C residues at that position will have increased binding efficiency due to the stronger hydrogen bonding of G/C residues. It also helps to improve the efficiency of priming by minimizing any 'breathing' that might occur. However, clamping the 3′ of the primer should be avoided because ambiguous binding of oligonucleotides to the target site may result in misprimed elongation ('slippage effect'). A reasonable compromise is to aim for two or three G/Cs within the five 3′ bases of the primer. Similarly, it is best to avoid long runs of a single base (i.e. more than three or four, especially G or C) since homopolymeric runs can also cause the 'slippage effect'.

Direct repeats in the target sequence should be avoided since stable hybridization to secondary binding sites in repetitive regions results in non-productive binding of primers to non-specific regions of the sequence so that efficiency of DNA amplification and detection decreases.

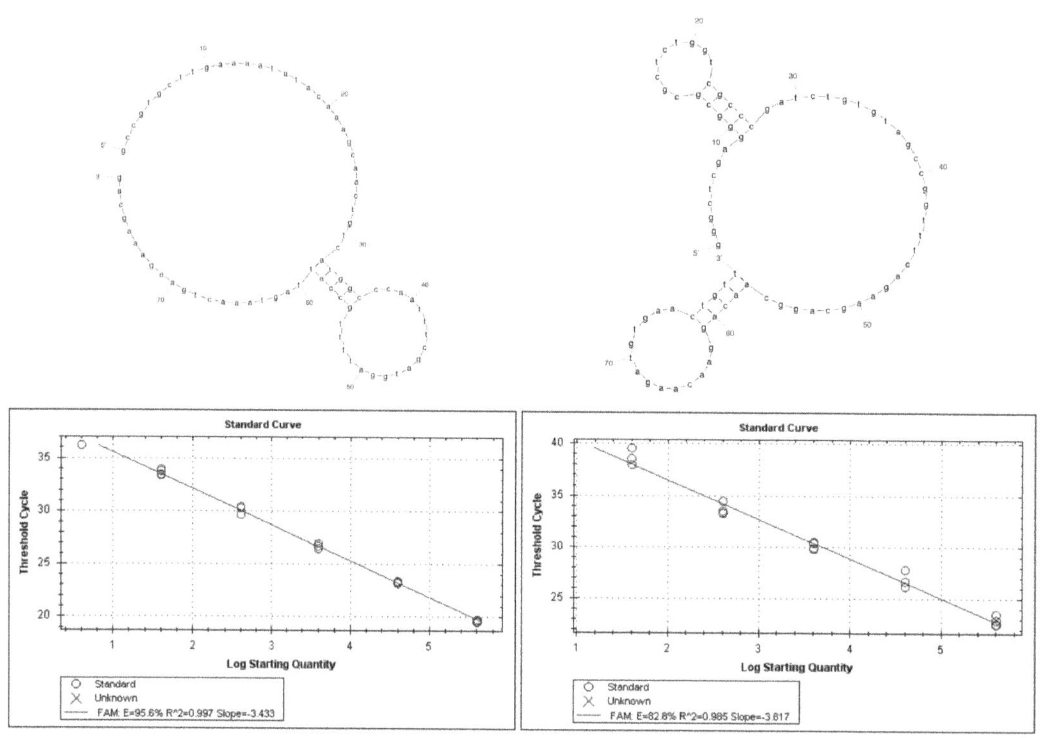

Figure 3.1 Importance of secondary structure at primer binding site. The amplicon on the left has a stem/loop structure within the amplicon, with both upstream and downstream (RT) primers able to hybridize to an open structure. A dilution curve results in an efficient qPCR assay (96%) and tight triplicate replicates (R^2=0.997). The amplicon on the right has stem loop structures at the primer binding sites, resulting in a significantly poorer amplification efficiency (82.8%) and more variation amongst the replicates (R^2=0.985).

Primers must have no intraprimer homology to avoid the formation of secondary structures that can interfere with annealing to the template. They are generally between 15 and 25 bases long when composed entirely of DNA, to maximize specificity, with a G/C content of around 50%. This design contributes to a higher yield of full-length product after synthesis and purification. For primers binding to very AT-rich sequences it is advantageous to substitute one or more of the bases with a locked nucleic acid LNA analogue to reduce the overall primer length while maintaining an appropriate T_m.

Primers should also not have any secondary structure (i.e. inverted repeats) or have sequence complementarity at the 3′ ends that could form dimers. Owing to competition between intermolecular (primer-template, probe-target) hybridization, and intramolecular hybridization, inverse repeats can cause inefficient priming and probing of the target sequence. PCR and/or probe binding can completely fail because of formation of stable hairpins at the binding region, or inside the amplicon in general. Primer dimers have a negative ΔG value, so primers should be chosen with a ΔG value no less negative than −10 kcal/mol.

Quantitative PCR assays using hydrolysis probes are usually carried out as two-step reactions, with a denaturation step followed by a combined annealing/polymerization step, during which fluorescence is measured. Most detection systems rely upon binding of the probe to the template before hybridization of the primers. It is for this reason that the probe T_m should be around 10°C higher than the T_m of the primers, usually in the 68–70°C range. The exception to this is the design of Scorpion primers (Whitcombe *et al.*, 1999) where the probe region hybridizes to the newly formed template and so has a T_m that is the same or 1°C lower than the primers. Probes should be designed with the 5′ end as close to the 3′ end of one of the primers as is possible without them overlapping. This ensures rapid cleavage by the DNA polymerase.

The probe should be no more than 30 nucleotides in length to maximize quenching, with a G/C content of around 50%. If the target sequence is AT-rich, incorporate analogues such as LNA or minor groove binders. Guanine (G) can quench fluorescence and so is avoided as the 5′ base. In this position the G would continue to quench the reporter even after cleavage, resulting in reduced fluorescence values (ΔR), which could result in reduced sensitivity. Any runs of four or more identical nucleotides, especially G, should be avoided because these can cause secondary structure of the probe and reduce hybridization efficiency. Where an alternative sequence cannot be selected, disruption of a series of Gs by the substitution of an inosine can significantly improve probe performance.

Oligonucleotide suppliers are best chosen based on consistent oligo quality, service and support. If using amplicon-specific oligonucleotides to generate standard curves, it is essential to order those oligonucleotides on a different date or to request that these are produced at a different location, so as not to permit any chance of contamination of the primers with that oligonucleotide. It is also advisable to order a number of potential primers for a single target, optimize the assay and then order the probe. This helps to ensure that the best assay is selected. Different batches of primer and probe can generate different results, hence it is essential to verify batch concentrations and compare results obtained with any new batch of oligonucleotide.

Artificially synthesized amplicon target oligonucleotides should be PAGE purified, probes should be HPLC purified, whereas primers can be supplied in a standard desalted or reverse phase cartridge purified format. It is essential to aliquot all oligonucleotides, as this prevents contamination of a whole batch and reduces the need to repeatedly freeze/thaw them. Forward and reverse primers can be stored at 10 μM or mixed together after optimization at the required working stock concentrations. Probes should be stored frozen at −70°C both when lyophilized or in solution (at 5 μM). The probe must be protected from light, but long-term stability is variable, with some probes stable for several years, and others hydrolysed after 6 months.

RNA assessment

The quality of the RNA template is an essential determinant of the reproducibility, reliability and biological relevance of any subsequent RT-qPCR assay. RNA extraction is easiest from fresh, or frozen material, with the main concern relating to the maintenance of RNA integrity. RNA is easily degraded and it is easy to co-purify inhibitors of the RT, or PCR steps, which will generate inaccurate results (Huggett *et al.*, 2008). The diversity of the different sources of mRNA poses a real practical problem: they include not just relatively simple *in vitro* samples such as tissue culture cells, but *in vivo* biopsies obtained during endoscopy, post-surgery, post-mortem and from archival materials. Naked RNA is extremely susceptible to degradation by endogenous ribonucleases (RNases) that are present in all living cells. Therefore, the key to the successful isolation of high-quality RNA and to the reliable and meaningful comparison of qRT-PCR data is to ensure that neither endogenous nor exogenous RNases are introduced during the extraction procedure. Two related problems are that different mRNAs display differential stability and that mRNA expression profiles can change rapidly after cells or tissue samples have been collected, but before they have been frozen. This implies that quantification of a single transcript or determination of relative transcript levels could be seriously affected by the sampling procedure adopted and makes it essential that careful consideration is given to how best to conserve the original expression profile during sample extraction and subsequent processing. This is a particular problem with cells extracted from whole blood (Keilholz, 1998) and mRNA profiles can change over several orders of magnitude even in the short time it can take from collecting the blood to processing it in the laboratory. The main causes of the observed differences are RNA degradation and the induction of certain genes, e.g. Cox-2, as well as the method of RNA preparation (de Vries *et al.*, 2000). Therefore, appropriate sample acquisition has a major influence on the quality of the RNA and subsequently on any result of qRT-PCR assays (Vlems *et al.*, 2002a,b) and once a biological sample has been obtained, immediate stabilization of RNA is the most important consideration (Grotzer *et al.*, 2000; Madejon *et al.*, 2000; Rainen *et al.*, 2002; Muller *et al.*, 2004).

The analysis of RNA extracted from archival material constitutes an exception from the highest quality rule. RNA extracted from formalin-fixed, paraffin-embedded material is a little more difficult to analyse, as extensive degradation can occur before (Mizuno *et al.*, 1998) or during (Klimecki *et al.*, 1994) the formalin fixation process. Furthermore, formalin fixation generates cross-links between nucleic acids and proteins and covalently modifies RNA, making subsequent RNA extraction, reverse transcription and quantification analysis problematic (Masuda *et al.*, 1999). Not surprisingly, fixatives are important (Goldsworthy *et al.*, 1999) and different tissue preparation methods will invariably lead to different results from different laboratories. There are numerous reports that describe the successful extraction and quantification of RNA and miRNA (Doleshal *et al.*, 2008) from formalin-fixed, paraffin-embedded (FFPE) archival materials (Coombs *et al.*, 1999; Godfrey *et al.*, 2000; Specht *et al.*, 2001; Macabeo-Ong *et al.*, 2002; von Smolinski *et al.*, 2005; Yang *et al.*, 2006; Li *et al.*, 2008). Formalin-fixed tissues are ideal for retrospective clinical studies of disease mechanisms and as the use of PCR technology has become more common in molecular testing, it has enhanced the clinical utility of these tissues. This is important, since such studies have the potential to enable the correlation of molecular findings with the response of the patient to treatment and eventual clinical outcome. RT-qPCR assays, with amplicon lengths of below 100 bp, are ideally placed to amplify the usually degraded RNA from these archival samples whose average size is 200–250 nucleotides. However, care must be taken when interpreting the results obtained from archival material, as gene expression profiles from FFPE samples do not exactly correlate with the profiles generated from the corresponding frozen samples; for example, in a study demonstrating that 64 genes were differentially expressed in matching fresh-frozen normal colon and cancer samples, only 38 genes were identified with differential expression in the corresponding FFPE samples. Furthermore, only 28 of these genes were in common (Bibikova *et al.*, 2004a). Thus, any results obtained using FFPE samples require independent experimental determination (Bibikova *et al.*, 2004b), but also may underestimate or report misleading changes in gene expression patterns. On the other hand, variation appears to be constant for each assay, allowing a standardized assessment of RT-qPCR assay efficiencies and the calculation of correction factors. These have the potential to make FFPE-derived expression data comparable to those obtained from fresh material as well as among each other (Koch *et al.*, 2006).

Just as real-time chemistries have revolutionized the PCR reaction, so the introduction of the Bioanalyser 2100 (Agilent), Experion (Biorad) and Lab901 (Agilent) has revolutionized the ability to analyse the quality of RNA samples. These instruments are rapidly becoming the tools of choice for analysing RNA preparations destined to be used in RT-qPCR assays. All three systems use small bench top instruments that integrate sample detection, quantification and quality assessment. They do this by using a combination of microfluidics, capillary electrophoresis and fluorescent dyes. Prior to RNA quality analysis on the BioRad or Agilent systems, a chip containing an interconnected network of microchannels fabricated in glass is filled with a sieving polymer and fluorescence dye. Following chip preparation, either 12 (NanoChip kit) or 11 (PicoChip kit), sample wells together with a marker ladder, are filled and the chip becomes an integrated electrical circuit. The chip is placed into its appropriate instrument and electrodes fit into the wells of the chip. A voltage gradient is applied and, similar to slab gel electrophoresis, RNA molecules are separated by size. During the chip run, a fluorescent dye intercalates directly with the RNA and all bands pass a detector and are detected by laser-induced fluorescence. RNA ladders containing six RNA fragments are used as a reference for data analysis and the unknown samples are compared to the ladder fragments to determine the concentration of the unknown samples and to identify the rRNA peaks. The main differences with the Lab901 ScreenTape System are that it consists of 16 individual separation channels that have been preloaded with reagents and that one RNA sample can be analysed at a time. There is broad correlation between the different formats, although the Lab901 system gave better indications of performance classification for some of the criteria measured (Wilkes *et al.*, 2010). These systems have a wide dynamic range and can quantify as little as 5 ng/µl, but are most accurate when working at concentrations above 50 ng/µl. An electropherogram of fluorescence versus time and a virtual gel image are generated and the software assesses RNA quality by using the areas under the 28S and 18S rRNA peaks to calculate their ratios. However, it is worth noting that, despite sophisticated algorithms, these profiles provide a measure of rRNA integrity, which may or may not reflect the quality of the (largely unmeasured) mRNA. An Agilent Bioanalyzer analysis of 24 colorectal cancer samples is shown in Fig. 3.2.

Accurate delineation of mRNA quality is likely to demand additional quality assessments similar to the 3′:5′ ratios used during microarray experiments. A method using the relative expression of the GAPDH transcript has been described (Nolan *et al.*, 2006a) and modified for the successful, large-scale screening of primary neuroblastoma tumour samples (Vermeulen *et al.*, 2011). The data obtained are independent of ribosomal RNA (rRNA) integrity and provide a quantifiable measure of the degradation of the transcripts of interest.

Although the microfluidic instruments can be used for quantification, more reproducible measurements are achieved by using either the Nanodrop (www.nanodrop.com) system or, for higher throughput applications, the Ribogreen RNA (www.invitrogen.com) staining dye that binds to RNA and produces a fluorescent signal. The concentration of test samples is determined by comparison of the fluorescent signal to that of serial dilutions of controls forming a standard curve series. This can be used in conjunction with a fluorescent plate reader or some qPCR instruments. Note that it is essential that a single method of RNA quantification is selected and that this is used exclusively throughout a study. It has been clearly demonstrated that different methods of quantification give widely differing results (Bustin, 2005).

RNA integrity

RNA quality measurements are defined by both purity (absence of protein and DNA contamination, absence of inhibitors) and integrity. Traditionally RNA quality has been determined by analysis of the A260/A280 ratio and/or analysis of the rRNA bands on agarose gels, an approach updated with the introduction of the Agilent Bioanalyser/ Lab901/BioRad Experion microfluidic capillary electrophoresis systems. In a report that claims that it is possible to calculate a more objective measure of RNA quality by measuring several characteristics of the electropherogram generated by the Bioanalyser 2100, the fraction of the area in the region of 18S and 28S RNA, the relative height of the 28S peak, the area containing molecules which are defined as having a 'fast' mobility and marker height, are used to calculate and assign an RNA integrity number (RIN) to each RNA sample (Schroeder *et al.*, 2006). However, an exhaustive analysis of the influence of RNA integrity on RT-qPCR assay performance reaches a different conclusion (Fleige and Pfaffl, 2006; Fleige *et al.*, 2006). The authors extracted RNA samples extracted from numerous tissue types, subjected them to controlled degradation and analysed them using the Agilent Bioanalyser. The samples had RINs between 10 (apparently intact material) and 4.0 (almost no evidence of rRNA bands). The quantity of individual transcripts in each of these samples was then determined using RT-qPCR assays. In some tissues the quantity of the measured transcripts detected was unaffected by the RIN whereas in others there was a linear relationship and in others a threshold response. Critically, the relationship between transcript quantity and RIN was different for different tissues and there was not a predictable relationship between these factors. The authors conclude that moderately degraded RNA samples can be reliably analysed and quantified, as long as the amplicons are kept short (< 250 bp) and expression

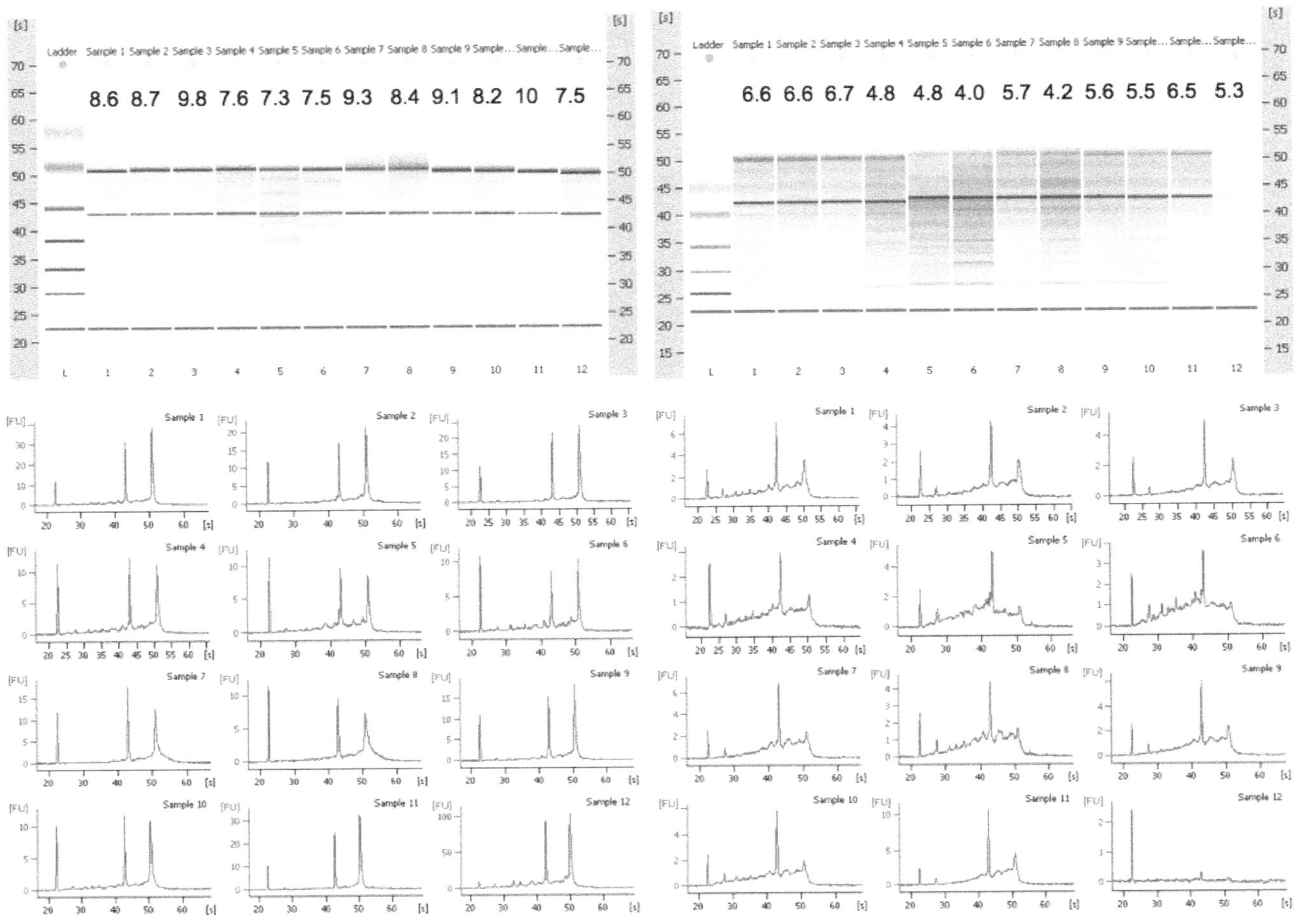

Figure 3.2 Analysis of RNA extracted from colorectal cancer samples using the Agilent 2100 Bioanalyzer. The gel view shows a representation of the RNA, with the 18S and 28S peaks prominent. The samples on the right have additional bands and the 28S band is less distinct, suggesting poorer integrity of the rRNA. The electropherogram uses the distinct differences between the samples to calculate the RIN numbers shown. The amount of small RNA (first peak), which can include 5.8S and 5S ribosomal peaks and transfer RNA, is highly dependent on the preparation protocol.

is normalized against an internal reference. The discrepancy between the two reports could be due to the relatively poor correlation coefficient (0.52) between RIN and expression values of the reference genes reported by the authors advocating the use of RINs. Similarly, a very recent study shows that the microfluidic analysis by the Experion instrument was a less reliable measure of sample quality than measurement of the 3′/5′ ratio of HPRT (Vermeulen *et al.*, 2011).

Inhibitors

Inhibitory components can result in a significant reduction in the sensitivity and kinetics of qPCR (Cone *et al.*, 1992; Liss, 2002; Guy *et al.*, 2003; Perch-Nielsen *et al.*, 2003; Lefevre *et al.*, 2004; Radstrom *et al.*, 2004; Sunen *et al.*, 2004; Jiang *et al.*, 2005; Huggett *et al.*, 2008). Inhibitors may be present in reagents used during nucleic acid extraction or co-purified components from the biological sample, for example bile salts, urea, heparin or IgG. Even the RT enzyme that remains in the cDNA sample and bound to cDNA also has an inhibitory effect on PCR (Chandler *et al.*, 1998). Blood is well-known for containing numerous inhibitors of the PCR reaction (Akane *et al.*, 1994; Fredricks and Relman, 1998; Al Soud and Radstrom, 2001) and we have observed that haem can affect qPCR reading by differentially enhancing FAM signals while reducing HEX fluorescence. The presence of inhibitors can produce inaccurate quantitative results and, at worst, a high degree of inhibition will create false-negative results. The most common procedure used to account for any differences in PCR efficiencies between samples is to amplify a reference gene in parallel to the reporter gene and relate their expression levels. However, this approach assumes that the two assays are inhibited to the same degree and it is clear that this is not the case (Huggett *et al.*, 2008). The problem is even more pronounced in 'absolute' quantification, where an external calibration curve is used to calculate the number of transcripts in the test samples, an approach that is commonly adopted for quantification of pathogens. Some, or all, of the biological samples may contain inhibitors that are not present in the nucleic acid samples used to construct the calibration curve, leading to an underestimation of the mRNA levels in the test samples (Stahlberg *et al.*, 2005). The increasing interest in extracting nucleic acids from formalin-fixed paraffin-embedded (FFPE) archival material will undoubtedly lead to an exacerbation of this problem. Obviously, such inhibitors are likely to distort any comparative quantitative data. However, a survey of practices adopted by qPCR researchers has revealed that only 6% of researchers test their nucleic acid samples for the presence of inhibitors (Bustin, 2005).

Various methods can be used to assess the presence of inhibitors within biological samples. The PCR efficiency in a test sample can be assessed by serial dilution of the sample (Stahlberg *et al.*, 2003), although this is impossible when using very small amounts of RNA extracted, for example, from single cells or from laser capture microdissected sections. When a large number of samples are to be analysed this approach becomes impractical. Furthermore, there are mathematical algorithms that provide a measure of PCR efficiency from analysis of the amplification response curves (Liu and Saint, 2002a,b; Ramakers *et al.*, 2003; Tichopad *et al.*, 2003; Rutledge, 2004; Rutledge and Stewart, 2008a,b) although these are generally suited to SYBR Green I dye based assays and are generally less effected when using probe based systems. We have observed that standard curve measurements are a simple and effective means for identification of samples that contain inhibitors, since both changes in efficiency (gradient) and shifts to higher cycles (efficiency is maintained but delays in initial amplification causes reduced sensitivity) are identified. Internal amplification controls (IAC) that co-purify and co-amplify with the target nucleic acid can be used to detect inhibitors as well as indicate template loss during processing (Pasloske *et al.*, 1998). Another approach utilizes a whole bacterial genome to detect inhibition from clinical samples (Cloud *et al.*, 2003). As an alternative, we have developed a method, which we have called the 'SPUD' assay, that is suitable for the detection of inhibitors in nucleic acids extracted from all tissue with the exception of potato (Nolan *et al.*, 2006b). This is a simple qPCR based assay that is amenable to high-throughput screening consisting of a predetermined, universal template sequence (SPUD template, based on a potato sequence), primers and the SPUD detecting probe. The sequences have been designed such that they are compatible with studies on all organisms (with the exception of potato). In the absence of RNA (or cDNA) sample template amplification of the SPUD target results in a quantification cycle, *Cq* (Bustin *et al.*, 2009) that reflects the optimal reaction conditions for the chosen SPUD template concentration. When RNA (or cDNA) sample is added, any observed deviation from this control *Cq* value indicates that the presence of some component within the sample has inhibited the qPCR reaction. The method can also be used with an artificial RNA template to monitor both the RT and qPCR steps.

Different polymerases react differently to inhibitors and it is worthwhile checking each template preparation for inhibition by testing several polymerases for their efficiency at amplifying the template. Sequence-dependent PCR amplification bias has been observed and results in incorrect or ambiguous quantification (Shanmugam *et al.*, 1993) and can even result in the preferential amplification of one allele over another (Weissensteiner and Lanchbury, 1996). AmpliTaq Gold and the *Taq* polymerases are totally inhibited in the presence of 0.004% (v/v) blood, whereas HotTub (*Tfl*) from *Thermus flavus*, *Pwo*, r*Tth*, and *Tfl* DNA polymerases are able to amplify DNA in the presence of 20% (v/v) blood without reduced amplification sensitivity; the DNA polymerase from *Thermotoga maritima* (Ultma) appears to be the most susceptible to a wide range of PCR inhibitors. HotTub and *Tth* are the most resistant to the inhibitory effect of K^+ and Na^+ (Al Soud and Radstrom, 1998) and biological samples (Poddar *et al.*, 1998). Thus, the PCR-inhibiting effect of various components in biological samples can, to some extent, be eliminated by the use of the appropriate thermostable DNA polymerase. One thing to bear in mind is that some enzymes are in fact enzyme mixes that combine a highly processive enzyme with a proof-reading enzyme to balance fidelity and yield (Barnes, 1994).

Reverse transcription

The most appropriate method for cDNA priming remains a contentious issue. Direct comparison of the different methods available reveals that there is no one universally best method, and that results are target and enzyme-dependent (Bustin and Nolan, 2004; Stahlberg *et al.*, 2004a,b; Tichopad *et al.*, 2009). Random or oligo-dT priming both allow a representative pool of cDNA to be produced during a single reaction. However, it has been shown that RT priming using random hexamer primers does not necessarily result in equal efficiencies of reverse transcription for all targets in the sample and that there is not a linear correlation between input target amount and cDNA yield when specific targets are measured (Bustin and Nolan, 2004; Lacey *et al.*, 2005). The linearity of cDNA synthesis may be restored by addition of PEG or linearized acrylamide (Gaillard *et al.*, 1999; Stahlberg *et al.*, 2004a) although we have observed that even after such additions, not all genes are detected at the appropriate dilution from the same cDNA series. A comparison of the efficiency of RT priming by random primers of varying lengths showed that 15-nucleotide-long random oligonucleotides consistently yielded at least twice the amount of cDNA as random hexamers (Stangegaard *et al.*, 2006). The 15-mers were more efficient at priming, resulting in reverse transcription of > 80%, while random hexamers induced reverse transcription of only 40% of the template. Not surprisingly, this resulted in the detection of one order of magnitude more genes in whole transcriptome DNA microarray experiments. Since the authors did not address the question of linearity it remains to be seen whether this change overcomes that particular limitation of the random hexamer approach. Unfortunately, we have not observed the same beneficial effect and found that specific priming, followed by random nonomers resulted in the most sensitive detection of targets from colon cancer RNA samples, with 15-mers the least efficient primers (Fig. 3.3).

The choice of oligo dT primers requires the use of non-degraded RNA. Even when intact mRNA is used, the cDNA molecules may be truncated, since the RT enzyme cannot proceed efficiently through highly structured regions; accordingly assays should be targeted towards the 3′ end of the transcript. This is an unsuitable choice for experiments that require examination of splice variants, sequences with long 3′ UTR regions or those without polyA sequences. Furthermore, oligo-dT priming is not recommended when using RNA extracted

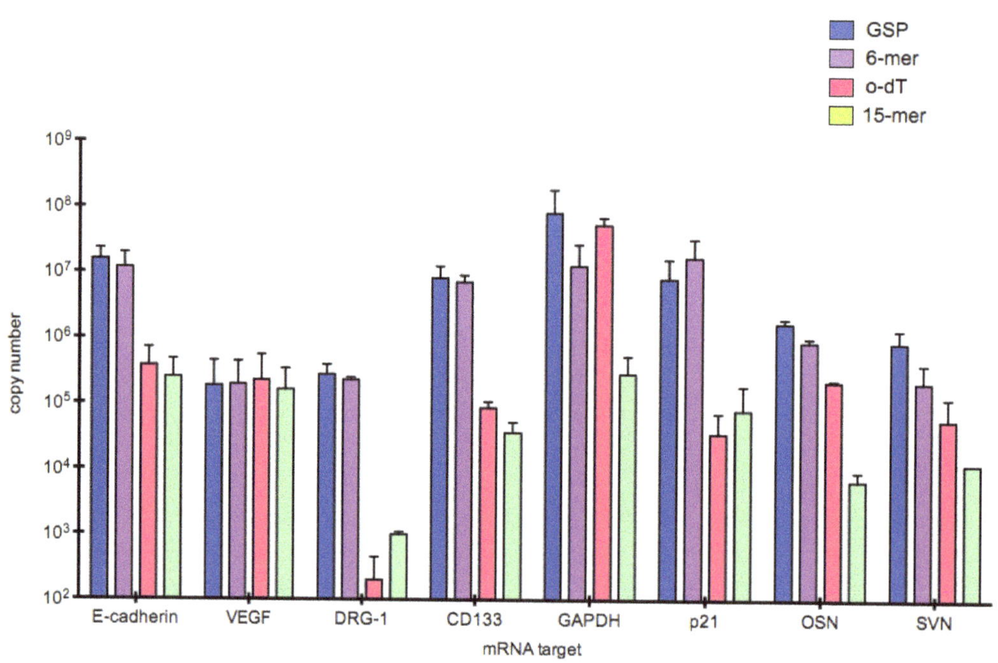

Figure 3.3 Choice of priming strategy affects RT-qPCR results. Four priming strategies were compared for the reverse transcription of eight target mRNAs. The cDNA from each was analysed using qPCR. In each case, gene-specific priming resulted in the most efficient reverse transcription, although the use of random hexamers generated very similar results. Reverse transcription from random 15-mers was consistently the least efficient priming method. Oligo-dT priming was less consistent, resulting in greater variability of quantification and relative quantity between targets.

from paraffin tissue sections, since formalin fixation results in the loss of the polyA tails on mRNA (Lewis and Maughan, 2004).

Target specific primers are the most unambiguous method for converting mRNA into cDNA (Lekanne Deprez *et al.*, 2002) and are the recommended choice when RNA quantity is not a limiting factor. In a comparison of priming methods, well-designed specific primers were also the most sensitive. Interestingly, it has been reported that specific primers can be applied in tandem for the reliable and specific amplification of 72 genes from limiting amounts of RNA using a multiplexed tandem PCR approach (Stanley and Szewczuk, 2005). Nevertheless, as with random priming there may be differences in the efficiencies at which individual RT reactions occur. These variations must be controlled for by reference of unknown samples to a calibrator sample (when using $\Delta\Delta Cq$ analysis) or to standard curves. Specific priming of RNA dilutions results in a linear response of target cDNA yield (Bustin and Nolan, 2004) and so a further advantage of using this priming method is that the efficiency of the combined RT-qPCR reaction can be confirmed by analysis of the gradient of the standard curve. The inclusion of a calibrator sample or standard curve on every plate is important since these serve as controls for measuring inter-assay variability that may occur when multiple samples are run on different plates.

Fortunately, whereas the RT step is highly variable, the qPCR segment of the assay is remarkably reproducible when run under optimal conditions (Kubista *et al.*, 2006), although unsurprisingly, greater variability is observed between replicates when the assay is run under suboptimal conditions (Hilscher *et al.*, 2005).

Choice of enzymes

The RT-qPCR assay can be performed either as a single RT and PCR enzyme procedure in a single tube or as a separate RT and PCR enzyme procedure in a single or a two tube format.

Single RT and PCR enzyme

A single enzyme such as *Tth* polymerase is able to function both as an RNA- and DNA-dependent DNA polymerase and can be used in a single tube without secondary additions to the reaction mix. The main advantages of the single enzyme approach are the reduced hands-on time and potential for contamination. There are several disadvantages:

1. Since all reagents are added to the reaction tube at the beginning of the reaction, separate optimization of the two reactions is not feasible.
2. The assay can only be carried out using target-specific primers.
3. The assay may be less sensitive, due possibly to the less efficient RT activity of *Tth* polymerase when compared with other RT enzymes (Cusi *et al.*, 1994; Easton *et al.*, 1994).
4. The most thorough study comparing the two procedures found that the one-step reaction was characterized by extensive accumulation of primer dimers, which could obscure the true results in quantitative assays (Vandesompele *et al.*, 2002b).

Two enzyme procedures: separate RT and PCR enzymes

Reactions can be either 'uncoupled' or 'coupled'. In the 'uncoupled' alternative the reverse transcriptase is used to synthesize cDNA in a first tube, under optimal conditions, using random, oligo-dT or sequence-specific primers. An aliquot of the RT reaction is then transferred to another tube (containing the thermostable DNA polymerase, DNA polymerase buffer and PCR primers) and PCR is then carried out under optimal conditions for the DNA polymerase. In the 'continuous or coupled' alternative the reverse transcriptase is used to synthesize cDNA in the presence of high concentrations of dNTPs and either target-specific or oligo-dT primers. Following the RT reaction, PCR buffer (without Mg^{2+}), a thermostable DNA polymerase and target-specific primers are added and the PCR is performed in the same tube. Interassay variation of two enzyme protocols can be very small when carried out properly, with correlation coefficients ranging between 0.974 and 0.988 (Vandesompele *et al.*, 2002b). The disadvantages of this approach are:

1 In two-enzyme/one-tube assays, a template switching activity of viral RTs can generate artefacts during transcription (Mader *et al.*, 2001).
2 Two-enzyme/two-tube reactions present additional opportunities for contamination.
3 The RT can inhibit the PCR assay, resulting in reduced amplification efficiency and an underestimation of target quantity (Suslov and Steindler, 2005).

On balance, the increased flexibility and potential for optimization makes the use of a two enzyme protocol preferable to the single enzyme procedure.

PCR

When selecting a PCR enzyme, there are two important considerations; first, an absolute requirement for a thermostable polymerase with a 5′–3′ nuclease activity when carrying out 5′-nuclease assays using hydrolysis probes (TaqMan) and second, selection of a hot start *Taq* polymerase, either complexed with an antibody or chemically modified to inhibit polymerase activity at low temperatures. This is particularly important when targeting low copy number targets using SYBR Green I dye chemistry. We have used enzymes and kits from all the major suppliers and have found no major, consistent differences in quality but note that antibody inactivated enzymes require a shorter period of activation. Again, it is important to use the buffer supplied with the enzyme, since the manufacturer has optimized it for that enzyme. Some manufacturers include PCR enhancer solutions in their kits. It is important to read the instructions carefully, since they can affect the reaction conditions. For example, if designed for amplification of G/C-rich regions, they may lower the DNA melting temperature, thus lowering the primer annealing temperature. Combined reagent mastermixes are increasingly popular. The advantage of using these is that there is a reduction in potential errors that could be introduced when combining components prior to reactions. In many cases these buffers are stabilized with components that will also disrupt secondary structures in templates and so can give improved results. In the case of difficult templates, a sequence specific effect with specific reagent blends will be observed and so trying a different blend may alleviate problems. We have also found that the 'fast' reagents offered by all suppliers are better at amplifying difficult templates or from samples showing inhibition. Blended buffers are provided in a convenient format and are suitable for routine or high-throughput assays. The disadvantage is that there is less flexibility for optimization of challenging simplex or multiplex assays. It is important to check the mixture components with particular attention to matching the $MgCl_2$ concentration to the detection chemistry used and to the presence and concentration of optical normalization reference dyes, such as ROX.

Protocols

PCR optimization

The thermodynamic stability (ΔG) of a duplexed primer/target structure differs for different primers and varies with primer concentration. Therefore, it is important to use primers at concentrations that result in optimal hybridization and priming. Although optimization of PCR reactions used to be an essential part of assay development, recent trends towards high throughput and rapid data reporting have resulted in the elimination of this step in many laboratories. Indeed, some manufacturers claim it is not necessary to optimize primer concentrations when using their particular mastermixes. However, it is clear that the rationale underlying the original recommendations remain valid and that optimization of primer concentration can significantly improve assay detection sensitivity (Fig. 3.4).

The protocol below utilizes SYBR Green I dye to detect amplified products and to visualize melting of the amplicons after the reaction. Post-amplification melting is performed by gradually increasing the temperature of the reaction and assessing the fluorescence due to amplicon-bound dye at each temperature. As the temperature is increased the fluorescence decreases due to separation of the double strands of the DNA. The temperature at which this occurs depends on amplicon size and sequence. In this way the melting characteristic of primer dimer and specific products can be distinguished, since primer dimers are generally shorter and therefore melt at lower temperatures than longer, target amplicon products (Fig. 3.5).

1 Selection of target sample to use during optimization. When using an artificial amplicon or linearized plasmid as the template to determine the PCR efficiency prepare a stock at around 10^4–10^5 copies/µl. While working with the concentrated, artificial template is it absolutely critical to find a working area that is well away from the qPCR set up area and to take all precautions necessary to ensure that reaction contamination does not occur. Test the amplification of the stock template using 200 nM of each primer to establish that amplification of the sample results in *Cq* values of close to 20. If the concentration is outside of this range assume that a 10-fold difference in concentration corresponds to 3.323 cycles and adjust the sample concentrations accordingly. Alternatively, if cDNA synthesized from universal RNA is to be used, prepare 6–8 fivefold serial dilutions. Test the amplification of the dilution series of template using 200 nM primers and select the diluted sample that results in *Cq* values closest to 20.

2 Prepare individual working dilutions of forward and reverse primers for the optimization matrix. Prepare 20 µl of stocks of 0.5 µM, 1 µM, 3 µM, 6 µM and 9 µM each primer.

3 Prepare a matrix of primer concentrations by adding 2.5 µl of the stock concentration stated in the appropriate column of primer to two qPCR reaction tubes, such that each reaction is prepared in duplicate. Hint: Refer to Table 3.1 for the final concentrations of primers. Concentrations are stated as forward primer/reverse primer in nM in the final qPCR reaction.

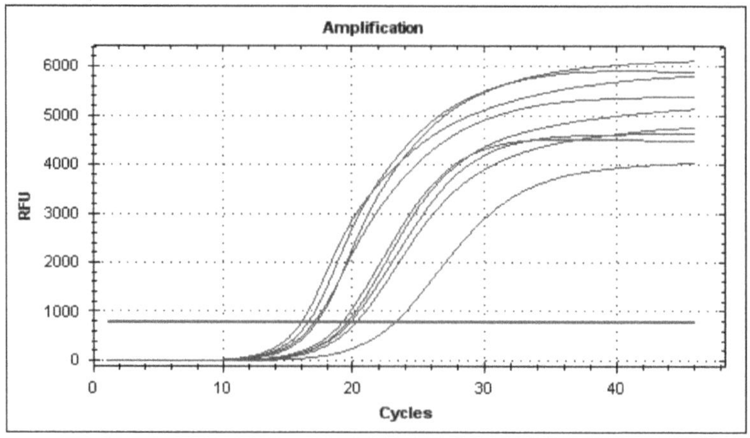

Figure 3.4 The difference in primer binding efficiency affects PCR kinetics. The amplification plots were generated by amplifying the same target using different primer concentrations between 100nM and 300nM as described in Table 3.1. In this example, the difference between the most and least efficient primer concentration combinations is 7 *Cq*s.

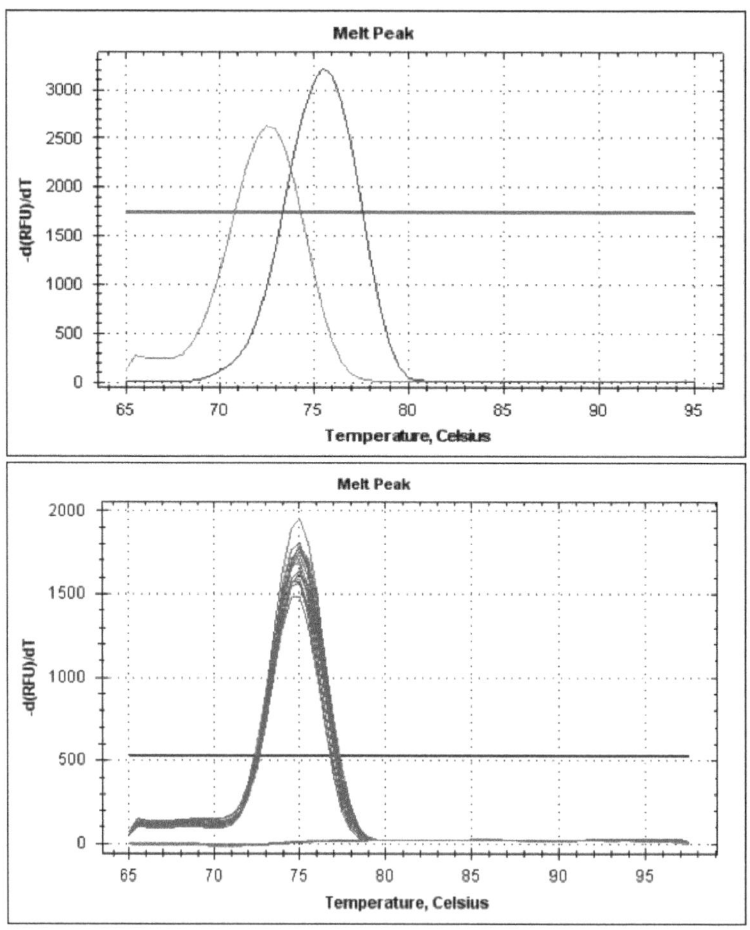

Figure 3.5 (A) SYBR Green I dye melt curve demonstrating the appearance of primer dimers in a poorly designed assay. The target amplicon has a T_m of around 76°C (right), whereas the primer dimers melt at 72°C (left). (B) SYBR Green I dye melt curve demonstrating the absence of primer dimers in a well-designed assay. The amplicon T_m is 75°C.

Table 3.1 Forward and reverse primer concentrations for optimization assay

Reverse primer (μM)	Forward primer (μM) 0.5	1	2	3	6
0.5	50/50	100/50	200/50	300/50	600/50
1	50/100	100/100	200/100	300/100	600/100
2	50/200	100/200	200/200	300/200	600/200
3	50/300	100/300	200/300	300/300	600/300
6	50/600	100/600	200/600	300/600	600/600

4 Prepare a mastermix for 55 reactions according to Table 3.2A.

Table 3.2A qPCR mastermix for primer optimization assay

	Stock concentration	Volume to add for single 25μl reaction
Water		2.125μl
SYBR Green I Dye Mastermix (qPCR)	2×	12.5μl
Reference dye (ROX)[a]	0.03μM (optional depending on instrument)	0.375μl
Sample	10^4 to 10^5 copies in 5μl diluted cDNA synthesis reaction or alternative template from step 1	5μl

[a]Note that the given protocol includes the use of the ROX reference dye; if this is not required, adjust water volume accordingly.

5 Add 20 μl mastermix to each primer pair in qPCR tubes. Mix, avoiding bubbles and briefly spin reaction tubes to ensure components are in the bottom of the tubes.
6 Repeat the set up but use a mastermix that does not contain template to create no template controls for each primer concentration.
7 Run in qPCR instrument using three step protocol as shown in Table 3.2B.

Table 3.2B qPCR three-step protocol

1 cycle	Activation	95°C, 2–5min, depending on mastermix
40 cycles	Denaturation	95°C, 30s
	Annealing/	60°C, 30s (collect data during each cycle)
	Extension	72°C, 30s
1 cycle	Melt curve	Between 55°C and 95 C (collect data at each temperature)

8 Refer to the data analysis section for general guidelines. Select the primer combination that fulfils the following criteria in the given order of priority:
9 Absence of primer dimers; use the melt curve profile.
10 Lowest *Cq*: This determines the primer combination resulting in the most sensitive (and efficient) reaction.

11 Highest end point fluorescence (the highest dRn). This is likely to signify the highest number of amplicon products being formed.
12 Absence of signal in the NTC or high $Cq > 35$.

Determination of assay sensitivity and reproducibility

1 Prepare a 10-fold serial dilution of the template described in step 1 of the PCR optimization above, using a minimum of five samples between 10^7 and 10^3 copies in 5 µl. It is possible to quantify a dynamic range of 10–12 logs on most systems (i.e. from 1 to between 10^{10} and 10^{12} target copies). Ensure that the range spans the Cq values anticipated for the experimental analysis, repeating the dilution series using different concentrations if required.
2 Add 5 µl of each dilution point to qPCR reaction tubes, including each sample in duplicate.
3 Prepare a qPCR mastermix according to Table 3.3A using the primer concentrations determined from the primer concentration optimization process.

Table 3.3A qPCR mastermix for determination of assay efficiency and sensitivity

	Stock concentration	Volume (µl) to add for single 25µl reaction
Water		2.125µl
SYBR Green I Dye Mastermix (qPCR)	2×	12.5µl
Reference dye (ROX)[a]	0.03µM (optional depending on instrument)	0.375µl
Primers	As determined during primer optimization process	5µl

[a]Note that the given protocol includes the use of the ROX reference dye, determine whether this is required for the instrument and omit, replacing with water, if desired.

4 Add 20 µl mastermix to each sample in qPCR tubes. Mix, avoiding bubbles and spin places briefly to gather mix in the bottom of the tubes.
5 Run in qPCR instrument using three-step protocol as shown in Table 3.3B.

Table 3.3B qPCR three-step protocol

1 cycle	Activation	95°C 10min (or 2min verify enzyme hot start requirement)
40 cycles	Denaturation	95°C 30s
	Annealing/	60°C 30s (collect data during each cycle)
	Extension	72°C 30s
1 cycle	Melt curve	Between 55°C and 95°C (collect data at each temperature)

6 Follow the steps outlined in the data analysis section.

7 Use the instrument software to plot *Cq* against log target concentration and determine the slope and R^2 values. Confirm that the slope lays between −3.2 and −3.5 and that the R^2 value is above 0.98, preferably above 0.99 as described for standard curve analysis.

Perform the qPCR as soon as possible after preparing the reaction mixes. Some assays are stable for 12 hours if stored at 4°C overnight; however others are more sensitive to storage. Reactions containing glycerol and DMSO or made from commercial master mixes can be stored at −20°C.

RT-qPCR assay

Reverse transcription

See flow chart for a summary of the steps (Fig. 3.6).

Variation between RT experiments

When using random primers, the efficiency of the RT reaction may vary between reactions. It also appears that the RT efficiency for different targets within the same reaction is variable (Lacey *et al.*, 2005) and it is important to control for these factors by following the steps below:

- When possible perform RT reactions on all samples at the same time.
- Use the same amount of RNA in all reactions.

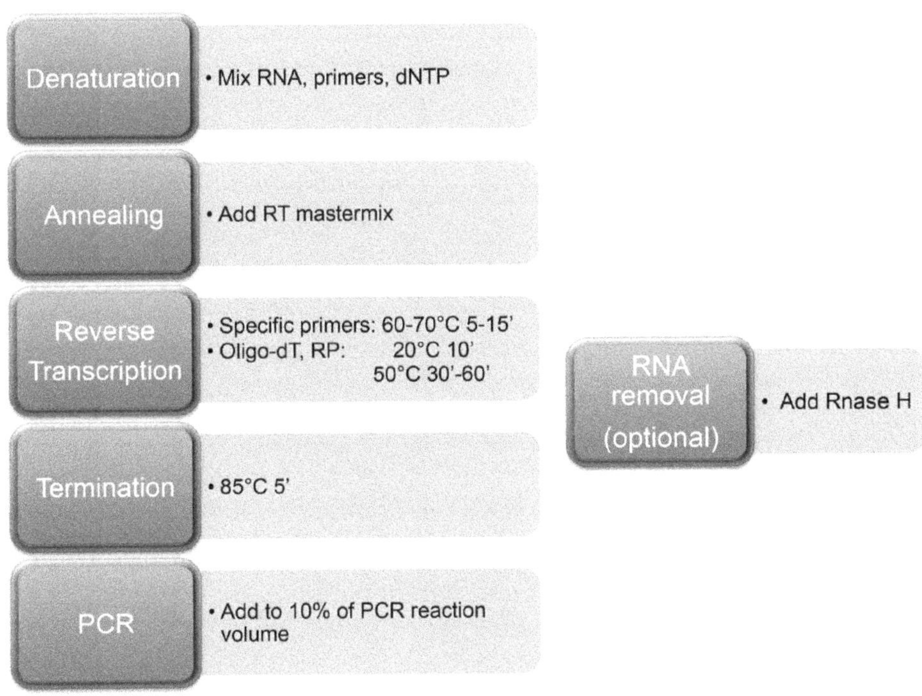

Figure 3.6 Outline of reverse transcription protocol.

- When multiple samples are processed in different batches, include a common positive reference sample alongside all batches.
- Quantify target genes relative to a standard curve including the target copy number in the positive control samples from each batch.
- Express quantity of target in samples relative to the quantity measured in the control sample included in the same RT batch (Lacey *et al.*, 2005).

1 Keep reagents and microfuge tubes on ice. Mix and briefly spin each component and prepare a mastermix in a microfuge tube by combining the reagents in the order listed in Table 3.4A. Note that this protocol is adapted for use with specific primers, random primers or oligo-dT priming. In some cases a combination of oligo dT and random primers may be appropriate.

Critical: Prepare no-RT controls in duplicate: To prepare each no-RT control, add 18 µl mastermix to 7 µl water prior to adding RT enzyme to the mastermix.

Table 3.4A RNA/primer mix components

Component	Volume per reaction
Total RNA 1–200ng (max. 500ng)	× (max 7µl)
3′ (antisense) primer (2µM) or	1µl
Random 15-mer (50ng/µl) *or*	1µl
Oligo-dT (500ng/µl)	1µl
Water	to make a total volume 10µl

2 Incubate at 65°C for 10 minutes, then snap cool on ice for 5 minutes.

3 Prepare an RT mastermix for the required number of reactions by combining the reagents as listed in Table 3.4B.

Table 3.4B RT mastermix components

Component	Volume per reaction
10×RT buffer	2.5µl
$MgCl_2$ (25µM)	5µl
DTT (0.1M)	2.5µl
RT enzyme (200U/µl)	1µl
Water	to 15µl

4 Add 15 µl of the RT mastermix to the RNA/primer mix. Gently mix the tube contents and briefly spin.

4a *Specific priming*: incubate at 60°C for 15 minutes.

4b Oligo dT priming or random priming: incubate at 20°C for 10 minutes followed by 50°C for 60 minutes.

5 Terminate the reaction at 85°C for 5 minutes, and then place on ice. Collect by brief centrifugation.

6 Optional step: add 1 µl (2 U) of *E. coli* RNase H and incubate at 37°C for 15 minutes.
7 Store at −20°C or proceed to qPCR.

PCR

Option A: qPCR using SYBR Green I dye detection

1a Make a master mix by adding the reagents in the order shown in Table 3.5A, mix gently by repeatedly pipetting up and down making sure that there are no bubbles, (add DNA polymerase after mastermix production if separate components are being used and mix gently again). Buffer, MgCl2, dNTPs, SYBR Green I dye and DNA polymerase are combined in commercial 2 × mastermix preparations.

Table 3.5A PCR mastermix components (SYBR Green I dye)

Master mix	Final concentration	Volume per reaction (µl)
dH_2O		6.625[a]
2×buffer	1×	12.5
Stock Primers (or 10µM for a final concentration of 100nM)	Refer to optimization results	(0.25µl each primer) 0.5
ROX diluted (optional)	Refer to instrument manufacturer's recommendations	0.375[a]

[a]Note that this protocol includes the use of the reference dye ROX; if this is not required, adjust water volume accordingly.

2a Add 5 µl template to the reaction tubes or wells.
3a Add 20 µl volume of reaction mix to each tube or well containing template. Cap carefully, spin briefly to ensure there are no bubbles, and place the tubes, strips, rotor or 96/384-well plate into the real time thermal cycler.
4a Perform a three-step PCR reaction according to the thermal profile in Table 3.5B.

Table 3.5B qPCR three-step protocol

1 cycle	Activation	95°C	10min (or 2min verify enzyme hot start requirement)	
40 cycles	Denaturation	95°C	30s	
	Annealing	60°C[b]	30s	Collect data
	Extension	72°C	30s	Collect data
Followed by a melting profile (as below or according to the instrument manufacturer's instructions)				
1 cycle	95°C		1min	Collect data
41 cycles	55°C		30s	
	repeat and increase temperature by 1°C per cycle			Collect data

[b]Annealing temperature is primer design dependent. A two step protocol with combined annealing and extension steps into a single 62°C incubation for 60 seconds can be used when primers are being evaluated to be used with a TaqMan probe.

Option B: qPCR using dual-labelled probe detection

1b. Make a master mix by adding the reagents in the order shown in Table 3.6A, mix gently by repeatedly pipetting up and down, making sure that there are no bubbles. (Add DNA polymerase after preparing the rest of the mastermix, if separate components are being used and mix gently again.)

Table 3.6A PCR mastermix components (TaqMan)

Master mix	Final concentration	Volume per reaction (µl)
dH_2O		5.625[a]
2×buffer	1×	12.5
Stock Primers (or 10µM for 100nM)	Refer to optimization results	(0.25 each primer) 0.5
ROX diluted	Refer to instrument manufacturer's recommendations	0.375[a]
5µM Probe	200nM	1

[a]Note that the given protocol includes the use of the reference dye ROX. For alternative volumes or to exclude, adjust water volume accordingly.

2b Add 5 µl template to the reaction tubes or wells.

3b Add 20 µl of reaction mix to each tube or well containing template. Cap carefully, spin briefly to ensure there are no bubbles, and place the tubes, strips, rotor or 96/384-well plate into the real time thermal cycler.

4b Perform a two step PCR reaction according to the thermal profile outlined in Table 3.6B.

Table 3.6B qPCR two step protocol

1 cycle	Activation	95°C	10min (or 2min verify enzyme hot start requirement)	
40 cycles	Denaturation	95°C	30s	
	Annealing/extension	60°C	1min	Collect data

Note: reactions containing hydrolysis probes are usually performed using a two-step PCR profile. During this reaction the double-stranded template is melted at 95°C and the annealing and extension steps both occur during a single incubation at 60°C. Although this is suboptimal for amplification by *Taq* polymerase it is believed to encourage more efficient cleavage of the internal probe, resulting in maximum fluorescent signal per cycle. Experiments including other probe based chemistries, e.g. Molecular Beacons, Scorpions, can be run under these conditions but a three step protocol is optimal (Table 3.6C).

Table 3.6C qPCR three-step protocol

1 cycle	Activation	95°C	10min (or 2min verify enzyme hot start requirement)	
40 cycles	Denaturation	95°C	30s	
	Annealing	60°C or as appropriate for primers	30s	Collect data
	Extension	72°C	30s	Collect data

Data analysis

The analysis of qPCR data can be highly subjective, since *Cq* values can be altered by changes to baselines and threshold levels. Therefore it is important to understand that *Cq* values, in themselves are not sufficient information. Although software default options can assuage some subjectivity, it is also true that these settings are not always appropriate. Clearly, it is important to understand the analysis procedure in order to be able to generate reliable data. The following steps constitute a basic data analysis workflow:

1 Examine amplification plots of standard or reference samples and identify abnormal plots. A normal amplification plot is defined as consisting of a horizontal baseline region, a log phase of amplification, then a linear amplification phase, followed by a plateau. Note samples that record a *Cq*, but where the amplification plot clearly does not show the appropriate characteristics (Fig. 3.7), these should be subjected to further investigation.

2 The threshold settings are subjective. Select a setting that is within the logarithmic region of amplification. This is best viewed as the linear region of the plot of log normalized fluorescence versus cycle (log dRn vs. cycle). The precise setting should be within a range where all the amplification plots to be analysed are parallel, hence the setting represents cycles during which the amplification efficiencies are equivalent for each reaction to be compared. Select the threshold setting where the replicates are closest to each other.

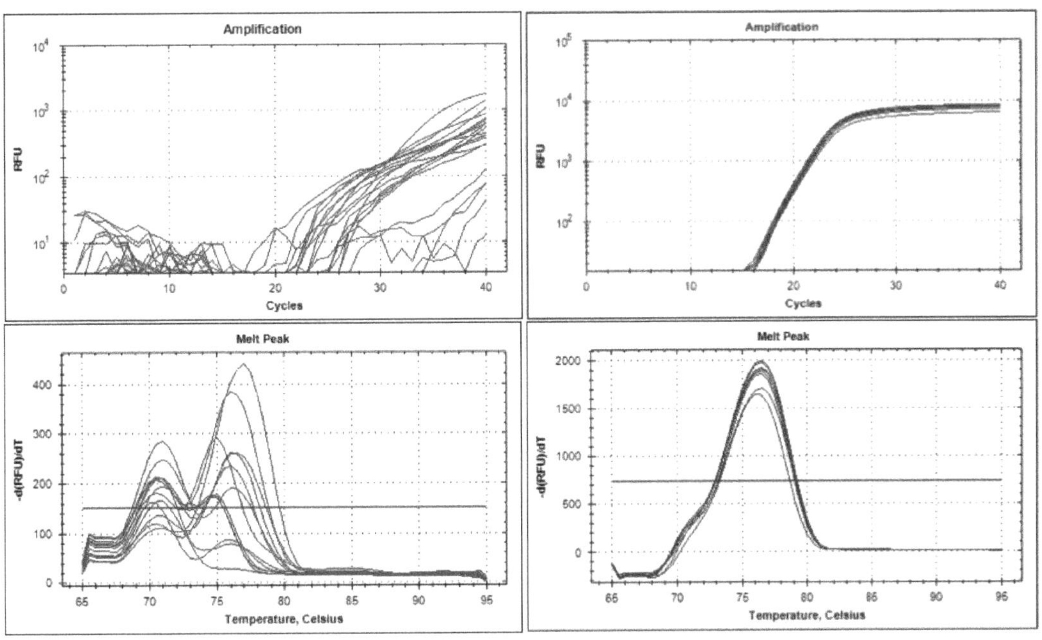

Figure 3.7 Example of problematic amplification reaction resulting in inappropriate *Cq* values. (A) Clearly, there is a problem with the amplification of the unknown samples, as the plots are of an unusual shape. (B) The melt curves are also indistinct, with more than one peak per sample, and different T_m values for the same amplicon. This is often due to the presence of inhibitors in the assay, and a 10-fold dilution of the RNA results in efficient amplification plots (C) and good melt curves (D) on the right.

3 Check slope is between −3.2 and −3.5 and reproducibility of replicates $R^2 > 0.98$.
4 Check controls for expected results.
5 Check melt curve profiles for SYBR Green I dye detection and ensure that products are specific.
6 Where possible, ensure that the *Cq* of the positive control (or dilutions of standard curve) is within one cycle of that recorded during a previous experiment and that the initial and final fluorescence levels are similar.
7 Ensure that all data points to be recorded are within the dynamic range defined by the standard curve.
8 Examine replicates. All replicates should be within 0.5 *Cq* of each other. At low *Cq* the tolerance should be lower than at high *Cq*. Above cycle 35 the variability will be greater and quantification may be unreliable.
9 Ensure a minimum *Cq* difference of 5 between any NTC signal and a sample data point (Bustin and Nolan, 2004) (Fig. 3.8). When using SYBR Green I dye take care with *Cq* values when primer dimer products are evident in the sample after melt curve analysis.

In general, the higher the fluorescence released per cycle, the more sensitive and accurate the measurement.

Normalization

All RT-qPCR assay results are subject to variability caused by technical, as well as biological, variation. It is essential that technical variability is kept at a minimum so as to optimize the chances of identifying biologically relevant changes in mRNA levels. Consequently, data normalization is an essential part of a meaningful RT-qPCR assay and requires standardization. Unfortunately, normalization is a rather problematic area (Vandesompele *et al.*, 2002a; Smith *et al.*, 2003; de Kok *et al.*, 2005; Rubie *et al.*, 2005; Argyropoulos *et al.*, 2006; Kanno *et al.*, 2006; Port *et al.*, 2007; Guenin *et al.*, 2009; Marullo *et al.*, 2010) and, as yet, there is no universally accepted method for data normalization that accounts for all variables encountered during the course of a RT-qPCR experiment. The use of RT-qPCR to quantify miRNAs adds to the concern as to which method is the most appropriate to ensure reliable normalization of non-coding RNAs (Peltier and Latham, 2008; Mestdagh *et al.*, 2009; Chang *et al.*, 2010). One report clearly demonstrates that the use of non-miRNA reference genes has a pronounced effect on the differential expression of tested miRNAs, an effect especially pronounced when the overall abundance of miRNAs varies, e.g. in experiments affecting the miRNA processing machinery, or in comparisons involving multiple tissues (Git *et al.*, 2010). All proposed methods represent a compromise and the selection depends upon the study aims and acceptable tolerance (Huggett *et al.*, 2005).

Normalization against input RNA amount is feasible and, if validated, perfectly acceptable (Bustin, 2002). However, there are problems when comparing tissues that are proliferating at different rates, since they will contain different amounts of rRNA and the mRNA/rRNA ratios are likely to be different. Furthermore, differences in cDNA synthesis are not taken into account and in some situations (e.g. embryology) may generate irrelevant results.

Normalizing to an endogenous reference gene (it is best not to use the term 'housekeeping gene') or genes is the most common method for internally controlling for error in RT-qPCR and is currently the preferred option, as long as it is carried out correctly (Goossens *et al.*, 2005) The inevitable consequence of inappropriate normalization is the acquisition of

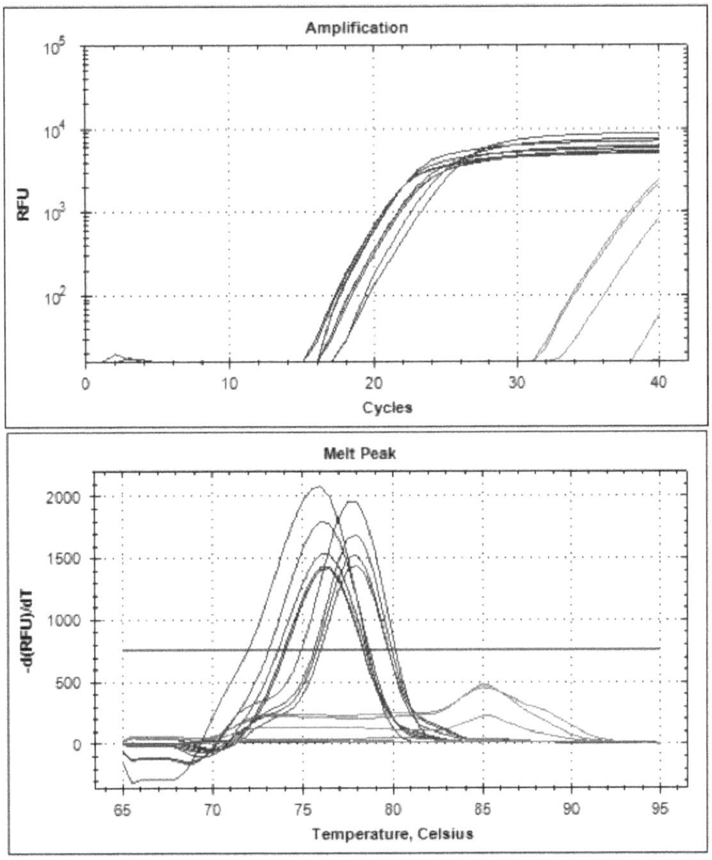

Figure 3.8 Analysis of a SYBR Green I dye reaction with positive NTCs. In this example, the replicate NTCs are recording *Cq* values of >33, whereas the lowest *Cq* generated by the unknowns is at 23, i.e. the d*Cq* is >5. The melt curves also clearly distinguishes between the two amplicons and the NTCs. Hence it is acceptable to use these data for quantification. The positive *Cq*s are caused by primer dimers, and if the aim of the experiment was to quantitate very low abundance targets, it would probably be necessary to redesign the primers.

biologically irrelevant data (Tricarico *et al.*, 2002; Dheda *et al.*, 2004, 2005). When using the endogenous reference strategy target gene levels are expressed relative to those of internal reference genes and so all of the steps required for the final PCR measurement are controlled. The procedure is simplified as both the gene of interest and the reference gene are measured using real-time RT-PCR. However, it is essential that reference gene expression in the target tissue is carefully analysed (Perez-Novo *et al.*, 2005) and the minimum variability is determined and reported (Vandesompele *et al.*, 2002a). The currently accepted method combines the evaluation of a panel of several reference genes together with a method for selecting reference genes with the most stable expression, e.g. GeNorm medgen.ugent.be/~jvdesomp/genorm/or Bestkeeper (Pfaffl *et al.*, 2004). A recent development of this strategy is the suggestion that transcript levels can be normalized to the expression of small interspersed nuclear elements (SINEs) such as expressed Alu repeats in primates (Marullo *et al.*, 2010) or B-elements in mouse models (Huggett, 2011, personal communication). In this system a single assay is designed to target the repetitive element but detection is representative of several genes. Fluctuations in mRNA levels of different genes should result in the overall level of SINEs remaining constant. This appears to be the closest system to a universal normalization method though care must be taken that the model does not cause expression of repetitive elements, such as heat shock responses.

Normalization against gDNA is another option (Kanno *et al.*, 2006), particularly when performing RT-qPCR analyses on RNA obtained using kits such as the Invitrogen CellsDirect kit. The main problems here are (1) differential stability of DNA and RNA may distort quantification; (2) sample cannot be DNase treated; (3), whilst being an internal control, there is no equivalent RT step; and (4) high relative concentrations of gDNA may inhibit the RT-PCR. Finally, it is also possible to normalize against area dissected when using laser capture microdissection, or cell number when extracting RNA from nucleated blood cells.

Standard curve

Analysis of the data from the standard curve can provide a substantial amount of information about the assay. For this reason, initially, assays should be validated on a serial dilution of high quality template, even in situations where data collection will not require reference to a standard curve. The template material used to generate a standard curve should accurately reflect the sample complexity and so this means that it is preferable to use a total RNA preparation, either from the tissue sample under investigation or from a commercially supplied reference RNA, or spiking a known amount of cDNA, plasmid DNA or oligonucleotides into a total RNA solution. Each concentration should be amplified in triplicate to allow a determination of reproducibility. The standard curve is constructed from a measure of *Cq* against log template quantity. It is realistic to expect a linear dynamic range of at least four logs with highly reproducible quantification, with five to nine logs typical (Fig. 3.9). However, it is important to emphasize that if all the unknowns are clustered between *Cqs* of 20 and 30, there is little point in subsequently using a standard curve that is linear all the way from 11 to 35. As with everything else, common sense must be applied to the use of this tool. This defines the working dynamic range for the assay. One measure of assay efficiency is made by comparison of the relative *Cq* values for subsequent dilutions of sample. The efficiency of the reaction can be calculated by the equation: $E = 10^{(-1/\text{slope})} - 1$. The efficiency of the PCR should be as close to 100% as possible, corresponding to a doubling of the target amplicon at each cycle. Using this measure, an assay of 100% efficiency will result in a standard curve with a gradient of −3.323 (also see http://www.gene-quantification.de/efficiency.html). An optimized assay will result in a standard curve with a slope between −3.2 and −3.5. Reproducibility of the replicate reactions also reflects assay stability, with R^2 values of 0.98 or above and preferably closer to 0.99 being indicative of a stable and reliable assay. The intercept on the *Cq* axis indicates the *Cq* at which a single unit of template concentration would be detected and is therefore an indication of the theoretical sensitivity of the assay.

Fig. 3.9 illustrates a typical result obtained for amplicons detected using SYBR Green I dye. Note that the slopes of all amplification plots are identical, indicating that the amplification efficiencies of every sample are the same. The plateau of one of the plots shows a significantly reduced ΔR_n value, but clearly this does not interfere with the ability to quantify accurately from that sample. The NTCs in the assay on the left are negative, whereas the NTC for the assay on the right shows a positive *Cq* value of 42. This phenomenon is fairly common with SYBR Green I dye based assays and is due to the formation of primer dimers in the absence of template. We would suggest that the presence of the primer dimers is reported, together with the ΔCq between the highest *Cq* recorded by an unknown and

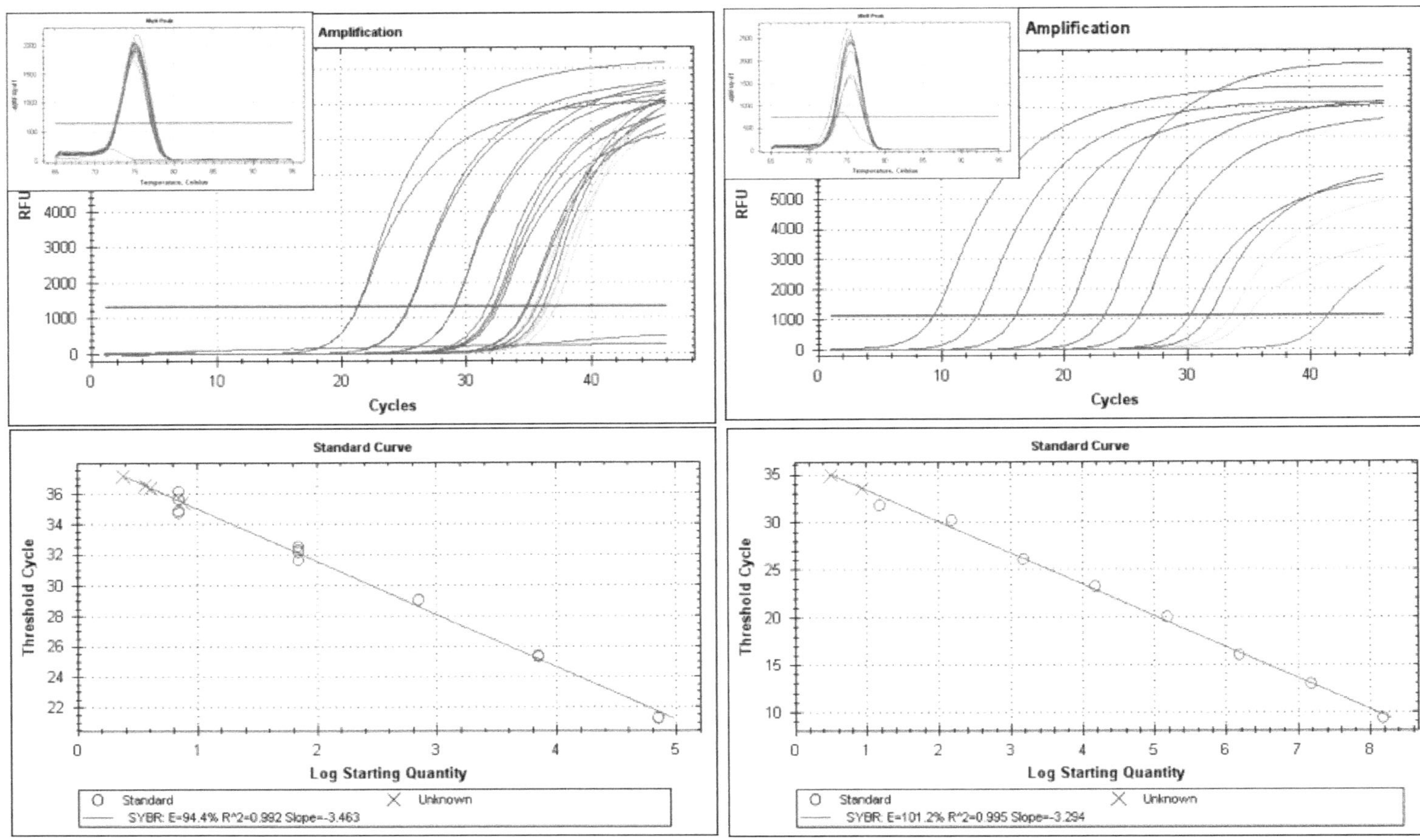

Figure 3.9 Standard curves are generated from a plot of *Cq* against the log copy number of a 10-fold serial dilution of template. They should be linear, with a slope of –3.323 and an R^2 of greater than 0.98 and preferably 0.99. In the example shown on the left, the slope of the SYBR Green I dye assay is –3.463 with an R^2 of 0.992, with a dynamic range of five logs. The NTC is negative even after 45 cycles. The example on the right is another SYBR Green assay that is also within the above specifications. In this case the NTC is positive, but the ΔCq between the lowest copy number standard (eight copies) and the NTC is >5.

the NTC. For probe-based assays, NTCs should always be negative, since the probe will not detect dimers and should not bind to non-specific amplicons.

Conclusions

qPCR is a remarkable technology that has enabled many of the advances made in our understanding of basic biological and disease processes and is increasingly applied to human and veterinary diagnostics, forensics and the detection of genetically modified organisms. Its applications are constantly expanding, posing new opportunities, but also new challenges. Analysis of expression profiles from single cells is one such challenge (Diercks *et al.*, 2009; Stahlberg and Bengtsson, 2010), nanolitre high-throughput RT-qPCR is another (Dixon *et al.*, 2009), proximity ligation assay (PLA) technology is expanding the range of qPCR applications to include the direct detection of proteins through the amplification of a surrogate DNA template after antibody binding (Swartzman *et al.*, 2010), even in single cells (Renfrow *et al.*, 2011) and qPCR is increasingly used to validate digital gene expression tag profiling data obtained using next generation sequencing technology (Wang *et al.*, 2010).

Unfortunately, the combination of ease of use and lack of rigorous standards of practice has resulted in widespread misinterpretation of RT-qPCR data and consequent publication of erroneous conclusions. Hence the need to make PCR-based assays more reliable, a Herculean task that requires both an appreciation and an understanding of numerous attributes that include biological concepts, statistics, mathematical modelling, technical know-how and a willingness to share that information. MIQE, discussed elsewhere (Bustin *et al.*, 2009), addresses a number of these issues. In addition, these problems are being addressed by a next generation of assays that aim to overcome the limitations that are inherent to any RT-PCR reaction. Developments in microfluidics, (Liu *et al.*, 2002), single cell analysis (Pierce and Wangh, 2011), digital PCR (McCaughan and Dear, 2010) and direct analysis of single molecules (Pushkarev *et al.*, 2009) are rapidly approaching a threshold that will make them practical and accessible to the wider scientific community. These advances this will eventually allow accurate and precise quantification without any of the drawbacks inherent in the need to convert RNA to DNA enzymatically, pre-amplify small amounts of target and utilize complex normalization procedures to obtain relative quantification data. That is the future. For now, qRT-PCR assays, when carried out appropriately, continue to be the method of choice for RNA detection and quantification.

References

Abruzzo, L.V., Wang, J., Kapoor, M., Jeffrey Medeiros, L., Keating, M.J., Edward Highsmith, W., Barron, L.L., Cromwell, C.C., and Coombes, K.R. (2005). Biological validation of differentially expressed genes in chronic lymphocytic leukemia identified by applying multiple statistical methods to oligonucleotide microarrays. J. Mol. Diagn. *7*, 337–345.

Abu Al-Soud, W., and Radstrom, P. (1998). Capacity of nine thermostable DNA polymerases To mediate DNA amplification in the presence of PCR-inhibiting samples. Appl. Environ. Microbiol. *64*, 3748–3753.

Akane, A., Matsubara, K., Nakamura, H., Takahashi, S., and Kimura, K. (1994). Identification of the heme compound copurified with deoxyribonucleic acid (DNA) from bloodstains, a major inhibitor of polymerase chain reaction (PCR) amplification. J. Forensic Sci. *39*, 362–372.

Akilesh, S., Shaffer, D.J., and Roopenian, D. (2003). Customized molecular phenotyping by quantitative gene expression and pattern recognition analysis. Genome Res. *13*, 1719–1727.

Al-Soud, W.A., and Radstrom, P. (2001). Purification and characterization of PCR-inhibitory components in blood cells. J. Clin. Microbiol. *39*, 485–493.

Andersen, C.L., Jensen, J.L., and Orntoft, T.F. (2004). Normalization of real-time quantitative reverse transcription-PCR data: a model-based variance estimation approach to identify genes suited for normalization, applied to bladder and colon cancer data sets. Cancer Res. *64*, 5245–5250.

Aradaib, I.E. (2009). PCR detection of African horse sickness virus serogroup based on genome segment three sequence analysis. J. Virol. Methods *159*, 1–5.

Argyropoulos, D., Psallida, C., and Spyropoulos, C.G. (2006). Generic normalization method for real-time PCR. Application for the analysis of the mannanase gene expressed in germinating tomato seed. FEBS J. *273*, 770–777.

Barnes, W.M. (1994). PCR amplification of up to 35-kb DNA with high fidelity and high yield from lambda bacteriophage templates. Proc. Natl. Acad. Sci. U.S.A. *91*, 2216–2220.

Batsch, A., Noetel, A., Fork, C., Urban, A., Lazic, D., Lucas, T., Pietsch, J., Lazar, A., Schomig, E., and Grundemann, D. (2008). Simultaneous fitting of real-time PCR data with efficiency of amplification modeled as Gaussian function of target fluorescence. BMC Bioinformatics *9*, 95.

Beck, E.T., Jurgens, L.A., Kehl, S.C., Bose, M.E., Patitucci, T., LaGue, E., Darga, P., Wilkinson, K., Witt, L.M., Fan, J., *et al.* (2010). Development of a rapid automated influenza A, influenza B, and respiratory syncytial virus A/B multiplex real-time RT-PCR assay and its use during the 2009 H1N1 swine-origin influenza virus epidemic in Milwaukee, Wisconsin. J. Mol. Diagn. *12*, 74–81.

Bengtsson, M., Hemberg, M., Rorsman, P., and Stahlberg, A. (2008). Quantification of mRNA in single cells and modelling of RT-qPCR induced noise. BMC Mol. Biol. *9*, 63.

Benoy, I.H., Elst, H., Van Dam, P., Scharpe, S., Van Marck, E., Vermeulen, P.B., and Dirix, L.Y. (2006). Detection of circulating tumour cells in blood by quantitative real-time RT-PCR: effect of pre-analytical time. Clin. Chem. Lab. Med. *44*, 1082–1087.

Brazeau, D.A. (2004). Combining genome-wide and targeted gene expression profiling in drug discovery: microarrays and real-time PCR. Drug Discov. Today *9*, 838–845.

Buh Gasparic, M., Tengs, T., La Paz, J.L., Holst-Jensen, A., Pla, M., Esteve, T., Zel, J., and Gruden, K. (2010). Comparison of nine different real-time PCR chemistries for qualitative and quantitative applications in GMO detection. Anal. Bioanal. Chem. *396*, 2023–2029.

Burns, M.J., Nixon, G.J., Foy, C.A., and Harris, N. (2005). Standardisation of data from real-time quantitative PCR methods – evaluation of outliers and comparison of calibration curves. BMC Biotechnol. *5*, 31.

Bustin, S.A. (2000). Absolute quantification of mRNA using real-time reverse transcription polymerase chain reaction assays. J. Mol. Endocrinol. *25*, 169–193.

Bustin, S.A. (2002). Quantification of mRNA using real-time reverse transcription PCR (RT-PCR): trends and problems. J. Mol. Endocrinol. *29*, 23–39.

Bustin, S.A. (2005). Real-time, fluorescence-based quantitative PCR: a snapshot of current procedures and preferences. Expert Rev. Mol. Diagn. *5*, 493–498.

Bustin, S.A. (2010). Why the need for qPCR publication guidelines? The case for MIQE. Methods *50*, 217–226.

Bustin, S.A., and Mueller, R. (2005). Real-time reverse transcription PCR (qRT-PCR) and its potential use in clinical diagnosis. Clin. Sci. (Lond.) *109*, 365–379.

Bustin, S.A., and Mueller, R. (2006). Real-time reverse transcription PCR and the detection of occult disease in colorectal cancer. Mol. Aspects Med. *27*, 192–223.

Bustin, S.A., and Nolan, T. (2004). Pitfalls of quantitative real-time reverse-transcription polymerase chain reaction. J. Biomol. Tech. *15*, 155–166.

Bustin, S.A., Benes, V., Nolan, T., and Pfaffl, M.W. (2005). Quantitative real-time RT-PCR – a perspective. J. Mol. Endocrinol. *34*, 597–601.

Bustin, S.A., Benes, V., Garson, J.A., Hellemans, J., Huggett, J., Kubista, M., Mueller, R., Nolan, T., Pfaffl, M.W., Shipley, G.L., *et al.* (2009). The MIQE guidelines: minimum information for publication of quantitative real-time PCR experiments. Clin. Chem. *55*, 611–622.

Bustin, S., Bergkvist, A., and Nolan, T. (2011a). In Silico Tools for qPCR assay design and data analysis. Methods Mol. Biol. *760*, 283–306.

Bustin, S.A., Benes, V., Garson, J.A., Hellemans, J., Huggett, J., Kubista, M., Mueller, R., Nolan, T., Pfaffl, M.W., Shipley, G.L., *et al.* (2011b). Primer sequence disclosure: a clarification of the MIQE guidelines. Clin. Chem. *57*, 919–921.

Chandler, D.P., Wagnon, C.A., and Bolton, H.J. (1998). Reverse transcriptase (RT) inhibition of PCR at low concentrations of template and its implications for quantitative RT-PCR. Appl. Environ. Microbiol. *64*, 669–677.

Chang, K.H., Mestdagh, P., Vandesompele, J., Kerin, M.J., and Miller, N. (2010). MicroRNA expression profiling to identify and validate reference genes for relative quantification in colorectal cancer. BMC Cancer *10*, 173.

Cloud, J.L., Hymas, W.C., Turlak, A., Croft, A., Reischl, U., Daly, J.A., and Carroll, K.C. (2003). Description of a multiplex Bordetella pertussis and Bordetella parapertussis LightCycler PCR assay with inhibition control. Diagn. Microbiol. Infect. Dis. *46*, 189–195.

Cone, R.W., Hobson, A.C., and Huang, M.L. (1992). Coamplified positive control detects inhibition of polymerase chain reactions. J. Clin. Microbiol. *30*, 3185–3189.

Coombs, N.J., Gough, A.C., and Primrose, J.N. (1999). Optimisation of DNA and RNA extraction from archival formalin-fixed tissue. Nucleic Acids Res. *27*, e12.

Cusi, M.G., Valassina, M., and Valensin, P.E. (1994). Comparison of M-MLV reverse transcriptase and Tth polymerase activity in RT-PCR of samples with low virus burden. Biotechniques *17*, 1034–1036.

Dheda, K., Huggett, J.F., Bustin, S.A., Johnson, M.A., Rook, G., and Zumla, A. (2004). Validation of housekeeping genes for normalizing RNA expression in real-time PCR. Biotechniques *37*, 112–119.

Dheda, K., Huggett, J.F., Chang, J.S., Kim, L.U., Bustin, S.A., Johnson, M.A., Rook, G.A., and Zumla, A. (2005). The implications of using an inappropriate reference gene for real-time reverse transcription PCR data normalization. Anal. Biochem. *344*, 141–143.

Diercks, A., Kostner, H., and Ozinsky, A. (2009). Resolving cell population heterogeneity: real-time PCR for simultaneous multiplexed gene detection in multiple single-cell samples. PLoS One *4*, e6326.

Dixon, J.M., Lubomirski, M., Amaratunga, D., Morrison, T.B., Brenan, C.J., and Ilyin, S.E. (2009). Nanoliter high-throughput RT-qPCR: a statistical analysis and assessment. Biotechniques *46*, ii-viii.

Doleshal, M., Magotra, A.A., Choudhury, B., Cannon, B.D., Labourier, E., and Szafranska, A.E. (2008). Evaluation and validation of total RNA extraction methods for microRNA expression analyses in formalin-fixed, paraffin-embedded tissues. J. Mol. Diagn. *10*, 203–211.

Easton, L.A., Vilcek, S., and Nettleton, P.F. (1994). Evaluation of a 'one tube' reverse transcription-polymerase chain reaction for the detection of ruminant pestiviruses. J. Virol. Methods *50*, 343–348.

Emmert-Buck, M.R., Bonner, R.F., Smith, P.D., Chuaqui, R.F., Zhuang, Z., Goldstein, S.R., Weiss, R.A., and Liotta, L.A. (1996). Laser capture microdissection. Science *274*, 998–1001.

Fink, L., Kohlhoff, S., Stein, M.M., Hanze, J., Weissmann, N., Rose, F., Akkayagil, E., Manz, D., Grimminger, F., Seeger, W., *et al.* (2002). cDNA array hybridization after laser-assisted microdissection from nonneoplastic tissue. Am. J. Pathol. *160*, 81–90.

Fleige, S., and Pfaffl, M.W. (2006). RNA integrity and the effect on the real-time qRT-PCR performance. Mol. Aspects Med. *27*, 126–139.

Fleige, S., Walf, V., Huch, S., Prgomet, C., Sehm, J., and Pfaffl, M.W. (2006). Comparison of relative mRNA quantification models and the impact of RNA integrity in quantitative real-time RT-PCR. Biotechnol. Lett. *28*, 1601–1613.

Fredricks, D.N., and Relman, D.A. (1998). Improved amplification of microbial DNA from blood cultures by removal of the PCR inhibitor sodium polyanetholesulfonate. J. Clin. Microbiol. *36*, 2810–2816.

Fu, W.J., Hu, J., Spencer, T., Carroll, R., and Wu, G. (2006). Statistical models in assessing fold change of gene expression in real-time RT-PCR experiments. Comput. Biol. Chem. *30*, 21–26.

Gaillard, C., Flavin, M., Woisard, A., and Strauss, F. (1999). Association of double-stranded DNA fragments into multistranded DNA structures. *50*, 679–689.

Gallup, J.M., and Ackermann, M.R. (2008). The 'PREXCEL-Q Method' for qPCR. Int. J. Biomed. Sci. *4*, 273–293.

Gibson, U.E., Heid, C.A., and Williams, P.M. (1996). A novel method for real time quantitative RT-PCR. Genome Res. *6*, 995–1001.

Ginzinger, D.G. (2002). Gene quantification using real-time quantitative PCR: an emerging technology hits the mainstream. Exp. Hematol. *30*, 503–512.

Git, A., Dvinge, H., Salmon-Divon, M., Osborne, M., Kutter, C., Hadfield, J., Bertone, P., and Caldas, C. (2010). Systematic comparison of microarray profiling, real-time PCR, and next-generation sequencing technologies for measuring differential microRNA expression. RNA *16*, 991–1006.

Godfrey, T.E., Kim, S.H., Chavira, M., Ruff, D.W., Warren, R.S., Gray, J.W., and Jensen, R.H. (2000). Quantitative mRNA expression analysis from formalin-fixed, paraffin-embedded tissues using 5' nuclease quantitative reverse transcription polymerase chain reaction. J. Mol. Diagn. *2*, 84–91.

Goldsworthy, S.M., Stockton, P.S., Trempus, C.S., Foley, J.F., and Maronpot, R.R. (1999). Effects of fixation on RNA extraction and amplification from laser capture microdissected tissue. Mol. Carcinog. *25*, 86–91.

Goossens, K., Van Poucke, M., Van Soom, A., Vandesompele, J., Van Zeveren, A., and Peelman, L.J. (2005). Selection of reference genes for quantitative real-time PCR in bovine preimplantation embryos. BMC Dev. Biol. *5*, 27.

Grisold, A.J., and Kessler, H.H. (2006). Use of hybridization probes in a real-time PCR assay on the LightCycler for the detection of methicillin-resistant Staphylococcus aureus. Methods Mol. Biol. *345*, 79–89.

Grotzer, M.A., Patti, R., Geoerger, B., Eggert, A., Chou, T.T., and Phillips, P.C. (2000). Biological stability of RNA isolated from RNAlater-treated brain tumor and neuroblastoma xenografts. *34*, 438–442.

Guenin, S., Mauriat, M., Pelloux, J., Van Wuytswinkel, O., Bellini, C., and Gutierrez, L. (2009). Normalization of qRT-PCR data: the necessity of adopting a systematic, experimental conditions-specific, validation of references. J. Exp. Bot. *60*, 487–493.

Gunson, R.N., Bennett, S., Maclean, A., and Carman, W.F. (2008). Using multiplex real time PCR in order to streamline a routine diagnostic service. J. Clin. Virol. *43*, 372–375.

Guy, R.A., Payment, P., Krull, U.J., and Horgen, P.A. (2003). Real-time PCR for quantification of Giardia and Cryptosporidium in environmental water samples and sewage. Appl. Environ. Microbiol. *69*, 5178–5185.

Haller, F., Kulle, B., Schwager, S., Gunawan, B., von Heydebreck, A., Sultmann, H., and Fuzesi, L. (2004). Equivalence test in quantitative reverse transcription polymerase chain reaction: confirmation of reference genes suitable for normalization. Anal. Biochem. *335*, 1–9.

Hartshorn, C., Anshelevich, A., and Wangh, L.J. (2005). Rapid, single-tube method for quantitative preparation and analysis of RNA and DNA in samples as small as one cell. BMC Biotechnol. *5*, 2.

Haupt, C., Tolner, E.A., Heinemann, U., Witte, O.W., and Frahm, C. (2006). The combined use of non-radioactive in situ hybridization and real-time RT-PCR to assess gene expression in cryosections. Brain Res. *1118*, 232–238.

Hellemans, J., Mortier, G., De Paepe, A., Speleman, F., and Vandesompele, J. (2007). qBase relative quantification framework and software for management and automated analysis of real-time quantitative PCR data. Genome Biol. *8*, R19.

Hilscher, C., Vahrson, W., and Dittmer, D.P. (2005). Faster quantitative real-time PCR protocols may lose sensitivity and show increased variability. Nucleic Acids Res. *33*, e182.

Huggett, J., and Bustin, S.A. (2011). Standardisation and reporting for nucleic acid quantification. Accredit. Qual. Assur. *16*, 399–405.

Huggett, J., Dheda, K., Bustin, S., and Zumla, A. (2005). Real-time RT-PCR normalisation; strategies and considerations. Genes Immun. *6*, 279–284.

Huggett, J.F., Novak, T., Garson, J.A., Green, C., Morris-Jones, S.D., Miller, R.F., and Zumla, A. (2008). Differential susceptibility of PCR reactions to inhibitors: an important and unrecognised phenomenon. BMC Res. Notes *1*, 70.

Jiang, J., Alderisio, K.A., Singh, A., and Xiao, L. (2005). Development of procedures for direct extraction of Cryptosporidium DNA from water concentrates and for relief of PCR inhibitors. Appl. Environ. Microbiol. *71*, 1135–1141.

Jin, N., He, K., and Liu, L. (2006). qPCR-DAMS: a database tool to analyze, manage, and store both relative and absolute quantitative real-time PCR data. Physiol. Genomics *25*, 525–527.

Kanno, J., Aisaki, K., Igarashi, K., Nakatsu, N., Ono, A., Kodama, Y., and Nagao, T. (2006). 'Per cell' normalization method for mRNA measurement by quantitative PCR and microarrays. BMC Genomics *7*, 64.

Keilholz, U. (1998). New prognostic factors in melanoma: mRNA tumour markers. Eur. Cancer, J. *34(Suppl. 3)*, S37-S41.

Klimecki, W.T., Futscher, B.W., and Dalton, W.S. (1994). Effects of ethanol and paraformaldehyde on RNA yield and quality. Biotechniques *16*, 1021–1023.

Koch, I., Slotta-Huspenina, J., Hollweck, R., Anastasov, N., Hofler, H., Quintanilla-Martinez, L., and Fend, F. (2006). Real-time quantitative RT-PCR shows variable, assay-dependent sensitivity to formalin fixation: implications for direct comparison of transcript levels in paraffin-embedded tissues. Diagn. Mol. Pathol. *15*, 149–156.

de Kok, J.B., Roelofs, R.W., Giesendorf, B.A., Pennings, J.L., Waas, E.T., Feuth, T., Swinkels, D.W., and Span, P.N. (2005). Normalization of gene expression measurements in tumor tissues: comparison of 13 endogenous control genes. Lab Invest *85*, 154–159.

Kubista, M., Andrade, J.M., Bengtsson, M., Forootan, A., Jonak, J., Lind, K., Sindelka, R., Sjoback, R., Sjogreen, B., Strombom, L., *et al.* (2006). The real-time polymerase chain reaction. Mol. Aspects Med. *27*, 95–125.

Lacey, H.A., Nolan, T., Greenwood, S.L., Glazier, J.D., and Sibley, C.P. (2005). Gestational profile of Na+/H+ exchanger and Cl-/HCO3- anion exchanger mRNA expression in placenta using real-time QPCR. Placenta *26*, 93–98.

Larionov, A., Krause, A., and Miller, W. (2005). A standard curve based method for relative real time PCR data processing. BMC Bioinformatics *6*, 62.

Lefevre, J., Hankins, C., Pourreaux, K., Voyer, H., and Coutlee, F. (2004). Prevalence of selective inhibition of HPV-16 DNA amplification in cervicovaginal lavages. J. Med. Virol. *72*, 132–137.

Lekanne Deprez, R.H., Fijnvandraat, A.C., Ruijter, J.M., and Moorman, A.F. (2002). Sensitivity and accuracy of quantitative real-time polymerase chain reaction using SYBR green I depends on cDNA synthesis conditions. Anal. Biochem. *307*, 63–69.

Lewis, F., and Maughan, N.J. (2004). Extraction of total RNA from formalin-fixed paraffin-embedded tissue. In A-Z of Quantitative PCR, Bustin, S.A., ed. (IUL, La Jolla, CA), pp. 591–603.

Li, J., Smyth, P., Cahill, S., Denning, K., Flavin, R., Aherne, S., Pirotta, M., Guenther, S.M., O'Leary, J.J., and Sheils, O. (2008). Improved RNA quality and TaqMan Pre-amplification method (PreAmp) to enhance expression analysis from formalin fixed paraffin embedded (FFPE) materials. BMC Biotechnol. *8*, 10.

Liss, B. (2002). Improved quantitative real-time RT-PCR for expression profiling of individual cells. *30*, E89.

Liu, J., Enzelberger, M., and Quake, S. (2002). A nanoliter rotary device for polymerase chain reaction. Electrophoresis *23*, 1531–1536.

Liu, W., and Saint, D.A. (2002a). A new quantitative method of real time reverse transcription polymerase chain reaction assay based on simulation of polymerase chain reaction kinetics. Anal. Biochem. *302*, 52–59.

Liu, W., and Saint, D.A. (2002b). Validation of a quantitative method for real time PCR kinetics. Biochem. Biophys. Res. Commun. *294*, 347–353.

Macabeo-Ong, M., Ginzinger, D.G., Dekker, N., McMillan, A., Regezi, J.A., Wong, D.T., and Jordan, R.C. (2002). Effect of duration of fixation on quantitative reverse transcription polymerase chain reaction analyses. *15*, 979–987.

McCaughan, F., and Dear, P.H. (2010). Single-molecule genomics. J. Pathol. *220*, 297–306.

Madejon, A., Manzano, M.L., Arocena, C., Castillo, I., and Carreno, V. (2000). Effects of delayed freezing of liver biopsies on the detection of hepatitis C virus RNA strands. J. Hepatol. *32*, 1019–1025.

Mader, R.M., Schmidt, W.M., Sedivy, R., Rizovski, B., Braun, J., Kalipciyan, M., Exner, M., Steger, G.G., and Mueller, M.W. (2001). Reverse transcriptase template switching during reverse transcriptase-polymerase chain reaction: artificial generation of deletions in ribonucleotide reductase mRNA. J. Lab. Clin. Med. *137*, 422–428.

Marcus, J.S., Anderson, W.F., and Quake, S.R. (2006). Parallel picoliter rt-PCR assays using microfluidics. Anal. Chem. *78*, 956–958.

Marshall, R., Chernesky, M., Jang, D., Hook, E.W., Cartwright, C.P., Howell-Adams, B., Ho, S., Welk, J., Lai-Zhang, J., Brashear, J., *et al.* (2007). Characteristics of the m2000 automated sample preparation and multiplex real-time PCR system for detection of *Chlamydia trachomatis* and *Neisseria gonorrhoeae*. J. Clin. Microbiol. *45*, 747–751.

Marubini, E., Verderio, P., Raggi, C.C., Pazzagli, M., and Orlando, C. (2004). Statistical diagnostics emerging from external quality control of real-time PCR. Int. J. Biol. Markers *19*, 141–146.

Marullo, M., Zuccato, C., Mariotti, C., Lahiri, N., Tabrizi, S.J., Di Donato, S., and Cattaneo, E. (2010). Expressed Alu repeats as a novel, reliable tool for normalization of real-time quantitative RT-PCR data. Genome Biol. *11*, R9.

Masuda, N., Ohnishi, T., Kawamoto, S., Monden, M., and Okubo, K. (1999). Analysis of chemical modification of RNA from formalin-fixed samples and optimization of molecular biology applications for such samples. Nucleic Acids Res. *27*, 4436–4443.

Mestdagh, P., Van Vlierberghe, P., De Weer, A., Muth, D., Westermann, F., Speleman, F., and Vandesompele, J. (2009). A novel and universal method for microRNA RT-qPCR data normalization. Genome Biol. *10*, R64.

Mizuno, T., Nagamura, H., Iwamoto, K.S., Ito, T., Fukuhara, T., Tokunaga, M., Tokuoka, S., Mabuchi, K., and Seyama, T. (1998). RNA from decades-old archival tissue blocks for retrospective studies. Diagn. Mol. Pathol. *7*, 202–208.

Morrison, T., Hurley, J., Garcia, J., Yoder, K., Katz, A., Roberts, D., Cho, J., Kanigan, T., Ilyin, S.E., Horowitz, D., *et al.* (2006). Nanoliter high throughput quantitative PCR. Nucleic Acids Res. *34*, e123.

Muller, M.C., Hordt, T., Paschka, P., Merx, K., La Rosee, P., Hehlmann, R., and Hochhaus, A. (2004). Standardization of preanalytical factors for minimal residual disease analysis in chronic myelogenous leukemia. Acta. Haematol. *112*, 30–33.

Neuzil, P., Zhang, C., Pipper, J., Oh, S., and Zhuo, L. (2006). Ultra fast miniaturized real-time PCR: 40 cycles in less than six minutes. Nucleic Acids Res. *34*, e77.

Nolan, T., Hands, R.E., and Bustin, S.A. (2006a). Quantification of mRNA using real-time RT-PCR. Nat. Protocols *1*, 1559–1582.

Nolan, T., Hands, R.E., Ogunkolade, B.W., and Bustin, S.A. (2006b). SPUD: a qPCR assay for the detection of inhibitors in nucleic acid preparations. Anal. Biochem. *351*, 308–310.

Nordgard, O., Kvaloy, J.T., Farmen, R.K., and Heikkila, R. (2006). Error propagation in relative real-time reverse transcription polymerase chain reaction quantification models: the balance between accuracy and precision. Anal. Biochem. *356*, 182–193.

O'Connor, W., and Runquist, E.A. (2008). Error minimization algorithm for comparative quantitative PCR analysis: Q-Anal. Anal. Biochem. *378*, 96–98.

Oji, Y., Yamamoto, H., Nomura, M., Nakano, Y., Ikeba, A., Nakatsuka, S., Abeno, S., Kiyotoh, E., Jomgeow, T., Sekimoto, M., *et al.* (2003). Overexpression of the Wilms' tumor gene WT1 in colorectal adenocarcinoma. Cancer Sci. *94*, 712–717.

Pasloske, B.L., Walkerpeach, C.R., Obermoeller, R.D., Winkler, M., and DuBois, D.B. (1998). Armored RNA technology for production of ribonuclease-resistant viral RNA controls and standards. J. Clin. Microbiol. *36*, 3590–3594.

Pattyn, F., Speleman, F., De Paepe, A., and Vandesompele, J. (2003). RTPrimerDB: the real-time PCR primer and probe database. Nucleic Acids Res. *31*, 122–123.

Pattyn, F., Robbrecht, P., De Paepe, A., Speleman, F., and Vandesompele, J. (2006). RTPrimerDB: the real-time PCR primer and probe database, major update 2006. Nucleic Acids Res. *34*, D684–D688.

Peixoto, A., Monteiro, M., Rocha, B., and Veiga-Fernandes, H. (2004). Quantification of multiple gene expression in individual cells. Genome Res. *14*, 1938–1947.

Peltier, H.J., and Latham, G.J. (2008). Normalization of microRNA expression levels in quantitative RT-PCR assays: identification of suitable reference RNA targets in normal and cancerous human solid tissues. RNA *14*, 844–852.

Perch-Nielsen, I.R., Bang, D.D., Poulsen, C.R., El-Ali, J., and Wolff, A. (2003). Removal of PCR inhibitors using dielectrophoresis as a selective filter in a microsystem. Lab. Chip *3*, 212–216.

Perez-Novo, C.A., Claeys, C., Speleman, F., Van Cauwenberge, P., Bachert, C., and Vandesompele, J. (2005). Impact of RNA quality on reference gene expression stability. Biotechniques *39*, 52, 54, 56.

Pfaffl, M.W., Horgan, G.W., and Dempfle, L. (2002). Relative expression software tool (REST) for group-wise comparison and statistical analysis of relative expression results in real-time PCR. Nucleic Acids Res. *30*, e36.

Pfaffl, M.W., Tichopad, A., Prgomet, C., and Neuvians, T.P. (2004). Determination of stable housekeeping genes, differentially regulated target genes and sample integrity: BestKeeper – Excel-based tool using pair-wise correlations. Biotechnol. Lett. *26*, 509–515.

Picard, C., Silvy, M., and Gabert, J. (2006). Overview of real-time RT-PCR strategies for quantification of gene rearrangements in the myeloid malignancies. Methods Mol. Med. *125*, 27–68.

Pierce, K.E., and Wangh, L.J. (2011). LATE-PCR and allied technologies: real-time detection strategies for rapid, reliable diagnosis from single cells. Methods Mol. Biol. *688*, 47–66.

Platts, A.E., Johnson, G.D., Linnemann, A.K., and Krawetz, S.A. (2008). Real-time PCR quantification using a variable reaction efficiency model. Anal. Biochem. *380*, 315–322.

Poddar, S.K., Sawyer, M.H., and Connor, J.D. (1998). Effect of inhibitors in clinical specimens on Taq and Tth DNA polymerase-based PCR amplification of influenza A virus. J. Med. Microbiol. *47*, 1131–1135.

Port, M., Schmelz, H.U., Stassen, T., Mueller, K., Stockinger, M., Obermair, R., and Abend, M. (2007). Correcting false gene expression measurements from degraded RNA using RTQ-PCR. Diagn. Mol. Pathol. *16*, 38–49.

Pushkarev, D., Neff, N.F., and Quake, S.R. (2009). Single-molecule sequencing of an individual human genome. Nat. Biotechnol. *27*, 847–852.

Radstrom, P., Knutsson, R., Wolffs, P., Lovenklev, M., and Lofstrom, C. (2004). Pre-PCR processing: strategies to generate PCR-compatible samples. Mol. Biotechnol. *26*, 133–146.

Raengsakulrach, B., Nisalak, A., Maneekarn, N., Yenchitsomanus, P.T., Limsomwong, C., Jairungsri, A., Thirawuth, V., Green, S., Kalayanarooj, S., Suntayakorn, S., *et al.* (2002). Comparison of four reverse transcription-polymerase chain reaction procedures for the detection of dengue virus in clinical specimens. J. Virol. Methods *105*, 219–232.

Raggi, C.C., Verderio, P., Pazzagli, M., Marubini, E., Simi, L., Pinzani, P., Paradiso, A., and Orlando, C. (2005). An Italian program of external quality control for quantitative assays based on real-time PCR with Taq-Man probes. Clin. Chem. Lab. Med. *43*, 542–548.

Rainen, L., Oelmueller, U., Jurgensen, S., Wyrich, R., Ballas, C., Schram, J., Herdman, C., Bankaitis-Davis, D., Nicholls, N., Trollinger, D., *et al.* (2002). Stabilization of mRNA expression in whole blood samples. Clin. Chem. *48*, 1883–1890.

Ramakers, C., Ruijter, J.M., Deprez, R.H., and Moorman, A.F. (2003). Assumption-free analysis of quantitative real-time polymerase chain reaction (PCR) data. Neurosci. Lett. *339*, 62–66.

Ramsden, S.C., Daly, S., Geilenkeuser, W.J., Duncan, G., Hermitte, F., Marubini, E., Neumaier, M., Orlando, C., Palicka, V., Paradiso, A., *et al.* (2006). EQUAL-quant: an international external quality assessment scheme for real-time PCR. Clin. Chem. *52*, 1584–1591.

Raymaekers, M., Smets, R., Maes, B., and Cartuyvels, R. (2009). Checklist for optimization and validation of real-time PCR assays. J. Clin. Lab. Anal. *23*, 145–151.

Renfrow, J.J., Scheck, A.C., Dhawan, N.S., Lukac, P.J., Vogel, H., Chandler, J.P., Raizer, J.J., Harsh, G.R., Chakravarti, A., and Bredel, M. (2011). Gene-protein correlation in single cells. Neuro. Oncol. *13*, 880–885.

Rubie, C., Kempf, K., Hans, J., Su, T., Tilton, B., Georg, T., Brittner, B., Ludwig, B., and Schilling, M. (2005). Housekeeping gene variability in normal and cancerous colorectal, pancreatic, esophageal, gastric and hepatic tissues. Mol. Cell. Probes *19*, 101–109.

Rutledge, R.G. (2004). Sigmoidal curve-fitting redefines quantitative real-time PCR with the prospective of developing automated high-throughput applications. Nucleic Acids Res. *32*, e178.

Rutledge, R.G., and Stewart, D. (2008a). A kinetic-based sigmoidal model for the polymerase chain reaction and its application to high-capacity absolute quantitative real-time PCR. BMC Biotechnol. *8*, 47.

Rutledge, R.G., and Stewart, D. (2008b). Critical evaluation of methods used to determine amplification efficiency refutes the exponential character of real-time PCR. BMC Mol. Biol. *9*, 96.

Schroeder, A., Mueller, O., Stocker, S., Salowsky, R., Leiber, M., Gassmann, M., Lightfoot, S., Menzel, W., Granzow, M., and Ragg, T. (2006). The RIN: an RNA integrity number for assigning integrity values to RNA measurements. BMC Mol. Biol. *7*, 3.

Shanmugam, V., Sell, K.W., and Saha, B.K. (1993). Mistyping ACE heterozygotes. PCR Methods Appl. *3*, 120–121.

Sindelka, R., Jonak, J., Hands, R., Bustin, S.A., and Kubista, M. (2008). Intracellular expression profiles measured by real-time PCR tomography in the Xenopus laevis oocyte. Nucleic Acids Res. *36*, 387–392.

Smith, M.V., Miller, C.R., Kohn, M., Walker, N.J., and Portier, C.J. (2007). Absolute estimation of initial concentrations of amplicon in a real-time RT-PCR process. BMC Bioinformatics *8*, 409.

Smith, R.D., Brown, B., Ikonomi, P., and Schechter, A.N. (2003). Exogenous reference RNA for normalization of real-time quantitative PCR. Biotechniques *34*, 88–91.

von Smolinski, D., Leverkoehne, I., von Samson-Himmelstjerna, G., and Gruber, A.D. (2005). Impact of formalin-fixation and paraffin-embedding on the ratio between mRNA copy numbers of differently expressed genes. Histochem. Cell Biol. *124*, 177–188.

Specht, K., Richter, T., Muller, U., Walch, A., Werner, M., and Hofler, H. (2001). Quantitative gene expression analysis in microdissected archival formalin-fixed and paraffin-embedded tumor tissue. Am. J. Pathol. *158*, 419–429.

Spiess, A.N., Feig, C., and Ritz, C. (2008). Highly accurate sigmoidal fitting of real-time PCR data by introducing a parameter for asymmetry. BMC Bioinformatics *9*, 221.

Stahlberg, A., and Bengtsson, M. (2010). Single-cell gene expression profiling using reverse transcription quantitative real-time PCR. Methods *50*, 282–288.

Stahlberg, A., Aman, P., Ridell, B., Mostad, P., and Kubista, M. (2003). Quantitative real-time PCR method for detection of B-lymphocyte monoclonality by comparison of kappa and lambda immunoglobulin light chain expression. Clin. Chem. *49*, 51–59.

Stahlberg, A., Hakansson, J., Xian, X., Semb, H., and Kubista, M. (2004a). Properties of the reverse transcription reaction in mRNA quantification. Clin. Chem. *50*, 509–515.

Stahlberg, A., Kubista, M., and Pfaffl, M. (2004b). Comparison of reverse transcriptases in gene expression analysis. Clin. Chem. *50*, 1678–1680.

Stahlberg, A., Zoric, N., Aman, P., and Kubista, M. (2005). Quantitative real-time PCR for cancer detection: the lymphoma case. Expert Rev. Mol. Diagn. *5*, 221–230.

Stahlberg, A., Elbing, K., Andrade-Garda, J.M., Sjogreen, B., Forootan, A., and Kubista, M. (2008). Multiway real-time PCR gene expression profiling in yeast Saccharomyces cerevisiae reveals altered transcriptional response of ADH-genes to glucose stimuli. BMC Genomics *9*, 170.

Stangegaard, M., Dufva, I.H., and Dufva, M. (2006). Reverse transcription using random pentadecamer primers increases yield and quality of resulting cDNA. Biotechniques *40*, 649–657.

Stanley, K.K., and Szewczuk, E. (2005). Multiplexed tandem PCR: gene profiling from small amounts of RNA using SYBR Green detection. Nucleic Acids Res. *33*, e180.

Sunen, E., Casas, N., Moreno, B., and Zigorraga, C. (2004). Comparison of two methods for the detection of hepatitis A virus in clam samples (Tapes spp.) by reverse transcription-nested PCR. Int. J. Food Microbiol. *91*, 147–154.

Suslov, O., and Steindler, D.A. (2005). PCR inhibition by reverse transcriptase leads to an overestimation of amplification efficiency. Nucleic Acids Res. *33*, e181.

Swartzman, E., Shannon, M., Lieu, P., Chen, S.M., Mooney, C., Wei, E., Kuykendall, J., Tan, R., Settineri, T., Egry, L., *et al.* (2010). Expanding applications of protein analysis using proximity ligation and qPCR. Methods *50*, S23–6.

Szabo, A., Boucher, K., Carroll, W.L., Klebanov, L.B., Tsodikov, A.D., and Yakovlev, A.Y. (2002). Variable selection and pattern recognition with gene expression data generated by the microarray technology. Math. Biosci. *176*, 71–98.

Tichopad, A., Dilger, M., Schwarz, G., and Pfaffl, M.W. (2003). Standardized determination of real-time PCR efficiency from a single reaction set-up. Nucleic Acids Res. *31*, e122.

Tichopad, A., Didier, A., and Pfaffl, M.W. (2004). Inhibition of real-time RT-PCR quantification due to tissue-specific contaminants. Mol. Cell. Probes *18*, 45–50.

Tichopad, A., Kitchen, R., Riedmaier, I., Becker, C., Stahlberg, A., and Kubista, M. (2009). Design and optimization of reverse-transcription quantitative PCR experiments. Clin. Chem. *55*, 1816–1823.

Tricarico, C., Pinzani, P., Bianchi, S., Paglierani, M., Distante, V., Pazzagli, M., Bustin, S.A., and Orlando, C. (2002). Quantitative real-time reverse transcription polymerase chain reaction: normalization to rRNA or single housekeeping genes is inappropriate for human tissue biopsies. Anal. Biochem. *309*, 293–300.

Valasek, M.A., and Repa, J.J. (2005). The power of real-time PCR. Adv. Physiol. Educ. *29*, 151–159.

Vandesompele, J., De Preter, K., Pattyn, F., Poppe, B., Van Roy, N., De Paepe, A., and Speleman, F. (2002a). Accurate normalization of real-time quantitative RT-PCR data by geometric averaging of multiple internal control genes. Genome Biol. *3*, 0034.1–0034.11.

Vandesompele, J., De Paepe, A., and Speleman, F. (2002b). Elimination of primer-dimer artifacts and genomic coamplification using a two-step SYBR green I real-time RT-PCR. Anal. Biochem. *303*, 95–98.

Verderio, P., Pizzamiglio, S., Gallo, F., and Ramsden, S.C. (2008a). FCI: an R-based algorithm for evaluating uncertainty of absolute real-time PCR quantification. BMC Bioinformatics *9*, 13.

Verderio, P., Ramsden, S.C., Orlando, C., Pizzamiglio, S., Paradiso, A., Neumaier, M., Pazzagli, M., and Marubini, E. (2008b). External quality assessment schemes for real-time PCR: a statistical procedure to corrective actions. Clin. Chem. Lab. Med. *46*, 717–721.

Vermeulen, J., De Preter, K., Lefever, S., Nuytens, J., De Vloed, F., Derveaux, S., Hellemans, J., Speleman, F., and Vandesompele, J. (2011). Measurable impact of RNA quality on gene expression results from quantitative PCR. Nucleic Acids Res. *39*, e63.

Vlems, F., Soong, R., Diepstra, H., Punt, C., Wobbes, T., Tabiti, K., and van Muijen, G. (2002a). Effect of blood sample handling and reverse transcriptase-polymerase chain reaction assay sensitivity on detection of CK20 expression in healthy donor blood. Diagn. Mol. Pathol. *11*, 90–97.

Vlems, F.A., Diepstra, J.H., Cornelissen, I.M., Ruers, T.J., Ligtenberg, M.J., Punt, C.J., van Krieken, J.H., Wobbes, T., and van Muijen, G.N. (2002b). Limitations of cytokeratin 20 RT-PCR to detect disseminated tumour cells in blood and bone marrow of patients with colorectal cancer: expression in controls and downregulation in tumour tissue. Mol. Pathol. *55*, 156–163.

de Vries, T.J., Fourkour, A., Punt, C.J., Ruiter, D.J., and van Muijen, G.N. (2000). Analysis of melanoma cells in peripheral blood by reverse transcription-polymerase chain reaction for tyrosinase and MART-1 after mononuclear cell collection with cell preparation tubes: a comparison with the whole blood guanidinium isothiocyanate RNA isolation method. Melanoma Res. *10*, 119–126.

Wagner, B.K., and Arany, Z. (2009). High-throughput real-time PCR for detection of gene-expression levels. Methods Mol. Biol. *486*, 167–175.

Wang, Z., Dong, D., Ru, B., Young, R.L., Han, N., Guo, T., and Zhang, S. (2010). Digital gene expression tag profiling of bat digits provides robust candidates contributing to wing formation. BMC Genomics *11*, 619.

Weissensteiner, T., and Lanchbury, J.S. (1996). Strategy for controlling preferential amplification and avoiding false-negatives in PCR typing. Biotechniques *21*, 1102–1108.

Westh, H., Lisby, G., Breysse, F., Boddinghaus, B., Chomarat, M., Gant, V., Goglio, A., Raglio, A., Schuster, H., Stuber, F., *et al.* (2009). Multiplex real-time PCR and blood culture for identification of bloodstream pathogens in patients with suspected sepsis. Clin. Microbiol. Infect. *15*, 544–551.
Whitcombe, D., Theaker, J., Guy, S.P., Brown, T., and Little, S. (1999). Detection of PCR products using self-probing amplicons and fluorescence. Nat. Biotechnol. *17*, 804–807.
Wilkes, T.M., Devonshire, A.S., Ellison, S.L., and Foy, C.A. (2010). Evaluation of a novel approach for the measurement of RNA quality. BMC Res. Notes *3*, 89.
Yang, W., Maqsodi, B., Ma, Y., Bui, S., Crawford, K.L., McMaster, G.K., Witney, F., and Luo, Y. (2006). Direct quantification of gene expression in homogenates of formalin-fixed, paraffin-embedded tissues. Biotechniques *40*, 481–486.
Zaayman, D., Human, S., and Venter, M. (2009). A highly sensitive method for the detection and genotyping of West Nile virus by real-time PCR. J. Virol. Methods *157*, 155–160.
Zhang, C., and Xing, D. (2007). Miniaturized PCR chips for nucleic acid amplification and analysis: latest advances and future trends. Nucleic Acids Res. *35*, 4223–4237.

DOI: https://doi.org/10.21775/9781912530243.14

Real-time PCR Arrays

Nick A. Saunders

14

Abstract

Real-time PCR arrays are tools that allow convenient testing of samples in many assays concurrently, parallel testing of many samples or testing of multiple samples and targets simultaneously. It is desirable to standardize and automate primer and probe selection due to the large number of assays that must be designed. Furthermore, it is useful to use probe selection techniques that increase the robustness of the individual assays since this will increase the level of compatibility between the assays and decrease the complexity of interpretation of the outputs.

A simple approach to creating real-time PCR arrays is to use microtitre plates which currently have capacities of 96, 384 or 1536 features. Such arrays can be populated with user designed assays or with tests selected from a menu of over one million that are commercially available. A primary application of such arrays has been to verify gene expression data obtained using hybridization.

Cramming additional features into a device of manageable scale has led to the introduction of nanolitre volume arrays that diverge from the microtitre plate pattern. Several thousand different reactions can now be included in a single real-time PCR array. The reduction in scale also has advantages in terms of the volumes of materials required.

As real-time arrays are miniaturised the number of pipetting steps required increases and it is often necessary to pre-configure them commercially leading to relative inflexibility. This limitation has prompted the development of arrays that include microfluidic channels and valves. These 'chips' can be loaded via relatively few liquid handling steps to create custom applications.

Introduction

In a conventional PCR application a collection of samples might be tested under identical reaction conditions, i.e. with the same primers. PCR arrays represent a simple inversion of this set-up in which samples are split between parallel reactions each with a different primer set. Using PCR arrays it is possible to obtain multiple data points from a sample in a single run. Although some non-real-time PCR array applications exist real-time detection is generally required.

The common applications of real-time PCR arrays can be broadly divided into typing and gene expression analyses. For example, typing by SNP profiling can now be performed conveniently using arrays comprised of libraries of individual assays brought together in a microtitre tray and one major supplier now has a collection of over 4.5 million pre-designed human SNP assays of which 170,000 have been validated (Abi website https://products.appliedbiosystems.com). These assays can be combined as required by the investigator. In the same way, assays designed to measure the levels of specific mRNAs may be brought together to give a transcript profile and the same supplier has over 1 million assays covering genes from 19 species.

The first real-time PCR arrays built up from these assays were designed to use standard instruments and 96 well microtitre trays. However, the power of arrays lies in the opportunity to combine parallel assays. Advances in instrumentation and the associated consumables for real-time PCR have led to the development of platforms that allow several thousand nanolitre sized reactions to be run concurrently. Developments in microfluidics are also contributing to the convenience of PCR array based assays and it is possible to run multiple samples on a single device with rapid automated reaction set-up.

Handling the design of large numbers of sets of primers and probes for use in arrays is greatly assisted by using software that applies layered rules to the available sequence data. Data handling tools are also helpful for analysis of the greater quantum of information from the array platforms. The utility of such software needs to increase in step with the growing requirements of the field.

This chapter is designed to present available and emerging technologies associated with real-time PCR arrays. Some of the applications, strengths and weaknesses of different PCR array formats will be discussed.

Assay design

The performance of real-time PCR assays depends upon the suitability of the primers and probes chosen. For sensitivity the primers must match sequences in all target genomes or transcriptomes sufficiently for priming to occur and for specificity the primers and probe(s) should mismatch all other sequences that might yield a signal. Additional requirements for sensitivity are that amplification should proceed with reasonable efficiency, that the formation of primer artefacts occurs infrequently and that the probe(s) have a decent fluorescent yield on interacting with the amplicon.

Real-time assays that include hydrolysis (TaqMan) probes are very extensively used in PCR array type applications mainly because of the relative ease of data interpretation (De la Vega *et al.*, 2005). These assays have well understood but somewhat strict design requirements that lend themselves to being configured using computer programmes that follow a simple set of thermodynamic and heuristic rules recommended by Applied Biosystems. PrimerExpress® (Applied Biosystems, Foster City, CA, USA) is the best known software intended for this task. The primary requirement is the availability of a sequence that is conserved throughout the taxa that the assay is being designed to detect. The software then provides possible primer/probe sets that can be configured to allow single nucleotide polymorphism (SNP) detection. A BLAST (Altschul *et al.*, 1990) search is performed to ensure that the primers are not hitting unintended targets. Finally the assay is tested in the laboratory although this may be minimal depending on the application. However, for microbial targets testing may involve analysis of nucleic acids from many strains.

Approaches designed to improve assay robustness applicable to arrays

Real-time PCR arrays involve the implementation of multiple assays that must be run in parallel. To facilitate this requirement it is important to use design techniques intended to provide probes and primers that give robust real-time PCR assays.

In general robust primer design for arrays has depended upon adherence to the generally accepted parameters governing the compatibility of primer pairs. It is then possible to use a single set of amplification conditions for all assays on the array although this involves some degree of compromise, i.e. a default annealing temperature and magnesium ion concentration.

Various methods exist for the detection of SNPs and other mutations by real-time PCR. A widely used strategy is allele-specific PCR in which one or both primers anneal at variant sites (Wu *et al.*,1989). Ideally, a primer matching a specific variant should amplify only that variant. However, in practice, amplification of related sequences occurs although the breakthrough rate varies over many orders of magnitude (Huang *et al.*, 1992). Commonly, the 3′ end of the mutant allele-specific primer is positioned directly over a mutant base in order to selectively amplify the mutant template. This strategy greatly reduces the breakthrough rate for amplification of non-matching sequences but the variability of non-specific amplification may necessitate empirical testing to find highly selective assays (Newton *et al.*,1989). Mismatch amplification mutation assay (MAMA) where an additional mismatch is introduced at the penultimate base is effective in decreasing the breakthrough rate (Li *et al.*, 2004) as is the use of locked nucleic acid residues at the SNP (Latorra *et al.*, 2003).

Other methods include allele-specific competitive blocker PCR and blocker–PCR (Seyama *et al.*, 1992; Orou *et al.*, 1995; Prix *et al.*, 2002; Amicarelli *et al.*, 2007; Morlan *et al.*, 2009) in which oligonucleotides that cannot be extended by the polymerase are used to inhibit amplification of a specific sequence variant by annealing to the primer site, the inter-primer sequence or both. Locked nucleic acid hydrolysis probes can also be used to enhance the impact of a SNP (Ugozzoli *et al.*, 2004).

Co-amplification at lower denaturation temperature-PCR (COLD-PCR) uses a modified protocol to enrich the chosen alleles (minority population) from a mixture so that they can be detected reliably (Li *et al.*, 2008). Additional post-denaturation steps are introduced into every cycle. The reaction is held at a temperature above that required for primer annealing to allow the amplicons to re-anneal. The temperature is then increased to an intermediate point set so that any heteroduplexes present are denatured but homoduplexes remain in double stranded form. At the end of this additional process, designed to enrich the proportion of minority population sequence in the single-stranded population, the cycle is resumed at the annealing step. The cumulative effect of many cycles with the additional steps is to increase the proportion of the minority sequence close to a 1:1 ratio.

Another approach to robust, sensitive detection of SNPs is to exploit the high template specificity of DNA ligase. Circularizable ligation probes (Padlock probes) developed and applied by Nilsson and colleagues (1994) have been adapted for use in real-time PCR (van Doorn *et al.*, 2007) and named PRI-lock probes (Fig. 9.1). The circularised probes are amplified via primer pairs that are specific to each probe sequence that acts like a barcode. Each PRI-lock probe consists of terminal regions that are complementary to the target separated by a linker. Within the linker are the opposed unique primer sequences, a generic sequence and a desthiobiotin modified base that allows capture and release. Following ligation of the terminal probe sequences in the presence of the target the priming sites are in the correct orientation to allow formation of an amplimer. Many PRI-lock probes can be combined in a single assay provided the primers are designed to amplify the amplimer under the same conditions. For a multiplex assay in which many SNPs are to be detected

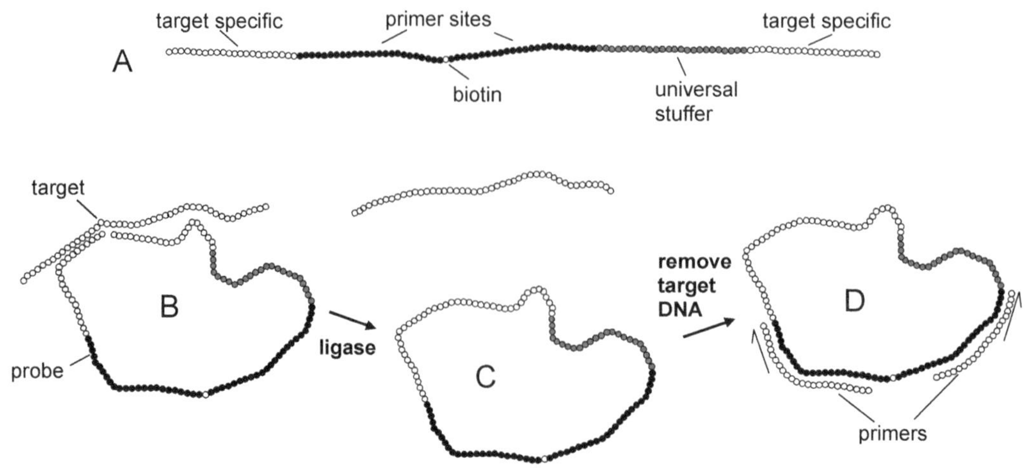

Figure 9.1 PRI-lock probes for real-time PCR. Each PRI-lock probe (A) consists of terminal regions that are complementary to the target separated by a linker. Within the linker are a set of primer sequences, a universal stuffer sequence and a desthiobiotin modified base to aid purification. Following ligation of the terminal probe sequences which requires that the target sequence hybridizes to both target specific regions of the probe (B) the target DNA is removed (C) by binding the probe to a streptavidin modified solid support via the biotin residue. The priming sites in the circular probe are in the correct orientation (D) to allow formation of an amplicon in a PCR with suitable primers. Each target sequence is coded to a different primer pair.

the probes are ligated on the test DNA recovered using the desthiobiotin moiety, treated with exonuclease to destroy non-ligated probe and assayed by real-time PCR on an array (OpenArray™).

Real-time PCR arrays in conventional plates

PCR arrays based on standard reaction plates (i.e. microtitre tray format; see Fig. 9.2) represent a straightforward application of readily available technology. Each well, or set of wells, is loaded with a different PCR but all wells receive an aliquot of the same sample. The primers and probes of each PCR reaction are generally designed to function with identical cycling conditions and reagents. This is a convenient but non-essential feature of plate real-time arrays which could, for example, be run on a gradient block.

Hydrolysis probe (TaqMan®) based assays are well suited for use in real-time PCR arrays partly because of the convenience of the format and partly because of the existence of large numbers of off-the-shelf designs and associated reagents. However, it is realistic to use lower cost assays based on SYBR Green detection and a wide range of off-the-shelf assays (SuperArray Biosciences now Qiagen) are available. SNP genotyping and expression profiling assays are available commercially and the individual tests are combined to form supplier designed off-the-shelf arrays or user specified custom arrays.

Potential applications of SNP arrays include their use for human genotyping for forensic purposes, the detection of human mutations associated with drug sensitivity, infectious agent genotyping and detection of markers for genetic modified foods (Mano *et al.*, 2009). Although, these applications have potential, TaqMan® arrays in conventional plates are not often the method of choice for work at this scale due to the availability of alternatives such

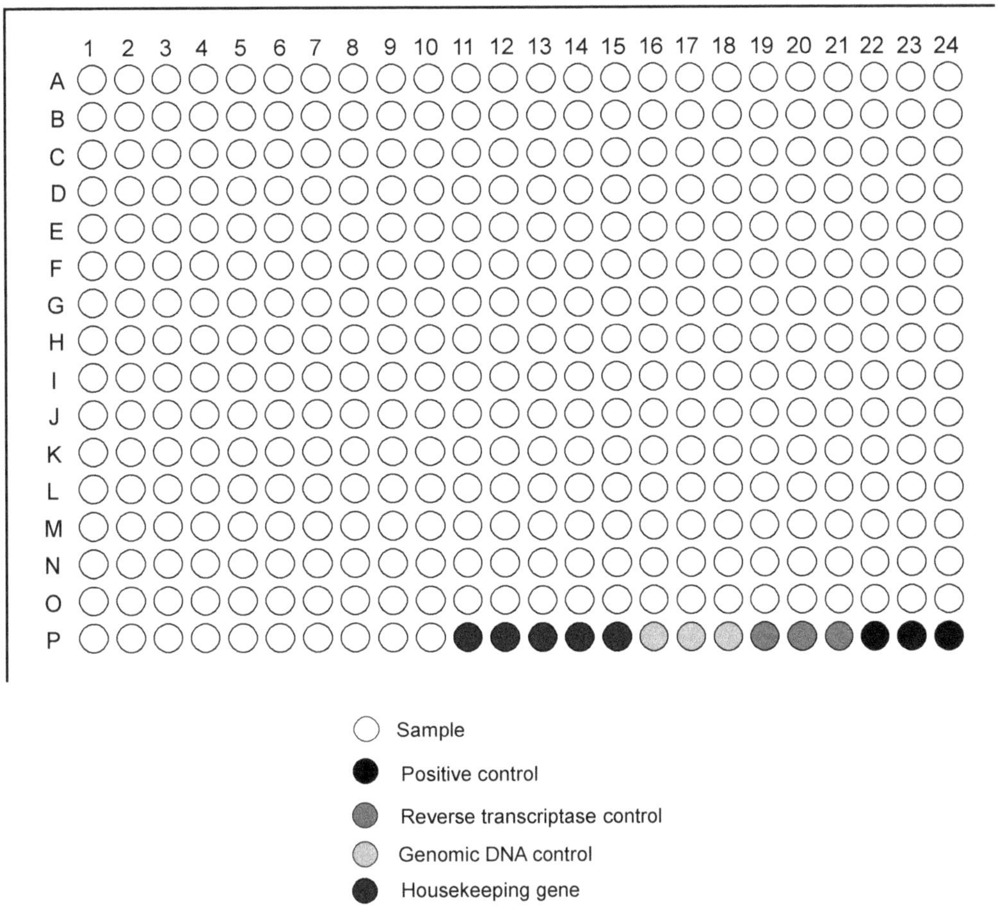

Figure 9.2 Real-time PCR on a plate array. The diagram shows a typical 384 well plate format. Here up to 370 PCR assays are included for measurement of the mRNA levels of target genes. A number of positive and negative controls can easily be accommodated. In this example assays for five housekeeping genes (i.e. genes expected to be expressed in the sample), three positive controls (i.e. a DNA target is spiked into the well) and three reverse transcriptase controls (i.e. a RNA target is spiked into the well) are included. The three negative genomic DNA controls detect genomic contamination (i.e. assay for a non-expressed sequence).

as multiplexing which can accommodate a larger number of samples (Tobler *et al.*, 2005; Tomas *et al.*, 2008).

Expression profiling arrays based on conventional plate formats have been more widely reported than SNP arrays. The key application has been validation of results obtained in microarray experiments. Validation is especially vital in the context that it may be very difficult to correlate any microarray results with real clinical or biological effects (Miklos and Maleszka, 2004). Reference RNA samples analysed in the MicroArray Quality Control study (The MicroArray Quality Control Study: http://www.fda.gov/nctr/science/centres/toxicoinformatics/maqc/index.htm) have been tested on a range of microarray formats (Shi *et al.*, 2006) and in real-time PCR arrays using TaqMan® probes (Shi *et al.*, 2006) or SYBR Green (Arikawa *et al.*, 2008). Although these studies have shown the general reliability of the microarray platforms in identifying genes that are differentially expressed it is also acknowledged that the PCR arrays provide the most sensitive and accurate data with a greater dynamic range than can currently be achieved using microarray formats (Wang *et al.*, 2006; Gaj *et al.*, 2008). Czechowski *et al.* (2004) used the additional sensitivity of the PCR array approach to overcome the detection and dynamic range limits inherent to hybridization based array methods of transcriptome analysis. The PCR array developed measured the levels of transcripts for 1465 Arabidopsis transcription factors (TFs) and was based on closely spaced gene-specific primer pairs and SYBR Green to monitor amplification. Transcript levels of 83% of all target genes could be measured with quantification over six orders of magnitude and a detection limit equivalent to one transcript molecule in 1000 cells. In comparison, Affymetrix chips detected less than 55% of TF transcripts in the same samples and the range of transcript levels was compressed by a factor more than 100. Thus the PCR array was able to reveal root-specific and shoot-specific TF genes that had not been identified as organ-specific previously.

Real-time PCR arrays are well suited for the measurement of the expression of a defined subset of genes. For example, SuperArray Biosciences supply a range of pre-formulated arrays. An example is the 'Human Toll-Like Receptor (TLR) Signalling Pathway RT^2 Profiler PCR Array', which can be used to measure the expression of 84 genes including pattern recognition receptors involved in TLR-mediated signal transduction and innate immunity to a wide range of bacteria, viruses, fungi and parasites. West and Damania (2008) used this array to demonstrate up-regulation of the TLR3 pathway by Kaposi's sarcoma-associated herpes virus during primary infection.

The determination of DNA methylation status at multiple sites is an interesting application of real-time DNA arrays. The technology is suited to the analysis of gene promoter sequences that become hypermethylated in defined biological settings. Hypermethylation may be correlated with biological phenotypes or disease outcomes. Restriction enzymes that either require or are inhibited by methylation are used to enrich a sample for target DNA with a defined methylation status (Herman *et al.*, 1996). The remaining DNA is then quantified in real-time PCR using specific gene PCRs. For example, the commercial Methyl-Profiler DNA Methylation qPCR array (SuperArray Biosciences Corp., Frederick, MD, USA) has been used to confirm results of a genome wide promoter CpG methylation array that uses the bisulfite conversion method (Kibriya *et al.*, 2011) It has also been used to analyse genomic DNA isolated from nasal epithelial cells of smokers and non-smokers for methylation in the promoter sites of human genes (Jaspers *et al.*, 2010). The relative concentrations of differentially methylated DNA are determined by comparing the amount of each digest with that of a mock digest.

Nanolitre plate arrays

Plate arrays that use nanolitre-scale reaction volumes are an interesting prospect for commercial development and some examples are now on the market. The advantages of moving to the nanolitre scale are the reductions in the use of samples and reagents together with an advance in the number of targets or samples that can be measured concurrently leading to potentially very high throughputs. An advantage of the array format is that if necessary each assay in a panel can be independently replaced without the need to re-optimize the entire assay set. However, current nanotitre plate array assay configurations are generally set by the commercial suppliers due to the need for specialized liquid handling and the use of large numbers of custom reagents (primers and probes) required. Array re-configuration is therefore likely to be relatively costly. Pre-formatted, high-throughput assays are likely to be very useful in a variety of applications. However, they lack the flexibility to address rapidly changing needs and cannot be expected to satisfy all requirements for niche assays. Consequently, systems with the facility for users to develop new applications hold a significant advantage.

As discussed in previous sections a significant advantage of the real-time PCR array approach compared to conventional arrays (i.e. those that depend on hybridization of labelled probes without prior amplification of the target) is high sensitivity. Further, at present qPCR provides a more reliable means of target quantification than conventional arrays and gives the potential for high sample throughput. However, compared to microarrays and macroarrays real-time arrays suffer the disadvantage that it is not feasible to achieve very high levels of multiplexing. Part of the motivation for introduction of nanolitre scale real-time PCR arrays is to reduce this relative performance deficit.

The OpenArray® Real-Time PCR System (Applied Biosystems) is designed to provide either genotyping or gene expression analysis. The arrays are constructed by micromachining holes into a stainless steel sheet with the dimensions of a standard microscope slide. The holes have a volume of 33 nl and the interior of each is treated to give a covalently bound hydrophilic coating of polyethylene glycol. The slide surface is given a hydrophobic fluoroalkyl coating. The OpenArray® holes are grouped in 48 subarrays each comprised of 64 wells giving a total number of 3072. For convenience in the robotic application of the reaction components from microtitre plates the subarray layout has the same pitch as a 384 well plate (Fig. 9.3). Primers are pre-loaded and dried down in user specified wells. The wells function to metre sample and master mix added by the loader via capillary action (i.e. the volume delivered, 33 nl, is determined by the hole capacity). The hydrophilic well and hydrophobic surface coating effectively ensures that the sample remains isolated when placed in a glass-walled cassette filled with an immiscible perfluorinated liquid (Fluorinert; 3M, St. Paul, MN, USA) hermetically sealed with epoxy.

Each well of the array can be used for an independent PCR assay although by design each subarray is used for analysis of a single sample in 64 PCRs. The system supports detection based on intercalation (SYBR Green) or hydrolysis probes. In addition, detection either in real-time or following cycling is possible.

Roberts and colleagues (2009) have reported use of OpenArrays® for high-throughput genotyping of human DNA based on assays for single nucleotide polymorphisms (SNPs) at 32 loci. SNPs represent the most convenient, abundant and accessible polymorphisms for recording genetic variation in a population. The attractions of SNP genotyping have increased with the emergence of a growing number of SNP databases, e.g. www.ncbi.nlm.nih.gov/projects/SNP. There are a number of significant applications of SNP genotyping including the diagnosis of human disease, pharmacogenetics, blood typing, forensics, paternity testing, fisheries management, pathogen identification and plant breeding (Brenan *et al.*, 2009). A panel of 90 DNA samples from the Centre d'Etude du Polymorphisme Humain (Fondation Jean Dausset-CEPH) was used for evaluation of the SNP panel on this platform. End-point analysis was used for the bi-allelic tests (most human SNPs are biallelic) based on pairs of hydrolysis probes labelled with VIC and FAM. Good rates of calling with high accuracy were obtained with from 0.1–1 ng human DNA per well although at the lower concentrations whole genome amplification (Dean *et al.*, 2001) was required to maintain the call rate for a proportion of the assays.

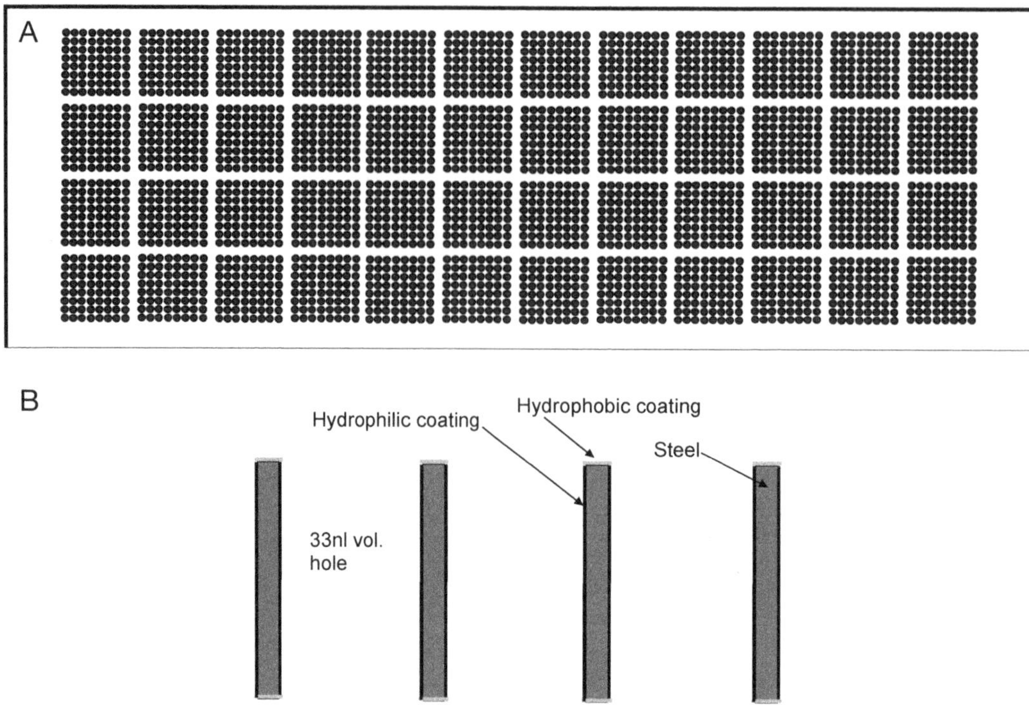

Figure 9.3 The OpenArray® slide. (A) Layout of 48 subarrays on an OpenArray® slide. The subarrays are spaced with the same pitch as a 384 well microtitre plate and each includes 64 holes (320µm diameter holes in an 8×8 array). (B) Cross-section of part of a slide with three complete holes included. The holes, formed by perforating a 0.3mm steel sheet the size of a microscope slide, have a 33nl capacity and are coated internally with a hydrophilic layer. The outer surface of the steel sheet has a hydrophobic coating.

OpenArrays® have also been exploited for the quantitative detection of multiple plant pathogens (van Doorn *et al.*, 2007). This paper introduced the concept of quantifiable PRI-lock probes (a variant of Padlock probes) as discussed in the section above 'Approaches designed to improve assay robustness applicable to arrays' and shows the potential application of the arrays to diagnosis of human and animal infection, food safety, biothreat detection and environmental pest management. The use of real-time arrays overcomes the major drawback of PCR multiplexing which is that the performance (i.e. particularly the sensitivity and accuracy of quantification) of the component assays is compromised by the number of target sequences included. In addition to the specificity conferred by the ligation based assay chemistry the use of the ligation probes confers the advantage that the array can be factory configured with a single set of primer pairs that can be applied universally.

The application to plant pathogens described by Van Doorn and colleagues (2007) provides a model for the development of similar detection systems offering high specificity and throughput with robust target quantification. In their study all targets were detected and no false-positives were recorded for non-target organisms with very similar ligation target sites, including alleles differing at a single nucleotide, indicating high specificity of the ligation reaction (Fig. 9.1). It was demonstrated that the presence of multiple targets had no statistically significant influence on either specificity or sensitivity. Indicating that this multiplex detection system could provide independent detection of multiple targets without inhibition due to competition.

The SmartChip™ Real-Time PCR System (WaferGen Biosystems) is a platform for rapid quantitative gene-expression analysis (Gelfand and Kastury, 2009). The SmartChip™ arrays consist of 5000–30,000 wells each of which can contain a different PCR reaction, in a volume as low as 100 nl. The arrays are run on a platform capable of rapid-cycling and real-time detection. The platform is capable of performing thousands of nanoscale quantitative PCR assays in a single run. This provides the capability to perform screens normally associated with microarrays but combined with the excellent sensitivity inherent in real-time PCR. The SmartChips can be configured either to measure a large number of target sequences on a single sample or in a multiarray format (i.e. with a smaller number of targets analysed for up to 384 samples on a single array). A dedicated dispenser is available to ensure reliable delivery of samples to the arrays. It is claimed that use of the system will allow the generation of expression results many times faster and at greatly reduced cost compared to conventional methods.

Current SmartChips are factory configured with panels of assays that target genes relevant to a particular application, e.g. cancer studies and toxicology. In addition, WaferGen offer custom arrays with user designed assay panels. Although this platform is promising it is yet to find significant levels of application. A problem is that its laboratory niche is currently occupied by platforms such as the high-density arrays manufactured by Affymetrix Inc and bead arrays such as those of Illumina Inc (Saunders, 2011).

Microfluidic arrays

The use of microfluidic devices is one solution to the problem of how to overcome one of the major potential bottlenecks hampering the use of real-time PCR arrays in high-throughput applications, i.e. the need to transfer samples into multiple test wells with minimal pipetting steps. Recent years have seen encouraging progress in the development of such devices and commercial products are now available.

The development of microfluidic devices for use in PCR has been an area of activity since the early 1990s and the developments have been reviewed by Zhang and colleagues (2006, 2007). For example, Belgrader and colleagues (1998) reported the development of a microchip PCR instrument with an array of ten silicon reaction chambers with thin-film resistive heaters and solid-state optics and showed that it could be applied to rapid pathogen detection. However, this approach to array construction, i.e. building an array from individual devices, is limited in its cost-effective scalability, compared with the alternative approach of building an array of reaction chambers into a single unit.

The TaqMan® Low Density Array (Applied Biosystems) is a consumable device consisting of 384 wells connected by a series of micro fluidic channels (Fig. 9.4) designed for custom or off-the-shelf configuration with TaqMan® assays. User-selected assays are factory loaded into each well. Up to eight cDNA samples per array are mixed with Master Mix and loaded into the eight ports. A single sample reaction mix is delivered to 48 wells via a centrifugal filling process. As for the plate arrays, low density microfluidic arrays are useful either as the primary tool for gene expression analysis of a limited number of genes (for example Rom *et al.*, 2009) or as a confirmatory assay for selected genes identified using high-density arrays (for example Hartman *et al.*, 2008). Although, these arrays were designed primarily as a

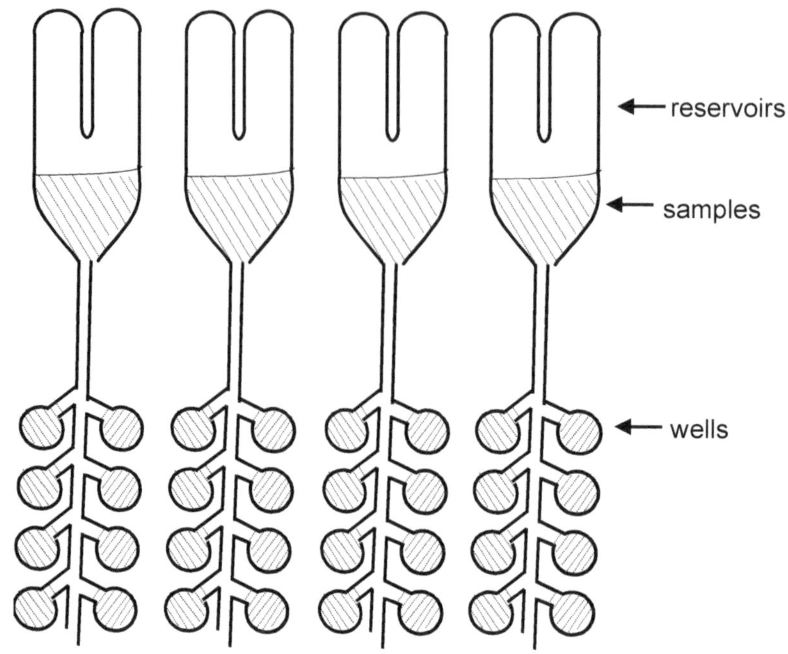

Figure 9.4 Low density microfluidic array. Each of the 384 wells is pre-loaded (Applied Biosystems) with dried primers and probes. Potentially each well can be loaded with a different assay though more usually each set of 48 wells is configured identically. For use each of the eight reservoirs is loaded with sample and this is distributed to the associated wells by centrifugation. After centrifugation each well is isolated. The wells of the card are configured as on a 384 well microtitre tray.

low-density alternative to microarrays for the analysis of gene expression they can also be employed for detection of multiple targets in parallel. For example, Kodani and colleagues (2011) have applied the array cards to the simultaneous detection of respiratory pathogens.

Fluidigm microfluidic chips are similar in capability to the TaqMan® Low Density Arrays. However, the array dimensions (i.e. the number of samples and different assays) of the devices are generally larger and all of the reaction components, including the probes and primers, may be added to the chip in the user laboratory so that the system is extremely flexible. A range of Fluidigm chips is now available which differ in the number of sample and assay (i.e. the probes and primers) inlets. The available formats are 192 samples × 24 assays, 48 samples × 48 assays and 96 samples × 96 assays. For most applications the chips are single use due to the difficulty of cleaning to the standard necessary to avoid carry-over. However, a reusable 48 samples × 48 assays chip is available that can be cleaned to the level necessary for use in SNP genotyping. The chips are mounted on a plastic carrier with the sample inlets and detector inlets together with interface and containment accumulators that can be pressurized to close the valves (i.e. NanoFlex™ valves) that are the central design feature of these arrays. The dimensions and positions of the inlet ports conform to those of a 384 well microtitre plate. A network of fluid lines, valves and reaction chambers are included on the chips. The valves deflect under pressure to seal and regulate the flow of liquids in the device. Each reaction chamber is connected to a sample and detector reservoir via a containment valve. The sample and detector sides of each reaction chamber are separated by a separately controllable interface valve (Fig. 9.5). For loading, pressure is applied to

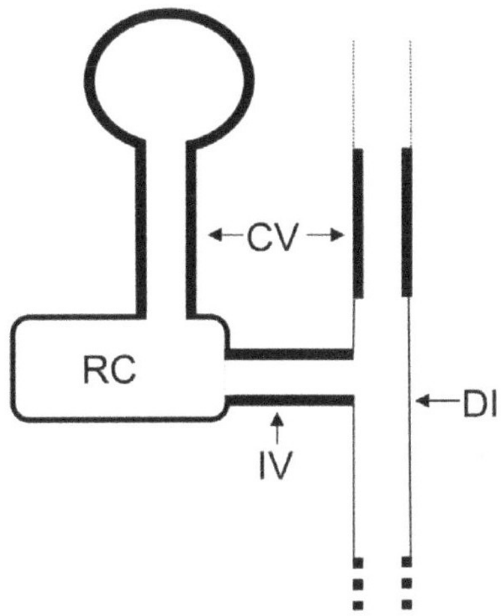

Figure 9.5 NanoFlex™ valves associated with a single reaction chamber. Each reaction chamber (RC) has two associated control valves (CV). One of the control valves is in the line leading to the detector and the other is in the sample line. The reactants are loaded down each line with the control valves open and the interface valve (IV) closed. This delivers sample to the reaction chamber and detector to the detector inlet (DI). For mixing the control valves are shut and the interface valve is opened allowing the detector and sample solutions to meet.

push the sample and detector fluids down separate fluid lines. Sample goes to the individual reaction chambers and detector solution collects in an inlet proximal to each well. At this point mixing of the two fluids is prevented by the closed interface valve. For mixing, the two containment valves associated with each reaction chamber are closed and then the interface valve is opened allowing the reagents in the detector inlet access to the reaction chamber so that mixing can occur. When mixing is complete the interface valve is closed and the chip is ready for PCR. The loading process can be completed in 1 hour.

Spurgeon and colleagues (2008) have evaluated the performance of the 48 × 48 Fluidigm array, which accommodates 2304 individual reactions. Gene expression data was compared with publicly available data (GeneNote database http://bioinfo2.weizmann.ac.il/cgi-bin/genenote/home_page.pl) obtained by microarray and conventional real time PCR. The reproducibility of the microfluidic array was good for replicates both within and between chips and the results were essentially identical in quality the conventional real-time RT PCR data and superior to the microarray data for RNA extracted from the same tissue type.

An interesting application of a 48×48 array has been described by Guo and colleagues (2010), who examined the levels of 48 gene transcripts in single-cell extracts from cells at various stages of progression from zygote to blastocyst. The cells were separated manually and lysed to release mRNA for reverse transcription and sequence-specific preamplification of the cDNA before loading onto the array. It was possible to show that in cells of embryos at the morula stage (consisting of about 30 cells) each expressed transcription factors at the same levels but that by the 64-cell stage three cell types were clearly distinguished by transcription factor expression levels. Fluidigm arrays have been further developed to enable single cells to be isolated, extracted and amplified within the same device (White *et al.*, 2011). These authors described an integrated microfluidic device for RT-qPCR analysis of 50 single cells from six independent samples (i.e. 300 cells in total). All steps including single-cell capture, lysis, reverse transcription, and qPCR can be executed on the device.

Digital PCR

PCR arrays are generally applicable to digital PCR in which DNA is partitioned to a level at which individual target molecules are isolated and can be individually amplified (Sykes *et al.*, 1992). The application of digital PCR to target quantification, the original driver of the concept, has now been rendered largely redundant by real-time PCR. However, digital arrays may be an alternative to real-time arrays in various applications. For example, haplotyping of individual samples can be achieved using digital arrays that probe individual sample DNA molecules. Fluidigm Digital Arrays™ have been used for this application since they allow individual target molecules to be isolated, amplified and detected within the device (Menzel *et al.*, 2010).

References

Altschul, S.F., Gish, W., Miller, W., Myers, E.W., and Lipman, D.J. (1990). Basic local alignment search tool. J. Mol. Biol. *215*, 403–410.

Amicarelli, G., Shehi, E., Makrigiorgos, G.M., and Adlerstein, D. (2007). FLAG assay as a novel method for real-time signal generation during PCR: application to detection and genotyping of KRAS codon 12 mutations. Nucleic Acids Res. *35*, e131.

Arikawa, E., Sun, Y., Wang, J., Zhou, Q., Ning, B., Dial, S., Guo, L., and Yang, J. (2008). Cross-platform comparison of SYBR(R) Green real-time PCR with TaqMan PCR, microarrays and other gene expression measurement technologies evaluated in the MicroArray Quality Control (MAQC) study. BMC Genomics *9*, 328.

Belgrader, P., Benett, W., Hadley, D., Long, G., Mariella, R., Jr., Milanovich, F., Nasarabadi, S., Nelson, W., Richards, J., and Stratton, P. (1998). Rapid pathogen detection using a microchip PCR array instrument. Clin. Chem. *44*, 2191–2194.

Brenan, C.J., Roberts, D., and Hurley, J. (2009). Nanoliter high-throughput PCR for DNA and RNA profiling. Methods Mol. Biol. *496*, 161–174.

Czechowski, T., Bari, R.P., Stitt, M., Scheible, W.R., and Udvardi, M.K. (2004). Real-time RT-PCR profiling of over 1400 Arabidopsis transcription factors: unprecedented sensitivity reveals novel root- and shoot-specific genes. Plant J. *38*, 366–379.

De la Vega, F.M., Lazaruk, K.D., Rhodes, M.D., and Wenz, M.H. (2005). Assessment of two flexible and compatible SNP genotyping platforms: TaqMan SNP Genotyping Assays and the SNPlex Genotyping System. Mutat. Res. *573*, 111–135.

Dean, F.B., Nelson, J.R., Giesler, T.L., and Lasken, R.S. (2001). Rapid amplification of plasmid and phage DNA using Phi 29 DNA polymerase and multiply-primed rolling circle amplification. Genome Res. *11*, 1095–1099.

van Doorn, R., Szemes, M., Bonants, P., Kowalchuk, G.A., Salles, J.F., Ortenberg, E., and Schoen, C.D. (2007). Quantitative multiplex detection of plant pathogens using a novel ligation probe-based system coupled with universal, high-throughput real-time PCR on OpenArrays™. BMC Genom. *8*, 276.

Gaj, S., Eijssen, L., Mensink, R.P., and Evelo, C.T. (2008). Validating nutrient-related gene expression changes from microarrays using RT(2) PCR-arrays. Genes Nutr. *3*, 153–157.

Gelfand, D., and Kumar, K. (2009). High-throughput nanovolume qPCR: next-generation platform combines microarray throughput with sensitivity. Genet. Eng. Biotechnol. News 42–43.

Guo, G., Huss, M., Tong, G.Q., Wang, C., Li Sun, L., Clarke, N.D., and Robson, P. (2010). Resolution of cell fate decisions revealed by single-cell gene expression analysis from zygote to blastocyst. Dev. Cell *18*, 675–685.

Hartman, A.L., Ling, L., Nichol, S.T., and Hibberd, M.L. (2008). Whole-genome expression profiling reveals that inhibition of host innate immune response pathways by Ebola virus can be reversed by a single amino acid change in the VP35 protein. J. Virol. *82*, 5348–5358.

Herman, J.G., Graff, J.R., Myöhänen, S., Nelkin, B.D., and Baylin, S.B. (1996). Methylation-specific PCR: a novel PCR assay for methylation status of CpG islands. Proc. Natl. Acad. Sci. U.S.A. *93*, 9821–9826.

Huang, M.M., Arnheim, N., and Goodman, M.F. (1992). Extension of base mispairs by Taq DNA polymerase: implications for single nucleotide discrimination in PCR. Nucleic Acids Res. *20*, 4567–4573.

Jaspers, I., Horvath, K.M., Zhang, W., Brighton, L.E., Carson, J.L., and Noah, T.L. (2010). Reduced expression of IRF7 in nasal epithelial cells from smokers after infection with influenza. Am. J. Respir. Cell Mol. Biol. *43*, 368–375.

Mano, J., Shigemitsu, N., Futo, S., Akiyama, H., Teshima, R., Hino, A., Furui, S., and Kitta, K. (2009). Real-time PCR array as a universal platform for the detection of genetically modified crops and its application in identifying unapproved genetically modified crops in Japan. *J. Agric. Food Chem.* *57*, 26–37.

Kodani, M., Yang, G., Conklin, L.M., Travis, T.C., Whitney, C.G., Anderson, L.J., Schrag, S.J., Taylor, T.H., Jr., Beall, B.W., Breiman, R.F., *et al.* (2011). Application of TaqMan low-density arrays for simultaneous detection of multiple respiratory pathogens. J. Clin. Microbiol. *49*, 2175–2182.

Kibriya, M.G., Raza, M., Jasmine, F., Roy, S., Paul-Brutus, R., Rahaman, R., Dodsworth, C., Rakibuz-Zaman, M., Kamal, M., and Ahsan, H. (2011). A genome-wide DNA methylation study in colorectal carcinoma. BMC Med. Genomics *4*, 50.

Latorra, D., Campbell, K., Wolter, A., and Hurley, J.M. (2003). Enhanced allele-specific PCR discrimination in SNP genotyping using 39 locked nucleic acid (LNA) primers. Hum. Mutat. *22*, 79–85.

Li, B., Kadura, I., Fu, D., and Watson, D.E. (2004). Genotyping with TaqMAMA. Genomics *83*, 311–320.

Li, J., Wang, L., Mamon, H., Kulke, M.H., Berbeco, R., and Makrigiorgos, G.M. (2008). Replacing PCR with COLD-PCR enriches variant DNA sequences and redefines the sensitivity of genetic testing. Nature Med. *14*, 579–584.

Menzel, S., Qin, J., Vasavda, N., Thein, S.L., and Ramakrishnan, R. (2010). Experimental generation of SNP haplotype signatures in patients with sickle cell anaemia. PLoS One 5, e13004.

Miklos, G.L., and Maleszka, R. (2004). Microarray reality checks in the context of a complex disease. Nat. Biotechnol. *22*, 615–621.

Morlan, J., Baker, J., and Sinicropi, D. (2009). Mutation detection by real-time PCR: a simple, robust and highly selective method. PLoS One *4*, e4584.

Newton, C.R., Graham, A., Heptinstall, L.E., Powell, S.J., Summers, C., Kalsheker, N., Smith, J.C., and Markhamet, A.F. (1989). Analysis of any point mutation in DNA. The amplification refractory mutation system (ARMS). Nucleic Acids Res. *17*, 2503–2516.

Nilsson, M., Malmgren, H., Samiotaki, M., Kwiatkowski, M., Chowdhary, B.P., and Landegren, U. (1994). Padlock probes: circularizing oligonucleotides for localized DNA detection. Science *265*, 2085–2088.

Orou, A., Fechner, B., Utermann, G., and Menzel, H.J. (1995). Allele-specific competitive blocker PCR: a one-step method with applicability to pool screening. Hum. Mutat. *6*, 163–169.

Prix, L., Uciechowski, P., Böckmann, B., Giesing, M., and Schuetz, A.J. (2002). Diagnostic biochip array for fast and sensitive detection of K-ras mutations in stool. Clin. Chem. *48*, 428–435.

Roberts, D.G., Morrison, T.B., Liu-Cordero, S.N., Cho, J., Garcia, J., Kanigan, T.S., Munnelly, K., and Brenan, C.J. (2009). A nanoliter fluidic platform for large-scale single nucleotide polymorphism genotyping. Biotechniques *46*, Suppl. ix–xiii.

Rom, J., von Minckwitz, G., Marmé, F., Ataseven, B., Kozian, D., Sievert, M., Schlehe, B., Schuetz, F., Scharf, A., Kaufmann, M., *et al.* (2009). Phase I study of apoptosis gene modulation with oblimersen within preoperative chemotherapy in patients with primary breast cancer. Ann. Oncol. *20*, 1829–1835.

Saunders, N.A. (2011). Applications of nanomaterials to arrays for infectious disease diagnosis. Nanomed. *6*, 271–280.

Seyama, T., Ito, T., Hayashi, T., Mizuno, T., Nakamura, N., and Akiyama, M. (1992). A novel blocker-PCR method for detection of rare mutant alleles in the presence of an excess amount of normal DNA. Nucleic Acids Res. *20*, 2493–2496.

Shi, L., Reid, L.H., Jones, W.D., Shippy, R., Warrington, J.A., Baker, S.C., Collins, P.J., de Longueville, F., Kawasaki, E.S., Lee, K.Y., *et al.* (2006). The MicroArray Quality Control (MAQC) project shows inter- and intraplatform reproducibility of gene expression measurements. Nat. Biotechnol. *24*, 1151–1161.

Spurgeon, S.L., Jones, R.C., and Ramakrishnan, R. (2008). High throughput gene expression measurement with real time PCR in a microfluidic dynamic array. PLoS One 3, e1662.

Sykes, P.J., Neoh, S.H., Brisco, M.J., Hughes, E., Condon, J., and Morley, A.A. (1992). Quantitation of targets for PCR by use of limiting dilution. Biotechniques *13*, 444–449.

Tobler, A.R., Short, S., Andersen, M.R., Paner, T.M., Briggs, J.C., Lambert, S.M., Wu, P.P., Wang, Y., Spoonde, A.Y., Koehler, R.T., *et al.* (2005). The SNPlex genotyping system: a flexible and scalable platform for SNP genotyping. J. Biomol. Tech. *16*, 398–406.

Tomas, C., Stangegaard, M., Børsting, C., Hansen, A.J., and Morling, N. (2008). SNPforID Consortium. Typing of 48 autosomal SNPs and amelogenin with GenPlex SNP genotyping system in forensic genetics. Forensic Sci. Int. Genet. 3, 1–6.

Ugozzoli, L.A., Latorra, D., Puckett, R., Pucket, R., Arar, K., and Hamby, K. (2004). Real-time genotyping with oligonucleotide probes containing locked nucleic acids. Anal. Biochem. *324*, 143–152.

Wang, Y., Barbacioru, C., Hyland, F., Xiao, W., Hunkapiller, K.L., Blake, J., Chan, F., Gonzalez, C., Zhang, L., and Samaha, R.R. (2006). Large scale real-time PCR validation on gene expression measurements from two commercial long-oligonucleotide microarrays. BMC Genom. *7*, 59.

White, A.K., Van Insberghe, M., Petriv, O.I., Hamidi, M., Sikorski, D., Marra, M.A., Piret, J., Aparicio, S., and Hansen, C.L. (2011). High-throughput microfluidic single-cell RT-qPCR. Proc. Natl. Acad. Sci. U.S.A. *108*, 13999–14004.

West, J., and Damania, B. (2008). Upregulation of the TLR3 pathway by Kaposi's sarcoma-associated herpesvirus during primary infection. J. Virol. *82*, 5440–5449.

Wu, D.Y., Ugozzoli, L., Pal, B.K., and Wallace, R.B. (1989). Allele-specific enzymatic amplification of beta-globin genomic DNA for diagnosis of sickle cell anemia. Proc. Natl. Acad. Sci. U.S.A. *86*, 2757–2760.

Zhang, C., and Xing, D. (2007). Miniaturized PCR chips for nucleic acid amplification and analysis: latest advances and future trends Nucl. Acids Res. *35*, 4223–4237.

Zhang, C., Xu, J., Ma, W., and Zheng, W. (2006). PCR microfluidic devices for DNA amplification. Biotechnol. Adv. *24*, 243–284.

www.ingramcontent.com/pod-product-compliance
Ingram Content Group UK Ltd.
Pitfield, Milton Keynes, MK11 3LW, UK
UKHW062003041026
15591UKWH00016B/1156

9 781912 530243